Philosophy of Engineering and Technology

Volume 19

The Philosophy of Engineering and Technology book series provides the multifaceted and rapidly growing discipline of philosophy of technology with a central overarching and integrative platform. Specifically it publishes edited volumes and monographs in: the phenomenology, anthropology and socio-politics of technology and engineering the emergent fields of the ontology and epistemology of artifacts, design, knowledge bases, and instrumentation engineering ethics and the ethics of specific technologies ranging from nuclear technologies to the converging nano-, bio-, information and cognitive technologies written from philosophical and practitioners perspectives and authored by philosophers and practitioners. The series also welcomes proposals that bring these fields together or advance philosophy of engineering and technology in other integrative ways. Proposals should include: A short synopsis of the work or the introduction chapter. The proposed Table of Contents The CV of the lead author(s). If available: one sample chapter. We aim to make a first decision within 1 month of submission. In case of a positive first decision the work will be provisionally contracted: the final decision about publication will depend upon the result of the anonymous peer review of the complete manuscript. We aim to have the complete work peer-reviewed within 3 months of submission. The series discourages the submission of manuscripts that contain reprints of previous published material and/or manuscripts that are below 150 pages / 75,000 words. For inquiries and submission of proposals authors can contact the editor-in-chief Pieter Vermaas via: p.e.vermaas@tudelft.nl, or contact one of the associate editors.

More information about this series at http://www.springer.com/series/8657

Andrea Bardin

Epistemology and Political Philosophy in Gilbert Simondon

Individuation, Technics, Social Systems

Springer

Andrea Bardin
Brunel University London
London, UK

Centre international des
études simondoniennes
MSH Paris-Nord

ISSN 1879-7202 ISSN 1879-7210 (electronic)
Philosophy of Engineering and Technology
ISBN 978-94-017-9830-3 ISBN 978-94-017-9831-0 (eBook)
DOI 10.1007/978-94-017-9831-0

Library of Congress Control Number: 2015934460

Springer Dordrecht Heidelberg New York London

Printed on acid-free paper

Springer Science+Business Media B.V. Dordrecht is part of Springer Science+Business Media (www.springer.com)

to Susanna

Acknowledgements

I am primarily indebted to Prof. Adone Brandalise, tireless master at the University of Padua. Many friends and colleagues between Padua, Paris and London have aided me – directly or incidentally – to manage the variegated topics this book aims to intertwine: Davide De Pretto, Bruno Karsenti, Enrico Manera, Giovanni Menegalle, Giangiorgio Pasqualotto, Gaetano Rametta, Alberto Zanon. In particular, I owe my thanks to Fabio Raimondi and Filippo Del Lucchese, whose sincere and straightforward friendship pushed me to continue, and Nathaniel Boyd, whose patient and intelligent editing in turbulent times provided a decisive contribution to the accomplishment of this book.

I cannot quantify the intellectual debt I contracted on occasions of structured discussions or informal chats with those who are currently part of the growing international network of Simondon's scholars, in particular the members of the *Centre international des études simondoniennes*, among them – last but not least – my friend Giovanni Carrozzini, always ready to share with intellectual generosity his incredibly detailed knowledge of Simondon's life and oeuvre.

This book is in fact the outcome of a 'metastable' milieu of friends, colleagues and students with whom I have lived, learned, thought, quarreled and sometimes written on science, technics, politics and even philosophy, during the lifetime we spent together: Massimo Bardin, Michele Faoro, Roberto Galvanin, Nicola Pagano, Sandro Pellarin, Barbara Plafoni, Filippo Ragazzo, Alessio Rizzo, Vittorio Saccardo, Mario Spezzapria, Robin Targon, Diego Vicenzutto and my unforgotten 'brother' Federico Sbordoni. Finally, I thank Mirella and Adriano for their unconditional trust, Bianca and Davide for having reinvented my life and Susanna for living it with me. To her – *la rosa bianca che fiorisce/nelle prime nebbie* – this book is dedicated.

Contents

Part I
Nature and Knowledge

> *Nulle part la confusion n'est aussi visible que dans les discussions sur l'individualité [...] Concluons donc que l'individualité n'est jamais parfaite, qu'il est souvent difficile, parfois impossible de dire ce qui est individu et ce qui ne l'est pas, mais que la vie n'en manifeste pas moins une recherche de l'individualité et qu'elle tend à constituer des systèmes naturels isolés, naturellement clos.*
>
> (Bergson, *L'évolution créatrice*)

> *The individuality of the body is that of a flame rather than that of a stone, of a form rather than of a bit of substance.*
>
> (Wiener, *The Human Use of Human Beings*)

Gilbert Simondon has been known as a philosopher of technics since his *Du mode d'existence des objets techniques* [On the Mode of Existence of Technical Objects] (MEOT) was published in 1958. Recently, after the posthumous first complete edition of his main work *L'individuation à la lumière des notions de forme et d'information* [Individuation in the Light of the Notions of Form and Information] (*Individuation*), Simondon's thought has become considered by scholars concerned with the connection between epistemology, ontology and political philosophy.[1] Taking into account the entire corpus of Simondon's oeuvre – the whole of his published works plus some unedited documents – and making substantial use of its sources, this book aims at showing the articulated interconnection between his philosophy of science and technology and his political philosophy. The book consists of three sections concerning different aspects of his research: (1) ontology and epistemology of individuation; (2) biological and social systems; (3) anthropology, technics and politics.

The first section analyses Simondon's attempt to re-configure the theoretical apparatus of philosophy according to some concepts he derived – following in the footsteps of his master Merleau-Ponty – from scientific and epistemological thought,

[1] At the time two theses were scheduled for a PhD, which preceded the entrance into French academia. While MEOT, the second thesis, directed by Georges Canguilhem, was immediately published in 1958, thus making Simondon known as a philosopher of technology, *Individuation*, the principal dissertation, directed by Jean Hyppolite, underwent a quite complicated editorial process (for a brief summary of it, see the note in the appendix to this volume). Simondon's texts will be quoted according to the list of abbreviations.

especially physics of *quanta*, thermodynamics and cybernetics. The second section shows the impact of biological concepts on the theorisation of the genesis and functioning of social systems, and the peculiar role played by technics in social dynamics. Simondon's main philosophical references in this field are Henri Bergson's biological and social theories, George Canguilhem's philosophy of life sciences and technics, Norbert Wiener's cybernetics of society and Leroi-Gourhan's palaeoanthropology. The third section of the book concerns the broader relationship between French epistemology and the conceptual renewal it elicited in the social and political field. I highlight Simondon's debts to the French sociological tradition, beginning with Mauss and Durkheim, and the way he posed the political problem outside of any positivistic faith in the power of technological progress and, at the same time, against the political regression inspired by Heidegger's anti-technological stance.

Simondon's view on the complex nature of social processes derives from his adoption of the paradigm of quantum physics to the study of social systems. Although he does not always make it explicit, a conception of human nature as a 'work in progress' is implicit in his epistemology. Hence his philosophy allows for a critique of the modern imagination – both ideological and scientific – of the contraposition between individuals and society, and can be a useful tool for questioning the contemporary relation between technological and social innovation in complex societies.

Chapter 1
Elements for a Philosophy of Individuation

In the title *Individuation in the Light of the Notions of Form and Information*, the concepts of form and information clearly indicate a theoretical progression toward the concept of individuation: 'form' and 'information' are a direct reference to the epistemological frameworks of *Gestalttheorie* and cybernetics respectively, in relation to which Simondon builds his own thesis. This chapter introduces the terms which constitute Simondon's jargon, showing how they are derived from and related to various fields of scientific research, and explaining their function in Simondon's philosophy. Crucial to his discourse are the *philosophical* notions of 'individual' and 'individuation' which, extended to every domain of being, Simondon proposes as the ontological foundations for a philosophical approach to what he calls a 'process of individuation' or 'ontogenesis'.

1.1 The Individual as a System: Structure and Operation

The centrality of the concept of individuation works as a counterpoint of Simondon's critique to the traditional concepts of form, matter, substance and cause.[1] *Individuation* begins with a critique of the Aristotelian distinction between matter and form, and aims to show the inadequacy of the conceptual apparatus of classical philosophy with regard to the results of twentieth century scientific thought. For this reason, if it is true that the term 'individual' spans all the domains that could be ascribed to 'being', it is also true that Simondon distances himself from its classical association to the concepts of 'substance' or 'essence'. And nevertheless, what is firstly noted is the extension of its use to all the different 'regimes of individuation'

[1] Concerning this topic see Bardin (2015), where part of this chapter has been developed.

A. Bardin, *Epistemology and Political Philosophy in Gilbert Simondon*,
Philosophy of Engineering and Technology 19, DOI 10.1007/978-94-017-9831-0_1

which Simondon analyses: physical, biological, psychic and collective.[2] An extension likely to prompt the further observation of Canguilhem, his *directeur de thèse* for MEOT, according to whom

> From the philosophical point of view, it would be a question of a new kind of Aristotelianism, on the condition, of course, that Aristotelian psychobiology and the modern technology of transmission would not be confused. (Canguilhem 1943: 277–278)[3]

Since 'individual' is a heavily layered term in the philosophical tradition, it should therefore be carefully redefined. I shall start from two successive and apparently contradictory definitions provided by Simondon, in order to set the conditions for their compatibility. The first definition appears in *Analyse des critères de l'individualité* [Analysis of Individuality Criteria], where Simondon states that '*there can be no science but of the individual*, this will be the epistemological consequence of our enquiry' (AI 553); the second appears in *Individuation*, where Simondon claims that 'to be rigorous, one should not speak of individual, but rather of individuation' (I 191). It is quite clear that only a redefinition of the concept of 'individual' could make sense of the above statements and allow the construction of a philosophy of the processes of individuation, although it is worth underlining immediately that Simondon could not avoid an equivocal use of the term 'individual' throughout his main work.

To formalize the complex status of the individual, Simondon uses, especially in the two programmatic texts, the terms 'structure' and 'operation'.[4] As for *structure*, the individual can always be considered a 'phase-shift' [*déphasé*] system; Simondon borrows the term 'phase' from physics and chemistry to indicate how different processes, parallel, divergent or convergent, are simultaneously taking place in a system. As for *operation*, the individual is necessarily involved in 'transductive' processes; the concept of 'transduction' has origins both biological (contamination) and technological (amplification), and refers to a mode of propagation – a non-deterministic sequence, presenting gaps and discontinuities. The individual is therefore defined in relation to both a phase-shift spatiality and a transductive temporality and, in addition, by the capability of producing further transformations in itself and in its own milieu. In Simondon's jargon, the individual is more or less 'metastable':

---

[2] If the entirety of Simondon scholars take for granted that from crystals to higher degrees of complexity one can speak of individuals, this is more problematic concerning the subatomic world. See for instance the discussion with Aspe and Bontems, thus concluded by Stengers quoting Simondon: 'It is unclear whether what we call a critical fissile mass is not an individual as such' (Stengers 2002: 318–19). Simondon is frankly ambiguous on this topic, but I assume consistent with his thought that there are no limits of order or magnitude for individuality: thus with the photon one can have, 'synthesised *in the same being*, and carried by the same carrier, both a structured and an amorphous measure, a pure potential' (I 102, italics added).

[3] In his quick reference Canguilhem links Ruyer (1954) and Simondon (IGPB), both involved in a similar project of re-elaboration and amplification of the cybernetic concept of information.

[4] These 'programmatic texts' (see the *Brief Note* in the appendix) will be seriously taken into account here, just as Garelli (2004) has done, and Barthélémy (2009) suggests. My final thesis concerning the respective dates of their production will result from the analysis carried on in this chapter.

the term 'metastability', derived from thermodynamics, defines a system not on the basis of its stable 'form', but in relation to the potential energy involved in its precarious but still lasting equilibrium.

I will return analytically to each of these points, since the terms Simondon derives from the natural sciences force a considerable effort on the reader in order to reconfigure his philosophical imagination. What I would like to stress at this point is how this 'double' status of the individual – as a structure and as a process – emphasises the crisis of the category of identity. In fact, the concept of a 'metastable system' forces us to reconsider the notion of individual in terms of individuation (i.e. a complex and discontinuous system of processes) and denies any possibility of referring to any identity of being in itself:

> The relation of being to itself is infinitely richer than identity. Identity, a poor relation, is the only relation of being to itself that one can conceive according to a doctrine which considers being as single phased. (I 318)

The alleged identity of being is nothing other than a purely fictive limit case, often philosophically translated with the term 'individual', while the real philosophical operation which Simondon tries to implement is precisely the disjunction of the concepts of individual and identity. Thus the individual can be considered the key term of *Individuation* only if it is radically reassessed in the light of the discoveries of the natural sciences and in particular, as I will explain, of quantum physics. In fact, the classical concept of the individual is absorbed by Simondon into a new concept of the individual developed in light of the notion of a 'metastable system'. The 'stable' individual becomes the impossible limit case of a perfectly static system, the fictive name for a completely accomplished process of individuation, while in actual fact one is always simply witnessing processes which deprive individuals of any fixed identity, since being is 'more than a unity and more than an identity' (I 26). Due to this double meaning, the use of the term 'individual' retains its ambiguity throughout the whole text of *Individuation*, where it primarily refers to the structured part of a process, but is also frequently used to name a system which can be individuated further.

This conception of the individual shapes Simondon's horizon when he is still aiming at a reformulation of the social sciences, countering their epistemology, which remains grounded on the concept of the individual. In fact, despite the pretension in the social sciences to assume the study of structures *and* of processes, according to Simondon they cannot avoid conceiving such processes outside of their interaction with stable, fixed and structured individuals. They are in the end sciences concerned with individuals in relation to other individuals or to the processes going beyond them, and they cannot consider individuals in themselves as processes and as relational structures.

On the contrary, Simondon's 'theory of individuation' seeks to overcome the conceptual deadlock that, firstly, conceives of interactions as only occurring between individuals and, secondly, sees the strict reduction of individuals to the processes out of which they emerge. According to Simondon, two complementary reductionisms are enacted by psychology and sociology: psychology reduces the individual to the ultimate constituent of social processes, while, *symmetrically*, sociology removes any status of reality from them by assuming that individuals are part of the social whole to which they belong, i.e. an 'individual' of higher scale by which they are entirely determined (I 295–96). The epistemological problem Simondon is facing here could be formulated in the following terms: What are the conditions for the possibility of a science of the individual conceived of as an identity neither entirely self-oriented nor strictly hetero-referential? Or: what conceptual apparatus can ground a science of the processes of individuation which constitute metastable systems?

In Simondon's view, a conception of the individual as a metastable system involves a complete overhaul of the methodology of the social sciences, and a task comparable to the one which the natural sciences seemed to be achieving through questioning the ontological status of their object:

> Could we do the same in the social sciences? Could we found a social science [*la Science humaine*] respecting, of course, multiple possibilities of application but having at least one common axiomatic applicable to different areas? (FIP 533)

Thus conceived, a science of the individual is in fact a science of individuation, of systemic relations and processes, which requires a method extended over each domain of individuation, in order to enable the analysis both of the structures and the processes composing a system, integrating synchronic and diachronic issues.

Taking up Simondon's words again, one can conclude that, since '*there can be no science but of the individual*' and that the individual is a structure of simultaneous processes, a science of the individual necessarily entails a philosophy of individuation of the systems which Simondon defines as 'metastable'.

1.2 Metastability, (Non)Identity and (Non)Causality

Although Simondon openly declares his debt towards Norbert Wiener,[5] the term 'metastability' also refers to physics and chemistry. In any case, it defines a condition of equilibrium in complex systems, the stability of which can be easily broken by the intake of a little bit of energy or information and, conversely, needs a continuative and regular energetic support to counter its tendency to entropy. What is important, according to Simondon, is that a 'metastable system' can be 'structurally' defined by an inhomogeneous distribution of potential energy, since it has no other 'substance' than the differential relations constituting it. The assumption that '*every true relation holds the rank of being*' (I 28–29) entails important consequences both on the epistemological and on the ontological level. Not only *is* any kind of

[5] 'With twentieth century industry, our society enters a new *evolutionary* phase or, according to Norbert Wiener's expression, "metastable"' (RPE). As I will explain in Chap. 2, Wiener, one of the fathers of cybernetics, is one of Simondon's main (polemical) references for the criticism of the concept of information.

knowledge, at any level – from perception to concept – essentially a system of relations,[6] but also the objects of knowledge undoubtedly are:

> The physical object is a beam [*faisceau*] of differential relations, and its perception as an individuated being depends on grasping the coherence of such a beam of relations. (I 239)

It is important to notice how 'to grasp' [*saisir*] does not merely refer here to the construction of an order, a mental image, a form – as it happens with *insight* in *Gestalttheorie* – while the physical object would independently have its own stable existence: here for Simondon 'to grasp' means rather to *produce* a relation (i.e. being), precisely beginning with the encounter of a subject-system and an object-system.

Within such a newly generated metastable system full of potentials, the act of perception is therefore part of a complex process of knowledge, as the 'mental image' and any other processes are, which is both logical and ontological.[7] Thus the 'logical' and 'ontological' structures of a system are but limit cases of the real differential structure (which one could call 'mixed', if the word would not presuppose the anteriority of the limit cases to the structure). What really exists is a system in which the process of knowledge, its subject and its object, acquire their partially stable structure. 'Real' is what resists any imaginary simplification and constitutes the basis of scientific knowledge, producing its object as well as its subject, in perfect coherence with the quantum scale model, in which the experimental observation is, in fact, a reconfiguration of a new complex system where both the subject-instrument and the object are included and eventually defined.[8]

As already noted, in the programmatic texts Simondon constantly turns to the concepts of 'structure' and 'operation' (the latter meaning 'process') in order to define such systems, i.e. being *as* relation. In *Allagmatique* [Allagmatics] the term structure delimits the field of 'a systematised set of particular forms of knowledge: astronomy, chemistry, biology' which Simondon calls the 'theory of structures' (A 559). According to Simondon the sciences of structures cannot consider structures as metastable systems, partial and provisional products of the 'operations' constituting them. Thus a 'theory of structures' is by definition unable to explain the operation 'that makes for the appearance of a structure or that modifies it' (A 559), and therefore to understand the becoming of a system. Nevertheless, any attempt to treat the problem of operations separately ends in particularly complicated results, since what Simondon calls 'operation' is – in evident consonance with a Bergsonian matrix – an actual process, inaccessible *as such* to objective knowledge.

---

[6] Simondon claims for instance that sensation (I 258, 313) and concept (I 245) have a relational and differential nature. But, more remarkably, one can refer to the French translation of the term 'cybernetics' as a 'science of relations [*science des relations*]' proposed by J. Loeb in his preface to De Broglie (1951: 1).

[7] About the peculiar conception of the 'cycle of the image' displayed by Simondon's course IMIN, see Sect. 9.1.

[8] Simondon's thought on this topic is strongly indebted to Gaston Bachelard, as clearly underlined by Barthélémy (2009: 230–33).

In the subsection dedicated to the *Théorie de l'acte analogique* [Theory of the Analogical Act] Simondon presents two 'basic intuitions' which should function as paradigms for the explication of the operation: crystallisation and modulation. Simondon's hypothesis is that each crystallisation is in fact a reversed modulation and vice-versa (A 566). This symmetric opposition between the two kinds of processes – which in fact poses a lot of hermeneutic problems – is nevertheless a good starting point to cross two fundamental and complementary themes traversing *Individuation*: the themes of non-deterministic causality and of non-substantial identity.

First of all, in both paradigms of the operation, the cause-effect relation cannot be reduced to a deterministic one. The process of crystallisation, the crystal growth, although displaying a mechanical sequence, begins with an aleatory encounter of the system with the singularity of a seed crystal[9]: an encounter which cannot be strictly reduced to the sequence it triggers, and therefore is not determinable within the system itself. On the other hand, the process of modulation consists of a coupling [*couplage*] of two different systems. Such a process could be considered determinable only at the level of the accomplished (macro)system, where in fact there would be no *emergence* of a new system, but merely the assemblage of two subsystems.

One must at all costs avoid any interpretation of the relations among different scale systems as a kind of Chinese box game culminating into a Nature-whole conceived as a System including all systems, since this is exactly what Simondon explicitly denies when challenging Kurt Goldstein's 'Parmenidean ontology'[10] and asserting his own theory of systems as metastable, phase-shift and 'in state of disparation',[11] therefore incomplete and not entirely determinate. According to

[9] Thus Simondon states that 'in a very remarkable study by M.P. Auger, it is said that a seed crystal can be replaced in certain cases by chance encounters, i.e. by a chance correlation between molecules' (FIP 550). Today we would better distinguish between processes in which the first crystal is introduced from without, already present within the system, or emerging from a random assembling of molecules.

[10] In *Individuation* Simondon attacks what he calls Kurt Goldstein's 'Parmenidean ontology' (I 229). Goldstein's book *The Organism* (in the German original: *Der Aufbau* – the Structure – *des Organismus* 1934) is a Gestaltic approach to organism through a joint study of biology, psychiatry and medicine, which had great relevance for an entire generation of French philosophers during and after the Second World War.

[11] The expressions 'phase-shift' [*déphasé*] and 'in the state of disparation' [*en état de disparation*] have different although strictly correlated meanings: they both refer to states of system related to processes. 'Phase-shift' refers to the actual presence of different phases within the same system, and better explains the system as the outcome of a process of individuation through 'phase-shifting' [*déphasage*] and the subsequent inclusion of previous phases of development within the resulting system. The 'disparation' of a system rather underlines that a system is actually capable of further individuations because of its internal tensions due to the 'disparation' of potentials. In short, a certain dose of disparation is a necessary precondition for 'phase-shifting', while phase-shift might determine a disparation of potentials or might not. While the term 'disparation' and the correlated adjective 'disparate' are used for the purpose of translating the French correlatives, I prefer to maintain the expression 'phase-shift' for both 'déphasé' and 'déphasage', according to the choice of the editors in De Boever et al. (2012). For a more accurate definition of the concepts of 'disparation' and 'phase', see respectively Sects. 2.3 and 3.1.

Simondon 'Nature' conceived as a macro-individual would be the silent and perfectly stable – dead – universe of maximum entropy; on the contrary, we are exclusively concerned with 'non-totalised' systems:

> Systems cannot be *totalised*, since the fact of considering them the sum of their elements spoils the awareness of what actually makes them systems: relative separation of the sets they contain, analogical structure, disparation in general, and the relational activity of information. (I 234, n. 1)[12]

My hypothesis is that the paradigms of modulation and crystallisation work as two different ways of understanding and describing the same processes at different levels, thus delineating a different representation of the individual depending on the level at which it is considered.[13]

In *crystallisation* the individual is understood as *a part* of a process which goes from the encounter between a simple individual (the seed crystal) and a milieu full of potentials (the supersaturated solution) producing a partially individuated system. In Simondon's terms, such an encounter is the trigger [*amorce*] of the system phase-shift into a complex individual (the crystal) and a milieu deprived of potentials (the low-concentration solution). On the contrary, in *modulation* the individual itself is understood as a metastable system, the result of a coupling of initially independent systems and processes, as in Simondon's examples of the moulding of a brick or the changing frequencies of coupled oscillators.[14]

The failed attempt to keep the two sides together in a single conceptual framework, probably pushed Simondon – between the programmatic texts and *Individuation* – towards the unification of the paradigms of crystallisation and modulation, thanks to what he names the 'transductive operation', or 'transductive process' or, more simply, 'transduction' (a concept which, in fact, does not appear in those earlier texts). In *Individuation* Simondon definitively abandoned the assumption that modulation and crystallisation could describe two different kinds of processes, and rather used the two notions to describe different and concurrent aspects of the same processes at different levels. Those who claim that in *Individuation* the paradigm of crystallisation exhausts the significance of transduction by quoting the paragraph *L'individuation comme genèse des formes cristallines à partir d'un état amorphe* [Individuation as a Genesis of Crystalline Forms from an Amorphous State], would end up in quite a bit of difficulty when trying to justify the following statement: 'individuation is a modulation' (I 220). As a matter of fact, in *Individuation* the concept of individuation seems to catalyse all the aporias displayed in Simondon's thought about the changing relations between structure and operation and the possibility of a science of such transformations. This would explain why the term

[12] See also MEOT 61–65, where Simondon proposes the same conception of the relations among the 'sub-sets' of the technical object, conceived as 'the theatre of a number of relations of reciprocal causality'.

[13] As Simondon's himself will admit: 'indeed, the action of the structural germ on a structurable field, in a metastable state that contains potential energy, is a modulation' (FIP 548).

[14] On the peculiar example of oscillators, see Sect. 2.2.

'individual' aporetically oscillates in the text between indicating the individual as a metastable system *and* the individual as the stable structure which results from *and* triggers further ('at the same time the result and the agent' I 191) processes of different orders of magnitude.

Simondon's attempt to provide a unified paradigm for the understanding of individuation will endure at least until the *Colloque* of Royaumont (1962).[15] But, before moving on to discuss that text, I will first analyse the 'transductive' mode of propagation and configuration of stable structures, assuming that Simondon basically maintained a consistent perspective on the nature of the processes implied. According to this perspective, the fact that the cause-effect relation is non-reducible to any deterministic formula (i.e. to the mechanical conception of nature which dates back to early-modern imaginary), entails dramatic consequences for a philosophical critique of substantialism. I will therefore assume that – both in the programmatic texts and in *Individuation* – Simondon aimed for a strictly connected reformulation of the concepts of cause and of individual:

> At the end of this double study [concerning the concepts of modulation and crystallisation], the philosophical notion of *causality* will be enriched and the notion of individual defined. (A 566)

1.3 Transduction, Singularity, Field

In *Individuation* – according to the inspiring methodological paradigm of quantum physics (Barthélémy 2005: 46; 2008: 66) – the process of transduction is defined by a fundamental discontinuity and by reiterated changes of the order of magnitude. On this topic it is worth recalling how Simondon's debt to the physicist Louis De Broglie, although not always evident, is constant and decisive throughout this text.[16] Although referring to microphysics, the discovery of the 'indeterminacy principle'[17] poses philosophical problems concerning not only the theoretical status of classical deterministic physics, but also the status of all sciences related to objects of a different magnitude in which, however invisible and non explicitly described, such factors still produce effects: 'its [microphysics] relevance is not limited to the domain

[15] I will further discuss the question by treating the concept of 'pre-individual' in Chap. 3.

[16] This is quite clear if one considers that, among the only 20 bibliographical references in *Individuation*, three are De Broglie's. According to F. Balibar, in order to criticise the classical conception of the individual, Simondon made explicit De Broglie's implicit philosophical stance concerning the wave-particle duality against Bohr's 'complementarity principle' (Balibar 1995).

[17] Even if the current English translation is 'uncertainty principle', the original term used by Heisenberg was *Unbestimmtheit*, which can also mean 'indeterminacy'. I will use the second term, since it better expresses an ontological lack of determinacy rather than an epistemological uncertainty of knowledge.

of physical sciences, it applies to the sciences studying life, human beings and human societies' (De Broglie 1947: 225).[18]

This indicates the direction of Simondon's quest for 'a foundation of individuation at its different levels' (I 35). As already explained, the concept of transduction aims to describe the processes of the destructuring/restructuring of a 'metastable system', which progressively amplify the singular origin of the process itself, i.e. the encounter between *that* singular structure and *that* field of potentials. Indeed, Simondon's use of the terms 'singular' or 'singularity' is a very restricted one, which exclusively refers to structured individuals when they are the 'germ' or the result of a process triggered from an aleatory encounter. But in a wider sense one could legitimately claim that, as far as such a 'singularity' can be the origin as much as the result of a *transductive* process, the process itself can be considered singular. In this sense I feel consistent with Simondon's philosophy in defining as 'singular' any transductive process.[19]

Thus the concept of transduction serves Simondon's reading of the problem of ontogenesis in terms of processes of individuation which cannot be reduced to any of the terms constituting the determinism/contingency antinomy. According to him, the process of individuation must be explained by referring to determinate structural conditions *and* to undetermined aleatory conditions, thus making the hypothesis of a 'theory of singularities' the possible basis of a unified 'transductive' theory:

> It is possible, in the last instance, to suppose that the theory of singularity can be ascribed neither to the framework of a deterministic physics nor to the framework of an indeterministic physics. The two would rather be considered the particular cases of a new representation of the real that one might call the theory of transductive time or theory of the phases of being. This completely innovative mode of thinking – which conceives determinism and indeterminism as mere limit-cases – can be applied to different domains of reality beyond the one of elementary particles. (I 144)

Each system can therefore be conceived as a 'centre of transductive activity', neither dominated by any superior necessity ('quantic operations seem to show that this operation works through steps and not continuously' I 143), nor characterised by any substantial essence ('the substance is not the model of being anymore' I 32).

From this perspective, as a main model for a unified theory of 'being *as* relation', Simondon often refers to the notion of 'field', 'a gift from social sciences to the sciences of nature' (FIP 538).[20] The fact that, as *Gestalttheorie* claims, the psychic and

---

[18] On the philosophical relevance of early twentieth century microphysics, see in particular Chap. 7 on *Les révélations de la microphysique*, and Chap. 11 on *Hasard et contingence en physique quantique*.

[19] As already explained, the concept of transduction has both a technological and a biological meaning. In both cases it refers to a process of amplification of information that Simondon conceives as endowed with 'a certain degree of indeterminacy' (MEOT 143).

[20] Thus Simondon continues explaining that the notion of field '*establishes a reciprocity of ontological status and operatory modality between the whole and the element*'. In effect, within a field – whether electrical, electromagnetical, of gravity, or of any other kind – the element has a double status and a double function: (1) as it receives the influence of the field, it is submitted to the forces of the field; it is situated at a certain point of the gradient representing the field; (2) it

physical fields are isomorphic, would in fact explain how it is possible to discover in both of them the same forms of organisation, and how the notion of field can be programmatically extended to social sciences. Nevertheless, this conception risks reducing the system to a complex net of causes and effects which, in the end, fails to undermine determinism.[21]

But this is precisely where the concept of transduction shows its strength, authorising Simondon to think individuation as – one could say – a partially aleatory relation. In conclusion – against the modern, deterministic, conception of nature, through and beyond the holistic conception of 'field' elaborated by the *Gestalttheorie* – Simondon understands 'being as relation', through quantum physics, in terms of a beam of transductive processes, the calculability of which is never complete, and the randomness of which is never absolute:

> In conclusion, one can advance an hypothesis analogous to physical quanta and to the relativity of the levels of potential energy [...] According to this hypothesis, it would be possible to *consider all true relations as being, and as developing within a new individuation* [...] This is the conception of being on which this study is grounded: the unity of being is not the identity of a stable state in which no transformation is possible; being is defined by a *transductive unity*, i.e. it can phase-shift in relation to itself, exceed *its own centre.* (I 28–31)

Thus for any process there are both determined conditions of state (i.e. possible effects and impossible ones), and indeterminacy margins excluding any uniform, linear and continuous relation between causes and effects. If processes tend to have a direction due to their irreversibility, nevertheless the actual processes can never be deduced from the initial state of the system.[22] And this, for Simondon, functions at any scale for any possible kind of science. This perspective excludes an exhaustively predictive science. It, on the contrary, necessitates a 'twofold' science, concerned on the one hand with conditions of state and structural tendencies and, on the other, with the ontogenesis of the singular operations of individuation. Simondon

actively intervenes in the field, by modifying its forces and the gradient distributions; one cannot define the gradient of the field omitting to define what appears at that point (FIP 538). When underlining the importance of the 'gift', Simondon is implicitly recalling Kurt Lewin's 'topological' psychology, derived from the physical notion of *field*: 'a totality of coexisting facts which are conceived of as mutually interdependent is called a *field* (Einstein 1933). Psychology has to view the life space, including the person and his environment, as one field' (Lewin 1946: 792. See also Lewin 1935). On Simondon's debt to and criticism of Lewin, see in particular Sects. 5.1 and 6.2.

[21] In his course IPM, Simondon will characterize the 'deterministic age' as the one which postulates the order of Nature is 'uniform, necessary, universal and analytical', i.e. eternal, deterministic, general and reducible to elementary elements. According to him the 'deterministic age' started collapsing at the end of nineteenth century, first attacked by evolutionary biology, then by holistic assumptions based on Maxwell's theory of fields, later integrated by *Gestalttheorie*, Goldstein and Merleau-Ponty (IPM 288–90). However, for Simondon the validity of the 'postulate of isomorphism' is rather to be found in morphogenetical processes (i.e. individuation processes), since the holism of 'form' does not in itself escape a deterministic horizon (IPM 298).

[22] This would correspond to the classical Laplacean definition of determinism: 'We ought then to regard the present state of the universe as the effect of its anterior state and as the cause of the one which is to follow' (Laplace 1814: 4).

refers continuously to such a science in the programmatic texts and in *Individuation* with the peculiar name 'allagmatics' [*allagmatique*]'.[23]

1.4 Allagmatics, Topology and Chronology

Simondon starts from the Bergsonian assumption that objective sciences are sciences of structures, incapable as such of grasping transductive processes. And yet he asserts that a theory of operations he defines as allagmatic, from the Greek *allagè* (change) and *màthema* (knowledge) is possible. In *Allagmatique* Simondon directly challenges the issue, not without incurring the potential risk of inconsistency patently exemplified by contrasting the following quotations:

> *Allagmatics* is the theory of operations. It is, at the level of sciences, symmetrical to the theory of structures, the systematised set of determinate fields of knowledge: astronomy, physics, chemistry, biology. (A 559)
>
> *Allagmatic Theory* is the study of the individuated being. It organises and defines the relation between the theory of operations (applied cybernetics) and the theory of structures (deterministic and analytic science). (A 565)

On the one hand allagmatics is conceived as a theory of operations complementary to the sciences of structures; on the other it is conceived as the study of the individual through the connection between the 'analytical' sciences of structures and the 'analogical' sciences of operations. The patent contradiction could be easily attributed to the schematic and provisional nature of the methodological writings, but I think it is worth going deeper in order to make sense of it and discuss the two possible interpretations it arouses.

In *Individuation* – where the methodological issue of a 'theory of operations' regularly returns – Simondon attacks certain approaches proposing a topological formalisation of processes which would reduce the dimension(s) of time to spatial coordinates. Such approaches can be only partially identified with Kurt Lewin's dynamical topology and René Thom's differential topology.[24] However, Simondon directly attacks what he calls topology:

---

[23] The term recurs in I 48–50; 61–62; 82; 111; 127; 228; 238; 328; NC 506; 523–24; A 558–59.

[24] On Lewin see above n. 20. René Thom's work on differential topology dates to the early 1950s, but Simondon only attended his seminar during the 1980s, possibly hoping to find in his theory of linguistic and biological systems the universalising power he previously had attributed to cybernetics. Thom criticised Simondon, claiming that – due to an inaccurate knowledge of topology – he had failed to develop an adequate analysis of the 'subject of transductive knowledge', and subsequently could not provide a satisfying theory of signification (Thom 1994: 105). On the contrary, Jean Petitot views Thom's morphodynamics and semiophysics among Simondon's most surprising scientific anticipations (Petitot 2004: 104–6). Unfortunately, in his essay Thom – who ironically declared he had read IGPB without understanding it – did not even mention the third part of *Individuation*, where instead Simondon, as I will show, provided a theory of signification which could be actually related to Thom's semiotics (e.g. Thom 1968). However, if it is true that Thom's

> What a topology lacks is the consideration of potentials. Precisely because they are not structures, potentials cannot be represented as graphical elements of the situation. (I 238)

Simondon consistently criticises this point of view in the paragraphs concluding the analysis of each 'domain' of *Individuation*: at the end of physical individuation, in *Topologie, chronologie et ordre de grandeur de l'individuation physique* [Topology, Chronology and Order of Magnitude of Physical Individuation], at the end of biological individuation, in *Topologie et ontogenèse* [Topology and Ontogenesis], and at the end of psychic and collective individuation, where he raises the issues of time and of emotion respectively.[25] He ends each 'movement' with a similar attempt to resume the type of relation which 'structures' and 'operations' entertain within the concerned domain – respectively dedicated to physical, biological or psychic-collective individuation – whether it is formulated as a 'relation between chronology and topology' or as a 'central operational zone'.

I will limit myself here to the paragraph *Topologie, chronologie et ordre de grandeur de l'individuation physique* (I 148–53), since it is – as the whole of physical individuation – paradigmatic, and because I shall refer to the other parts when treating the themes of biological and psychic-collective individuation. For Simondon, topology and chronology – such as structure and operation, discontinuous and continuous, matter and energy – are complementary features of all systems and at the same time complementary ways of understanding the individuation of systems. The physical individual is a 'chrono-topological set, the complex becoming of which is made of subsequent crises of individuation' (I 149),[26] and topology and chronology are directly related to the 'limit cases' of knowledge which determinism and indeterminism properly are: 'determinism and indeterminism are nothing but limit-cases, since there is a becoming of systems: this becoming is the one of their individuation' (I 148).

The problem is both ontological and epistemological, and the solution lies, according to Simondon, in a science of discontinuous processes, in which the coupled-notions related to the opposition between structure and operation inevitably reveal their heuristic limitations:

> From this point of view it seems possible to understand why the opposite representations of continuity and discontinuity, of matter and energy, of structure and operation, can only be used as complementary couples. This is because these notions define the opposed and extreme characteristics of the different realities in which individuation takes place. But the operation of individuation is the active centre of this relation. (I 150–151)

---

topology is 'a structuralism including dynamics' which 'develops, in fact, into a neo-mechanism' (Petitot 1975: 146), then Simondon's criticism of determinism could be easily extended to it.

[25] It is worth noting that in Simondon's original thesis the 'macro-domains' of individuation are only two, since the psychic-collective one is part of the individuation *'au niveau des êtres vivants'*. But this does not foil my argument, because it is true that the descriptions of individuation at every domain *and* sub-domain follow the same pattern. For a detailed explanation of the different structures of the original thesis and its published version, see below n. 28.

[26] 'The individuated physical being is not completely simultaneous to itself; its topology and chronology are separated by a certain gap' (I 149).

In conclusion, this accounts for Simondon's exclusion of the possibility of a purely topological science as well as of a 'pure' science of the operations of individuation; since both concern limit cases, they abstract from the real process and are therefore incapable of giving an adequate account of it. Thus, in *Individuation* allagmatics appears as a 'theory of metastabilities', i.e. of 'the exchange processes between spatial configurations and temporal sequences' (I 238), and no more as a science of operations only.

But the two different kinds of knowledge – of structures or of operations – cannot be considered symmetrical. In *Individuation*, the contradiction which opposed an allagmatic theory of operations to an allagmatic theory of 'exchanges' between structure and operation, seems to be eventually overcome. And nevertheless, challenging the theme of individuation, Simondon presents a theory which, despite treating *both* structure *and* operation, cannot consider the two 'sides' of its object strictly equivalent or symmetrical, if only because the sciences of structures already exist while, on the contrary, sciences of operations do not. In the end, the operational 'side' seems to remain the prevalent one of allagmatics, if we limit ourselves to the text of *Individuation*:

> At the foundations of the ontogenesis of physical individuals, there is a general theory of exchanges and of state modifications, which one might call *allagmatics*. (I 328)

Let's return then to the double hypothesis previously formulated in order to read it in the light of *Individuation*, i.e. of a philosophy in which structures are always conceived in relation to their ontogenesis. According to what Simondon writes in *Allagmatique*, if a theory of operations cannot directly refer to objective domains, as sciences of structures do, it must be concerned with what such sciences leave open as 'gaps' within and between them. In this way Simondon himself states that a science of operations 'can be achieved only if the science of structures experiences, from within, the limits of its own domain' (A 560–561). Thus, the very possibility of an allagmatics is dependent on the prerequisite of a 'systemic theory of structures' (A 561).

In short, Simondon's thesis can be expressed as follows: (1) there have been, since seventeenth century, sciences studying structures; (2) starting from that, it is now possible to build sciences which study ontogenetic processes by linking the epistemological domains those sciences are concerned with (otherwise, we would run the risk of accepting an inadequate representation of the processes from which structures emerge); (3) perhaps it will be possible to build a science of operations, but only after a system of sciences of structures will be produced. Simondon leaves the conclusion hanging in the air by questioning the nature of a fundamental 'theory of operations': should it define and classify 'the great categories of operations, the different kinds of transforming dynamisms that objective studies discover', or should it rather define

> one fundamental kind of operation, from which all particular operations would derive as simpler cases. These different degrees of simplicity would define then a hierarchy, a rigorous principle of classification. (A 559)

It is possible to argue that, by crossing quantum physics with other different fields of scientific research, Simondon was seeking a paradigm for the 'fundamental kind of operation' which could definitely depose 'hylomorphic'[27] substantialism from the throne of the whole of occidental metaphysics, from Aristotle to modern science. In that sense, *Individuation* can be interpreted as an attempt to extend – through the notions of metastability and transduction – such revolutionary power to every domain of being; however, the implementation of a fundamental model always remained open and problematic. Simondon's oeuvre is in fact characterised by the persistent experimentation of conceptual invention, a philosophical 'operation' he always intended to be grounded on the technical and scientific research of his time:

> Such a representation of being necessitates a conceptual reforming which depends on a radical revision of some basic schemas, some new paradigms which would replace the hylomorphic schema imposed by culture [...] We have tried to derive a paradigm from physical sciences. (I 319)

Thus, the genuinely philosophical performance of *Individuation* lies in the attempt to enable different 'schemas' – modulation, crystallisation, phase, metastability, transduction, and many other conceptual tools – for the analysis of the different domains of being, of their structural conditions and operational status: tools for defining the thresholds between different domains in order to make a problem of them, rather than to fix them. Simondon's concepts, in fact, do not define any separate 'realms' – matter, living beings, psyche, society – traversed by individuals or any kind of substances of which individuals would be composed. On the contrary, they indicate 'phases', processes, whose dynamic composition continuously constitutes and modifies the configuration of individuals, as it happens within a magnetic and gravitational field, in which different forces and processes constitute an irregular and unstable space, full of potentials, which can modify or be modified by whatever – matter or energy – becomes part of it. No ontological guarantee, then, of a stable and secure domain, and no science capable of defining the specific processes characterising a domain, is possible without a prior enquiry of the singular ontogenesis and functioning of the different structures which constitute it.

It becomes clear then that Simondon's philosophy of individuation is based on the critical analysis of the results of the 'structural' sciences, aiming to criticize rather than to confirm the alleged 'identity' of their objects, thus reactivating the ontogenetic hypothesis in order to discover the actual tensions which render each structure metastable, i.e. an individual undergoing individuation. This explains the path followed by his analysis: the two original sections of *Individuation* display the process of individuation through three different subsections (physical, biological, psychic and collective); and each of them concludes – as already said – with the

[27] About Simondon's criticism of the Aristotelian notion of 'hylomorphism', see Sect. 2.1.

demonstration of the insufficiency of any structural and topological definition of the individual which would not adequately consider the problem of time.[28]

The same pattern brings us to the same question: is a knowledge of 'operations' possible? The answer should be: yes, but only as a surplus knowledge always characterising the ontogenesis of structure in relation to structure itself. Sciences of structures, then, must be conduced to reveal the irreducible singularity of the structures they are concerned with. Once the individual is understood as such a structure, one can eventually make sense of Simondon's above mentioned statement:

> According to the doctrine I am going to present [...] *there can be science only of the individual*, this would be the epistemological consequence of this enquiry. (AI 553)

We can finally choose between the two different definitions of allagmatics Simondon put forward in *Allagmatique* and in *Individuation*. The first ('allagmatics is the theory of operations') appears to be provisional: correct, since what the sciences of structures fail to take into account is precisely the aleatory factors of ontogenesis, but insufficient because it leaves to the imagination a 'pure' theory of operations. Instead, a 'science of the individual', conceived as a science of processes of individuation, is a theory of thresholds and transitions between structures, the knowledge of which *presupposes* a science of structures but, necessarily, in the direction of a science of the ontogenesis of these structures. For these reasons we are compelled to choose the definition of allagmatics Simondon offers at the conclusion of *Individuation*: 'a general theory of exchanges and modifications of states' (I 328).

As it should be apparent, my theoretical reconstruction entails a hypothetical anteriority of the programmatic texts in relation to *Individuation*. Although not dated, in my hypothesis they express Simondon's need for an overall view on the project of *Individuation* before or during its elaboration.[29] Their theoretical conclusion is that the epistemological conditions for the possibility of a 'theory of operations' resides in the amplification of the allagmatic approach to every field of knowledge, through the aforementioned paradigms of crystallisation and modulation:

---

[28] *Individuation* was originally divided into two sections, the first concerned *L'individuation au niveau physique* [Individuation at the Physical Level], the second *L'individuation au niveau des êtres vivants* [Individuation at the Level of Living Beings]. The second section had three subsections, respectively devoted to *Information et ontogenèse: l'individuation vitale* [Individuation and Ontogenesis: Vital Individuation] *L'individuation psychique* [Psychic Individuation] and *Fondements du transindividuel* [Foundations of the Transindividual]. From here on I will speak of the two *sections* (and the relative subsections) when referring to the original partition, and to the three *parts* when referring to the commonly accepted tripartition (physic, biological, psychic-collective). Simondon's original thesis can be consulted at the *Archives de Georges Canguilhem* (Canguilhem GC: 40.2.1).

[29] They could also be a further revision of its outcomes, in view of the 'general theory of social sciences' Simondon exposes in his lecture at the *Société Philosophique* in 1960. However, in this case the absence of the concept of 'transductive operation', central in Simondon's lecture (FIP 531), could hardly be explained.

We still have to clarify how the act of crystallisation and modulation are intertwined in the becoming of physical, biological, psychological and social systems. This will be the function of the *allagmatic hypothesis on the nature of becoming*. (A 566)

According to his previous project, in *Individuation* Simondon conducted his quest for an universal key which – on the basis of a new conception of the individual *and* of causality – would translate this whole series of paradigmatic metamorphoses into what he considered at that moment the more accurate and innovative methodological tool: the concept of information.

References[30]

Balibar, F. 1995. Note sur le Chapitre III, I Partie. In *L'individu et sa genèse physico-biologique*, ed. Simondon, G. 1995, 249–252. Grenoble: Million.

Bardin, A. 2015. On substances and causes again: simondon's philosophy of individuation and the critique of the metaphysical roots of determinism. In *Morphogenesis and individuation*, ed. A. Sarti et al. Cham: Springer International Publishing.

Barthélémy, J.-H. 2005. *Penser la connaissance et la technique après Simondon*. Paris: L'Harmattan.

Barthélémy, J.-H. 2008. *Simondon ou l'encyclopédisme génétique*. Paris: PUF.

Barthélémy, J.-H. 2009. D'une rencontre fertile de Bergson et Bachelard: l'ontologie génétique de Simondon. In *Bachelard & Bergson, continuité et discontinuité?* ed. F. Worms and J.-J. Wunenburger, 223–238. Paris: PUF.

Canguilhem, G. 1943. *Le normal et le pathologique*. Paris: PUF, 1966 (English trans. *The Normal and the Pathological*. New York: Zone Books, 1998; *First edition: Essai sur quelques problèmes concernant le normal et le pathologique*. Paris: Les Belles Lettres).

Canguilhem, G. (GC). *Archives de Georges Canguilhem*. Paris: CAPHES, ENS.

De Boever, A., et al. (eds.). 2012. *Gilbert Simondon. Being and technology*. Edinburgh: Edinburgh University Press.

De Broglie, L. 1947. *Physique et microphysique*. Paris: Albin Michel.

De Broglie, L. 1951. La cybernétique. Théorie du signal et de l'information. (With a preface of J. Loeb). *Revue d'optique théorique et instrumental*. Paris.

Einstein, A. 1933. *On the method of theoretical physics*. New York: Oxford University Press.

Garelli, J. 2004. *De l'entité à l'événement*. Paris/Milano: Mimesis.

Goldstein, K. 1934. *Der Aufbau des Organismus*. Haag: Nijhoff (English trans. *The organism: A holistic approach to biology derived from pathological data in man*. New York: Zone Books, 1995).

Laplace, P.S. 1814. *Essai philosophique sur les probabilités*, 1986. Paris: Bourgois.

Lewin, K. 1935. *Principles of topological psychology*. New York: McGraw-Hill.

Lewin, K. 1946. Chapter 16: Behavior and development as a function of the total situation. In *Manual of child psychology*, ed. L. Carmichael, 791–844. New York/London: Wiley.

Petitot, J. 1975. Identité et catastrophes (Topologie de la différence). In *L'identité. Séminaire interdisciplinaire dirigé par Claude Lévi-Strauss professeur au Collège de France 1974-75*, ed. Lévi-Strauss, C. 1983, 109–156. Paris: PUF.

Petitot, J. 2004. Elogio della modernità. Il reincanto tecnico-scientifico del mondo nel pensiero di Gilbert Simondon. In *Realismo, illuminismo ed ermeneutica: percorsi della ricerca filosofica attuale*, ed. F. Minazzi, 89–107. Milano: FrancoAngeli.

[30] *Simondon's complete bibliography and a list of abbreviations are provided in the* Appendix.

Ruyer, R. 1954. *La cybernétique et l'origine de l'information*. Paris: Flammarion.
Stengers, I. 2002. Comment hériter de Simondon? In *Gilbert Simondon: une pensée opérative*, ed. Roux, J. 2002, 300–323. Saint-Etienne: Publications de l'Université de Saint-Etienne.
Thom, R. 1968. Topologie et signification. *L'âge de la science* 4: 1–24.
Thom, R. 1994. Morphologie et individuation. In *Gilbert Simondon, Une pensée de l'individuation et de la technique, Bibliothèque du Collège international de philosophie*, 100–112. Paris: Albin Michel.

Chapter 2
Reforming the Concepts of Form and Information

Simondon inherited the concept of information from one of the fathers of cybernetics, Norbert Wiener. Although the structure of DNA was discovered in 1953, during the 1950s 'information' was not yet considered *the* fundamental paradigm for biology – and certainly not in France.[1] However, from the beginning cybernetics conceived the technological concept of information as a paradigm which could be at least in principle extended to all the fields of scientific research: biology, psychology, psychopathology, sociology and political economy. In that period Simondon was adopting it as a key methodological tool for his philosophy, as the entry *La psychologie moderne* [Modern Psychology], edited by Simondon for the *Encyclopédie de la Pléiade* [ENC] shortly before writing *Individuation*, clearly shows:

> The language of cybernetics, already applicable to nervous system physiology, could prove to be suitable for describing the relations between the human being and his natural and social milieu, overcoming the alternative between liberty and determinism, which seems to be the major obstacle for any psychological science. (ENC 1701)[2]

---

[1] Even Watson, Crick and Wilkins did not win the Nobel prize for the discovery of the structure of DNA before 1962. Born in the USA in the 1940s, during the 1950s cybernetics started to spread in France too. The conference *Les machines à calculer et la pensée humaine* of January 8th 1951 marks the initiation of a European audience to cybernetics (Guchet 2001: 231, n. 3; see also Guchet 2005, who quotes Simondon's unpublished texts on cybernetics; Geoghegan and Hayward 2012: 4–8). It is worth noticing that already in 1950 a series of conferences organised by L. De Broglie had taken place, later published as *La cybernétique. Théorie du signal et de l'information* (1951), and the text appears in the bibliography of *Individuation*. The astonishingly short bibliography of *Individuation* (Canguilhem GC: 40.2.1) presents only twenty references, four of which were on cybernetics, eight on quantum physics, three on biology, and five on the human psyche. These texts are particularly important, because they are the only 'official' table of Simondon's non philosophical sources for the book. In my interpretation I shall make substantial use of them.

[2] Simondon wrote the text in collaboration with F. Le Terrier. A letter he sent to Canguilhem in January 14th 1989, just a month before his death, not only gives some evidence of his mental illness at the time, but also testifies the value he attributed to this text about which he was still asking for advice from his former master (Canguilhem GC: 40.2.2). In France the relationship between social sciences and cybernetics was becoming quite *à la page*, thanks to the merging of structural

A. Bardin, *Epistemology and Political Philosophy in Gilbert Simondon*,
Philosophy of Engineering and Technology 19, DOI 10.1007/978-94-017-9831-0_2

After the two theses,[3] terms like ‘transduction’ ‘allagmatics’ or even ‘individuation’ almost disappeared from Simondon’s vocabulary while, throughout his entire intellectual life, he never abandoned the term ‘information’. From this perspective *Individuation* can be considered an experimental work in progress in which all the problems Simondon had previously discussed, converge. The notion of information, because of its ‘purely operative character, not linked to a specific matter, and defining itself only in relation to an energetic and structural regime’ (I 220), perfectly responds to the above expressed exigency of an allagmatic theory, which therefore ‘must be in relation to the theory of information, which concerns the translation of temporal sequences into spatial organisations’ (I 238).

But, in order to reach an adequate elaboration of the concept, a few problems related to the notion of ‘form’ had to be settled:

> Cut off from the hylomorphic schema, the notion of form can become adequate to the polyphasic character of being structuring itself in a relational way. This accords with the research direction of theoreticians of Form. This relational meaning of form is better grasped through the notion of information, provided that one understands the concept of information as the relational signification of a disparation. (I 318)

In this chapter I shall follow the route traced by Simondon: after displaying his double criticism to the concept of form – both Aristotelian and *Gestalt*-like –, I will delve into what he conceived as a ‘reform’ of the cybernetic concept of information.

2.1 Criticism of the ‘Hylomorphic’ Concept of Form (*Gestalttheorie*)

The first conceptual enemy Simondon chooses to challenge in *Individuation* has a venerable name: hylomorphism. The Aristotelian ‘hylomorphic schema’ – he says – has prevented an ontogenetic approach to the question of being, and subsequently to knowledge, by maintaining a latent but undisputed dominion over both common sense and philosophical and scientific thought:

> The meaning of the present research is that in order to think individuation the hylomorphic schema must be abandoned [… as it] abusively replaces the knowledge of the genesis of a real; [that is,] it prevents knowledge of *ontogenesis*. (I 312)

The ‘hylomorphic schema’ derives from an ordinary conception of the technical operation as the shaping of some formless matter. This conception substantialises

---

linguistics, anthropology and psychoanalysis (in particular through the works of Lévi-Strauss and Lacan). The texts to which Simondon frequently refers are of course Wiener (1948) and (1950), plus the famous ‘Macy conferences’ held in New York from ‘46 to ‘53 (Pias 1946–53; three of them appear in the bibliography of *Individuation* and four in MEOT’s). The theme is also dominant in the text Simondon seems to rely on in order to build his argument: De Broglie (1951). As I will explain in Sect. 2.3, Simondon intends to answer the questions posed by Raymond Ruyer in *La cybernétique et l’origine de l’information* (1954).

[3] As said, *Individuation* and MEOT were Simondon’s *two* PhD theses (see p. 1, n. 1).

matter and form, hiding their constant becoming: it presents them as already individuated at the very beginning of the process of formation. Simondon develops a double objection. On the one hand, no inert and amorphous matter is present in nature: as it is evident in the case of wood venation or stone stratification, what we call matter always presents the implicit results of an earlier formation, and therefore is always partially individuated. On the other hand, no accomplished form exists in nature, neither is it a perfect idea in the artisan's mind: it is instead an operative sequence, a complex process with a determinate history and, in this sense, a characterising 'form'. In short, rather than correctly representing the actual dynamics involved in the technical operation, the hylomorphic conception of matter and form is definitely unfit to describe any real and singular process of 'formation' [*prise de forme*].

Simondon provides a sociological hypothesis concerning the historical success of the hylomorphic schema:

> If only the individual and technical operation existed, the hylomorphic schema could not emerge [...] What the hylomorphic schema first of all shows, is a socialised representation of work and an equally socialised representation of the individual living being. (I 50–51)

He develops part of his argument referring to the different relationship established by the master [*maître*] and the artisan [*artisan*] in regards to the technical operation. The master's abstract relation is that of property, while the artisan's is the concrete immersion in matter through the technical process of its (trans)formation. This 'evocation' of the master–slave dialectic is less a Marxian debt than a Hegelian reference to the abstraction of the master's knowledge in front of the artisan's ability to grasp the singularity, the 'implicit forms' of worked matter (I 57–60). But, in the end, this sociological hypothesis is insufficient:

> The psycho-social conditioning of thought, even if it can explain the vicissitudes of the hylomorphic schema, cannot at all explain its permanence and its universality within reflection. (I 52)

Thus, he concludes, the problem can be solved only at a deeper level: the level of the physical analysis of the 'process of formation', i.e. of individuation.

In fact, according to Simondon, only a theory of individuation can give adequate reason for the structural inherence of the hylomorphic paradigm to knowledge, since the operation of knowledge itself is an operation of individuation which – *as such* – works according to the hylomorphic schema. In fact, knowledge *normally* proceeds through binary oppositions of symmetrically polarised terms, instituting and rendering compatible couples of clear and distinct ideas enclosing (and thus hiding) their relation. Against this tendency Simondon attempts to grasp being in its relational and active centre, or 'central operative zone' [*zone opérative centrale*], as he calls it, acknowledging that the *milieu* of a relation cannot be considered less important than its limit cases (I 313).[4] In this sense, in *Individuation* hylomorphism

---

[4] On the contrary, 'The hylomorphic schema entails and accepts an obscure zone: the central operational zone. It is the example and the model of all logical processes through which one attributes a key role to limit-cases, to extreme terms of a reality organised as a series' (I 312).

becomes a synonym of a 'substantialist dualism' which causes knowledge to be knowledge of individuals, instead of knowledge of the processes of individuation.

Although *Gestalttheorie* gave some useful indications for an alternative approach to the 'central operative zone', it failed to subtract the Aristotelian concept of form from its subjection to the dominant philosophical and scientific tradition. According to Simondon, *Gestalttheorie*'s limits derive from the 'psychologism' implicit in its central hypothesis, the stability of the 'good form', which prevents a valid application of it in the different fields of knowledge. In the psychology of perception the law of 'good form' (or *Prägnanz*) should explain the definition and stability of the figure: the way it imposes itself on attention and thus to perception through a dialectical relation with the background (hierarchical superiority), and its permanence in memory. Now, after denying the validity of the 'good form' hypothesis also in the psychology of perception,[5] Simondon states that – in general – the hierarchical superiority of a form and its stability can neither *ontologically nor logically* coincide. Let's see how he proceeds.

His argument is based on an original conception and evaluation of systemic stability. For Simondon true stability characterizes systems deprived of potentials and therefore incapable of any further transformation. Such systems are difficult to understand precisely because of their high degree of stability, since knowledge – according to the thesis Bachelard (1951) derived from quantum physics – requires a perturbation of the system. In short, the stability of a system prevents its knowledge, therefore duration entails inferior evidence of form (*Prägnanz*). On the contrary, the superior evidence of form derives from the fact that the system is full of potentials, thus still becoming, and therefore capable of becoming involved in the further processes of formation, including knowledge processes. In this sense, by conceiving stability as the fixation of a 'good form' into a long lasting, clear identity, *Gestalttheorie* ends up presenting as the genesis of a 'good form' what is in fact a process of slow degradation (i.e. reduction of potentials) into a system the main feature of which is a long lasting sterility (FIP 540–41).

In other terms, what the concept of form lacks is precisely the possibility of conceiving the actual metastability of systems, their tendency towards producing transductive amplification, rather than (apparently) ensuring a long duration with no effects. For this reason Simondon turns his attention to the emergent concept of information, since it allows for the understanding of 'formation' as a process concerning a dynamical system. Also for cybernetics a system (whether physical, biological, social) is a complex system, each element of which is related to the others and with the system as a whole, but it is also characterised by self-regulatory processes. Thus the system is conceived as permanently active, its equilibrium 'dynamic' rather than 'stable'.

---

[5] According to Simondon, *Gestalttheorie*'s 'static' conception of form fails to explain the dynamic character of the background and the differential nature of the figure: in fact, only a stimulus variation, not its 'good form', can produce information (I 236).

2.2 Criticism of the 'Technological' Concept of Information (Cybernetics)

If the notion of 'form' is conceived in terms of identity and 'structure', 'information' instead can be conceived in terms of a differential relation and 'operation'. Thus Simondon feels confident that a critical enquiry into the concept of information could supply a paradigm to direct his own quest for a science of the relations between structure and operation. Indeed, he thinks that the notion of information elaborated by cybernetics in connection with the concept of homeostasis,[6] remains insufficient in explaining the operating of complex systems, and therefore it must also be reformulated.

The cybernetic paradigm for the understanding of information is derived from engineering problems related to cable communication technologies, such as the telegraph or telephone. The basic schema consists of a linear energetic exchange between a Sender and a Receiver connected by a channel through which low potential energy transports information. Such a schema has different technological fields of application and is extendible to biology and society. But what is necessary for the process to take place, is the presence of the same code in the Sender and in the Receiver. The code ensures that the initial and final information really *is* the same. In other words, the identity of the code preserves the identity of the piece of information going through the whole process, from the Sender to the Receiver. In addition, the elementary process is complicated by a *feedback* cycle in which the roles of Sender and Receiver are inverted.

Let us look at Simondon's basic example, however, since it aims to effectuate a change of paradigm. Two electronic oscillators[7] with different frequencies, if close enough for their magnetic fields to overlap, end up stabilising their frequencies on a value which corresponds to the magnetic field which results from their merging. In the proposed example there is neither an 'ontological' nor 'logical' identification of a system-Sender and a system-Receiver, since the two systems A and B actually fulfil both functions. Furthermore, there is no univocal transmission, nor a one-to-one correspondence (as it happens in a *feedback* cycle) between the systems, but rather we have a concurrent reciprocal influence, and therefore a macro-system composed by A, B and their interaction. Thus we have a newly constituted macro-system where the difference between the frequencies of the two sub-systems originates as an information flux which modifies both of them, and therefore the macro-system itself from within. In fact, from the moment in which a relation between the two oscillators is set (and their fields overlap), the unique differential relation between the frequencies is a single signal which generates two different

[6] The concept of 'homeostasis' refers to the tendency of some systems (notably organisms) to maintain stable functioning and constant properties. The notions of *homeostasis* and *entropy* play a central role in Simondon's argument against cybernetics. For a wider discussion of the topic, see Chap. 7.

[7] The example of oscillators recurs frequently in Simondon's writings, e.g. I 222–24, MEOT 134–37 and, notably linked to the notion of 'field', FIP 534, 539.

pieces of information in A and B, according to their respective frequencies, thus determining different modifications within them. The process goes on until the two frequencies of A and B coincide, and a system in dynamic equilibrium will be structured.

Simondon's example has the merit of highlighting what the schema derived from the 'cybernetic paradigm' has the tendency to hide: the dual oscillators schema, in fact, subverts a few classical assumptions which, according to Simondon, still 'infect' the cybernetic concept of information.

1. *Active/Passive*. There is such a perfect reciprocity between the Sender and the Receiver that, logically speaking, it is impossible to differentiate the two functions. Furthermore, since there are no isolated linear sequences in the systemic relations, one should not speak of feedback mechanism, but rather of a *simultaneity* of transmission-reception.
2. *Internal/External*. In an oscillator what is internal (oscillation) and what is external (magnetic field) are regimes of functioning which correspond to each other and influence one another. Thus it is not possible to conceive the second as the effect of the first, nor vice versa.
3. *Information/Relation*. From the moment the process starts (when the two fields overlap) it no longer makes sense to distinguish the relation between the two systems and the circulating information, since information *is* precisely the (differential) relation between the two oscillations, i.e. what drives the sequence during which information progressively emerges and the relations between the systems progressively change.

In Simondon's example, what results particularly questioned is the nature of the code. In cybernetics' technological paradigm the codes of Sender and Receiver must coincide in order to allow a correct exchange of information, which is a process independent of the code permanently inscribed in the system's structure. On the contrary, for Simondon the code and functioning of the system depend on each other. Therefore, on the one hand the functioning of a system according to the code entails an emission of signals which can be transformed into different information by other systems and, on the other hand, each signal which actually modifies the operating of a system in fact can modify its code. In short, the code is both producer-of and produced-by information exchange, i.e. it can generate and be modified by signals. And this explains how systems with completely different codes can in principle (and they actually do) communicate, such as human being and machine or machine and animal, but also a human being and virus, orchid and wasp.[8]

[8] The last examples are in fact to be ascribed to Deleuze rather than to Simondon. But also Simondon, as I will explain, is particularly concerned with code in organisms: 'the content becomes the code', 'the living being transforms information into forms, *a posteriori* into *a priori*; but this *a priori* is always oriented towards the reception of information' (MEOT 123, 137). A probable common reference – through Canguilhem 1952: 144 ff. – is J. von Uexküll (1934), the German ethologist who provides the well-known example of the *milieu* of the tick (Sect. 9.4).

As a result, Simondon's example of oscillators – although not completely flawless – enables him to shift towards a relational and non-deterministic point of view, focusing on the way information exchange continuously modifies the relations between systems and *therefore* their identity. According to Simondon, the cybernetic conception of information, affected by its technological origins, proves to be tied to a double fetishism of 'identity' and 'determinism', a symptom of which is the confusion between signal and information. Of course, the transmitted energy has not only a quantity, but also a form, a 'quality' derived from its frequency and tension, or just from its distribution in time – as happens with the Morse code. Of course, the signal *is* this energy modulated in order to be converted into something else, such as the possible beginning of a procedure (if received by a machine) or a meaning (if received by a human being) (I 221–222). But the signal *is not* to be considered information, unless it encounters and modifies a system (or a subsystem) with a proper code. Therefore one should not properly call 'information' what emerges from the natural expression of a code, but exclusively what produces the *interruption* in the continuity of communication processes, a crisis in the self-regulatory functioning of systems, and can trigger, after all, the structural reconfiguration of the system.

On the basis of this conceptual disjunction of signal and information, Simondon attacks the contradiction between the ordering function and the operational efficacy cybernetics attributes to the signal. For Simondon dynamic *order* depends on the transmission of *signals* expressed by the code for the normal functioning of the system, while *efficacy* concerns the disorganising impact of new *information* on the same functioning. They are two radically different processes – the first deterministic, the second partially aleatory – which must not be confused. On the contrary, by identifying signal and information (I 224), cybernetics reduces information exchange to a unique deterministic process which leaves substantially untouched the identity of the systems involved, reducing them to subsets of the macro-system they are supposed to entirely depend on.

2.3 Reforming the Concept of Information

It is now possible to understand how Simondon can 'reform' the concept of information both in terms of systems' metastability and of processes' transductivity. As I will show, this theoretical framework allows him to avoid the cybernetic assimilation of information and negentropy, and subsequently solve the problem of the origin of information which was posed – precisely against cybernetics – by Raymond Ruyer in *La cybernétique et l'origine de l'information* [Cybernetics and the Origin of Information] (1954).

On the one hand, a system is not variable by just any signal, its changing is submitted to differential conditions of possibility or 'disparation' (in fact the ideal condition of information exchange corresponds to a 'relative maximum' of 'disparation', a threshold over which there would be no relation at all). This entails the abolition of the ontological distinction Sender/Receiver, as much as the abolition of the Aristotelian distinction Form/Matter. These distinctions are no longer valid, since the system conditions do not depend on the supposedly 'stable' initial condition of the Receiver on which the metastability of the Sender would produce its effects. In fact, also the Receiver's metastability is needed for the information exchange to take place: 'the metastability of the receiver is the condition of efficacy of the actual information' (API 159).

On the other hand the production or exchange of information cannot be the necessary outcome of processes which could be entirely calculated on the basis of the initial metastability conditions of the two coupled systems (Sender and Receiver). The shape of a system resulting from processes of information exchange can be only approximately foreseen, as such processes are transductive, i.e. discontinuous. And the more the system is phase-shift, the less forecasting is possible, because the relation among different phases of different systems follows different rhythms and modalities. That is why, for instance, the development of a social system – which is made of physical, biological and psychic-collective phases functioning according to different regimes of individuation and communicating among them at different levels – is highly unpredictable.

Furthermore it is important to notice that information can be treated as a process both internal and external to the system, since for Simondon there is no difference in considering the exchange of information as a relation between systems or – at a larger scale – as an internal relation between different parts of a system: 'there's only information when what emits signals and what receives them form a system' (I 223, n. 30). Now, to point out that in any system there is *always* an internal exchange of information between different scales, Simondon speaks of the 'internal resonance' of systems as an actual condition of their functioning.[9] By 'internal resonance' he means, in fact, the discontinuous relations between different parts of a system which produces quantic structural changes and therefore prevents any determined knowledge of 'the becoming of this system according to a theory of continuity or to the laws of great numbers, as thermodynamic does' (I 148–49). This leads him to conceive the different individuals-systems as closely connected through processes of energetic exchange which simply happen through the mediation of the respective oscillations (FIP 532).

Consistently with this theoretical framework Simondon rejects the cybernetic equation information = negentropy, presented by Wiener.[10] In Wiener's terms, *information* is the unit of measurement of order, the contrary of which is *entropy*, as the unit of measurement of disorder: it follows that information is by definition negentropic, i.e. opposed to the system's energetic process of degradation (Wiener 1950: 28 ff.). From the same engineering example, Simondon arrives at the opposite

[9] In physics the 'internal resonance' of a system is the progressive widening of its oscillation, due to the application of an external force with compatible frequency. Simondon's usage of the concept is wider, including the actual functioning of any system. For a deeper discussion of the scale problems entailed by the concept of 'internal resonance', see Sect. 4.4.

[10] Wiener's expression is 'negative entropy', later abbreviated as 'negentropy'.

conclusion. To transmit information it is necessary to input some energy (a signal) into the system. Now, in order to avoid signal degradation and improve the transmission of information, two paths can be followed: on the one hand one can increase the signal energy (thus increasing the total amount of energy in the system), on the other one can decrease the background noise. In the second case, through a diminution of the total amount of energy in the system one improves the transmission of information, thanks to a different *distribution* of energy within the system. What is fundamental to note is that, in this case, a diminution of energy increases order (I 222–23).

For Simondon this is enough to prove that there is no constant mathematical relation (direct or inverse) between the quantity of energy input into a system and the quantity of information transmitted. On the contrary, it is the actual distribution of energy in a system, i.e. its 'form' or 'quality', which determines the quantity of information that can be transmitted. Simondon speaks also of the *ecceitas* of information (I 223), but in conclusion he rejects all terms incapable of expressing the 'relational attitude' of a system. What actually produces/transmits information by differentiating information from background noise, is in fact the relation between the code and an energetic variation. The singularity of this encounter can be reduced neither to structured form, nor to pure chance:

> Information is halfway between pure chance and absolute regularity [...] information is not a kind of form, neither a set of forms, it is the variability of forms, the intake of a variation upon a given form. It is the unpredictability of a variation, not the pure unpredictability of any variation. We shall distinguish three terms, then: pure chance, form and information. (MEOT 137)

According to Simondon, Wiener's identification of information and 'negentropic order' must therefore be rejected, since it explains information only in quantitative terms, hiding its relational, differential value. He claims there is *no* univocal relation between information and energy, since the *quantity* of information actually transmitted depends also on the relation between that quantity and the 'form' of energy, the asymmetrical distribution of potentials within a metastable system. In short, information is relatively independent of the calculus of the quantity of energy present in a system, and its transmission is the result of a differential relation between systems or parts of a system, which cannot be expressed by a scalar measure.

In Simondon's intentions, this constitutes also the solution to Ruyer's question on the origins of information. Ruyer underlines that the postulates of cybernetics can explain how information circulates, but not, in general, the way it emerges:

> The paradox clearly results from two of Wiener's theses. The first states that informational machines [*machines à information*] cannot increase information [...] the second that brain and nervous systems are informational machines [...] let us combine the two theses: it becomes impossible to conceive the origin of information. (Ruyer 1954: 13)

In order to overcome the determinism of cybernetics, Ruyer introduces the indeterminacy issues of microphysics into the chinks made on classical physics by the entropic evidence discovered with thermodynamics:

> Despite its unquestionably "modern" spirit, cybernetics exclusively borrows concepts from classical physics, not from microphysics [...] thermodynamics, although deterministic in its postulates, has been compelled by technical reasons to pose the question of origins. (Ruyer 1954: 25–26)

Now, precisely because Simondon will draw on the schema of Ruyer's criticism, it is necessary to underline immediately what clearly differentiates the two positions: Ruyer calls 'consciousness' the operation which orders a domain, thus generating information. Only consciousness, 'anti-causality' *par excellence* (127), can give a form to a structure, i.e. transform it into signification, information (11). His entire discourse aims to demonstrate how a fundamental mechanistic approach compels cybernetics, in order to keep an internal coherence, to be involved in a sort of dialectical antinomy (a quite classical one, indeed) which would reveal as genuinely original what was supposed to be explained at be beginning of the argument, i.e. consciousness (136). Ruyer's assumption eventually becomes explicit when he not only uses the concept of organism to explain the elementary features of matter, but he also goes so far as to expand the phenomenological paradigm of the 'absolute overview' [*survol*] – the precedence of consciousness over microphysical systems.[11]

Simondon's perspective is completely different, and in a way it reduces the problem of the origins of information to a false one. As explained above, the transmission of information does not necessarily entail any 'external' intervention (neither a physical operator – whether a human being or a machine – nor 'consciousness') to introduce a supplementary piece of information in the system. In fact, due to their constitutive disparation and metastability, systems continuously emit signals which *can* be converted into information, *if* only they encounter another metastable system with a 'compatible' code. Of course, this works for any kind of information exchanges among systems (physical, biological or social), since the actual encounter of partial indeterminacies of different systems is what really originates information, independently of the typology, scale and regime of their operating.

In this sense Simondon claims that not just any signal emitted by the Sender is information, but only the one which 'comes through the test', i.e. enters a structuring relation with the code (here 'form') of the Receiver, thus being implemented in its functioning:

> One can distinguish the *signal* transmitted, the *form* through which the signal is received by the receiver, and the *information* properly named, which becomes actually integrated in the functioning of the receptor after the test of disparation carried on the extrinsic signal and the intrinsic form. (I 224)

[11] 'Also an organism partially functions *according* to its structure. But its structuring is manifestly not an operation depending on a existing structure [...] in this sense the fundamental beings of microphysics resemble organisms' (139). In the domain of microphysics 'where the individuality of the constituents is partially scattered within the individuality of the system, experience reveals similar behaviours to those induced, in the psychic-organic individualities, by the existence of fields of consciousness of absolute overview [*à survol absolu*]' (139–40). On Ruyer's project of expanding the concept of form-structure, in order to overcome the opposition determinism/contingency, see Ruyer (1930).

If the program of cybernetics consists in expanding a technological paradigm to the biological and social systems, Simondon's attempt seems rather the opposite. He aims to expand a biological and/or psycho-social paradigm of communication to the physical and technological fields, relying on what quantum physics associated to thermodynamics allows him to think, i.e. the quantic nature of all systems and the non-deterministic characterisation of all processes, against the essentially *deterministic nature* of cybernetics' concept of information.

This partly explains the fact that in *Individuation* Simondon develops his criticism against the 'technological concept of information' precisely when closing the part devoted to the individuation of living beings, at the threshold of the psychic-collective individuation. The study of the organism, in fact, highlights many problems related to the identity of the 'code' in the relation between the organism and its environment, which is consistently read by Simondon as a transmission of information *within* an individual-milieu system. Similarly, the psyche-body relation shows at a macroscopic level its relative metastability within the larger system of society. The consequence is that human communication itself – if not highly formalised, as happens in logic and mathematics (but these are limit-cases, not the essence of communication) – cannot be explained in terms of information exchange between stable systems with stable codes.

Thus, the 'reformed' concept of information is consistent with: (a) the 'oscillating' structure of the individual as a metastable system, the 'internal resonance' of which derives from quantum gaps which keep the system in tension (I 330); (b) the discontinuous transductive operation, which is relatively non-deterministic, i.e. dependent on incalculable events based on calculable conditions of state. Finally, it appears to solve a problem within Simondon's research: the 'allagmatic' problem of explaining how 'structure' and 'operation' can be repeatedly converted one into another, making of this 'conversion' the central core of information exchange itself. Simondon's theory of information is thus intended to be 'non probabilistic' and 'non deterministic' (FIP 549–50), and therefore apt to describe metastable systems by highlighting what in their transductive-operational functioning exceeds any 'coded' homeostatic functioning: this surplus is precisely the indefinite re-emergence of information within the systems which, at the same time, it *constitutes*. And this is true at any scale one would consider the individual as a system.

2.4 Royaumont: All the Paradigms of Information

In July 1962 the prestigious *Colloque international de philosophie* traditionally held at the Royaumont Abbey was dedicated to *Le concept d'information dans la science contemporaine* [The Concept of Information in Contemporary Science]. Simondon was not only responsible for introducing Wiener's paper on *L'homme et la machine* [Man and Machine], but was also indeed, in the words of Martial Gueroult, 'the soul of the conference' (RO 157). Many of his interventions offer evidence of his attempts to orient the discussion towards questions he was particularly concerned with. One cannot but notice, for instance, the 'traces' of his allagmatic theory in what Wiener recalls, by referring to the previous day's discussion 'with a small group', on 'how to transform the function into a structure and vice-versa' (RO 131).[12]

Simondon's paper concerned *L'amplification dans les processus d'information* [Amplification in Information Processes] and was still focused, perhaps for the last time, on the research of a fundamental paradigm. However, when editing the acts of the conference, Simondon decided to substitute it with a brief abstract, in which he intended to summarize 'continuous modulation' and 'discontinuous transduction' in a unique paradigm of biological derivation which he calls 'organizing amplification'.[13] It is worth quoting it in full:

> There are three main typologies of amplification: transductive propagation, modulation, organisation. The first does not have limits in itself; it is discontinuous, proceeds by all-or-nothing and does not entail gradation; it is irreversible; its energetic performance is quite high. The second, which is continuous and progressive, presupposes a reduction of the energetic performance of the system; it corresponds to the operation of technical modulators used for treating the information signal. Lastly, organisation, which is manifest in biological processes, is a synthesis of the former two; it corresponds to a quantum regime and functions through consecutive waves, mainly during growth processes. The three typologies provide paradigms for the understanding of complex situations. They share the primordial condition of any process of information: the existence of a metastable state and of a quasi-system [*quasi-système*] capable of effectively receiving an incident signal which modifies the equilibrium of the system rich in potential energy. (RO 417)

Although a few years had passed, Simondon's paper at Royaumont was still inspired by the same goal: provide a paradigmatic unification of scientific research. As shown, *Allagmatique* was the attempt to derive fundamental paradigms from the processes of modulation and crystallisation, while in *Individuation* the same problem was solved with a partial convergence of the two in explaining the process of individuation. The Royaumont paper clearly represented for Simondon a further occasion to reformulate an old problem. The theory of information, such as he elaborated it in *Individuation* as a relation between metastable systems and the 'incident signal', underwent no substantial modifications at Royaumont, where it provided the common basis for the three paradigms of 'amplification'.[14]

---

[12] Wiener continues: 'I think it is worth considering the relation between structure and function by means of a general theory of synthesis and analysis of machines' (RO 131). Among the participants at the conference we can find scientists (N. Wiener, B. Mandelbrot, D. MacKay, A. Lwoff, L. Couffignal) and philosophers (J. Hyppolite, L. Goldmann, G. Granger, L. Sebag, M. Gueroult). Gueroult (president of the Royaumont International Philosophy Conferences Committee) declares he expects from the conference a contribution to the regeneration of Cartesian philosophy. In his paper *Is Information Theory still Useful?* the mathematician and IBM researcher Mandelbrot argues against the hypothesis of a future unification of sciences and the excessive publicity of a theory which, according to him, has already exhausted its 'historical function' (RO 98).

[13] The entire paper has been recently issued as API, but Simondon did not publish it during his lifetime.

[14] The concept of a 'transductive' amplification is used here by Simondon to cover the semantic field of 'crystallisation' as the counterpart of modulation. It is in a way the model of the primitive

In conclusion, at Royaumont Simondon still adopted the same perspective of unification for *all* sciences, definitively focusing on the field of 'human sciences'[15] through a 'biological' reform of the technological concept of information. This constant tendency in Simondon's thought not only is confirmed by his notes to the Royaumont paper (174–76), it is also evident in what he states when closing the conference:

> The idea of organising this conference derives from the fact that the notion of information originally elaborated within the fields of some exact sciences and of the technology for submarine cable transmission, has now some *fringes*. It is now used out of its original context, sometimes metaphorically, sometimes abusively. However, what the borrowing [*emprunt*] clearly shows is the presence of a need. The usage for an emerging function pre-exists the fully formed instrument. Put differently, we wished we could demonstrate – starting from a usage which is perhaps abusive but in fact reveals an actual tendency – a possible research path towards the widening of the notions of information and organisation, starting from the awareness [*prise de conscience*] of existing needs in exact sciences and, probably, in less exact sciences such as the social sciences which are now organising themselves. What we tried to do to be precise is to generalize this notion of information. (RO 157)

The whole argument refers to the lack of legitimation characterising social sciences, which was in that period a central issue *also* concerning the epistemological status of exact or 'hard' sciences and the debated nature of life sciences. Even if after Royaumont Simondon abandoned the issue, concentrating on his academic career and focusing on the teaching of psychology and of technics, in his speculation he always kept questioning the political significance of a science of society inspired by biology and technology. But, before tackling this topic, it is worth delving again into *Individuation*, where Simondon's attempt to merge different paradigms was still haunted by a notion which seemed to concentrate all the unsolved problems concerning the relation between structure and operation: the notion of 'pre-individual'. This notion, in fact, depends on a profound intellectual debt, which Simondon revealed at Royaumont by precisely summarising – in his own jargon – the sense of the whole conference:

---

amplification generating structures: 'transduction is precisely capable of creating structures starting from a homogeneous metastable milieu' (API 174). In *Individuation* Simondon already referred to 'a process of amplifying communication, the most primitive modality of which is transduction, which already exists in physical individuation' (I 33, n. 10). Although the notion of 'amplification' often recurs in Simondon's texts, it will never gain the epistemological centrality which characterizes the notion of transduction in *Individuation*. The same applies to the concept of 'organising amplification', although still present, for instance, in the course *Formes et niveaux de l'information* (1970–71), where Simondon proposes again the three typologies of amplification (Bontems 2006: 323).

[15] It is worth noting the French use of the expression '*sciences humaines*', which approximately corresponds to the English 'social sciences', with the premise that it bears a major importance to the theory – and an implicit contraposition – of the natural and human domains. In what follows, if not strictly necessary, I will keep the English expression. The reader must therefore assume that, in the quotations, the expression 'social sciences' corresponds to the French 'human sciences'.

> In order to explain how difficulties and possible encounters can emerge, it is necessary to go back to the ontogenesis of this conference [...] The notion of *fringes* to the concept of information was suggested one year ago by the late Merleau-Ponty, precisely when we were organizing this conference. (RO 157–58)[16]

References[17]

Bachelard, G. 1951. *L'activité rationaliste de la physique contemporaine*. Paris: PUF.

Bontems, V. 2006. Encyclopédisme et crise de la culture. *Revue philosophique de la France et de l'Étranger* "Gilbert Simondon" 2006/3: 311–324.

Canguilhem, G. (GC). *Archives de Georges Canguilhem*. Paris: CAPHES, ENS.

Canguilhem, G. 1952. *La connaissance de la vie*. Paris: Vrin (English trans. *Knowledge of life*. New York: Ordham University Press, 2008).

De Broglie, L. 1951. La cybernétique. Théorie du signal et de l'information. (With a preface of J. Loeb). *Revue d'optique théorique et instrumental*. Paris.

Geoghegan, B.D., and M. Hayward. 2012. Catching up with Simondon. *SubStance* 129(41/3): 3–15.

Guchet, X. 2001. Théorie du lien social, technologie et philosophie: Simondon lecteur de Merleau-Ponty. *Les Études Philosophiques* 50(2): 219–237.

Guchet, X. 2005. Simondon, la cybernétique et les sciences humaines. Genèse de l'ontologie simondonienne dans deux manuscrits sur la cybernétique. *Chiasmi International 7: Merleau-Ponty. Vie et Individuation*: 187–206.

Pias, C. (ed.). 1946–53. *Cybernetics – Kybernetik: The Macy-Conferences 1946–1953*, 2003. Zürich: Diaphanes.

Ruyer, R. 1930. *Esquisse d'une théorie de la structure*. Paris: Alcan.

Ruyer, R. 1954. *La cybernétique et l'origine de l'information*. Paris: Flammarion.

Von Uexküll, J. 1934. *Streifzüge durch die Umwellen von Tieren und Menschen – Bedeutungslehre*. Berlin: Springer (French trans. *Mondes animaux et monde humain. Suivi de La théorie de la signification*. Paris: Denoël, 1965).

Wiener, N. 1948. *Cybernetics: Or control of communication in the animal and the machine*. New York: John Wiley & Sons.

Wiener, N. 1950. *The human use of human beings. Cybernetics and Society*. London: Sphere Books, 1968.

---

[16] Merleau-Ponty died a few months before the conference, in May 1961. Three years later Simondon dedicated IGPB to Merleau-Ponty. Direct references to the latter are quite rare in Simondon's works (e.g. in CSP) and indeed not particularly relevant. I shall not measure here Simondon's unquestionable debt towards Merleau-Ponty as I will limit myself to the use of some of his theses in order to challenge a few problems in Simondon's philosophy.

[17] *Simondon's complete bibliography and a list of abbreviations are provided in the* Appendix.

Chapter 3
The Object of a Philosophy of Individuation

While in the first section of *Individuation* Simondon derived from the hard sciences all the paradigms he needed to found his philosophy of individuation, in the second he traversed the whole field of the individuation of living beings.[1] There, in the subsection concerning vital individuation, Simondon established the notion of information as a point of methodological convergence for his project. In the other two subsections, he elaborated the premises for a philosophy of psychic and collective individuation. Before moving to the second section of Simondon's original plan, it is worth clarifying the object and the method of Simondon's philosophy of individuation, which I shall do by tackling the problem of the 'pre-individual'.

The concept of the pre-individual Simondon develops in *Individuation* is both an outcome of his ontological approach inspired by quantum physics and the mark of a persisting debt to Merleau-Ponty's phenomenology, in particular to the courses on *Nature* he delivered while Simondon was writing *Individuation*. In this sense the choice of the term 'pre-individual' is quite revealing: on the one hand it indicates the attempt to abandon the theme of 'perception' as an alleged solution for the problem of the transcendental horizon, on the other hand it entails the reformulation – not the disappearance, in fact – of the problem of the subject.[2] As the analysis of the debate following the lecture Simondon delivered at the *Société française de philosophie* will show, this perspective allows us to measure how far Simondon's theory of information is influenced by the phenomenological concepts of perception, sense, and consciousness.

[1] On the original partition of *Individuation* in two sections, see Chap. 1, n. 28.

[2] In underlining Simondon's debt towards Merleau-Ponty, I do not mean to reduce the former to an epigone of the latter, but rather to grasp his phenomenological background *together with* the originality of his philosophy of individuation. However, following what Descombes says about Merleau-Ponty ('to connect thing and consciousness it was necessary to write a philosophy of nature' Descombes 1979: 73), the hypothesis that *Individuation* could be the continuation of a legacy is at very least to be taken seriously.

A. Bardin, *Epistemology and Political Philosophy in Gilbert Simondon*,
Philosophy of Engineering and Technology 19, DOI 10.1007/978-94-017-9831-0_3

3.1 The Pre-individual. Quantum Discontinuity and Phases of Being

Simondon's epistemological enquiry into what he calls 'pre-individual' is developed at length in the introduction and conclusions of *Individuation*. There, the concept of the pre-individual catalyses a whole series of problems that the diverse attempts for classifying and defining different typologies of 'operations' had left unsolved.

The introduction is an attack against the philosophies which put the knowledge of the individual before the knowledge of individuation, thus contributing to 'obscure' ontogenesis. This is what atomism and hylomorphism did in classical philosophy. Atom, form and matter are the keywords of a thought based on the principle of individuation [*principium individuationis*], a conceptual tool which precisely raises the problems it was meant to solve. In fact, atom, form and matter are the products of individuation processes that remain unexplained: in atomism and in hylomorphism individuation is rather considered the (individual) 'thing to be explained' than the very process which would explain the emergence of the individual. On the contrary, it is necessary to '*know the individual through individuation rather than individuation through the individual*'. And this means to transform a search for the definitive principle of individuation into the study of what is '*only ontogenesis*', i.e. the non individuated reality which accompanies individuation (I 24). This non individuated reality is what Simondon calls the 'pre-individual'.

The notion of the pre-individual can be derived from a twofold definition:

> The individual would be thus conceived as a relative reality, a phase of being presupposing a pre-individual reality, and which, also after individuation, does not exist on its own. In effect, individuation does not exhaust the potentials of pre-individual reality, and what emerges from individuation is not simply the individual, but the couple individual-milieu. (I 24–25)

In short, the pre-individual is both the reality preceding the genesis of the individual and, *at the same time*, the milieu full of potentials 'associated' to the individual, once the latter has been generated. Here Simondon's problem is to avoid reducing the pre-individual to a 'part' of the resulting system: in other words he is pushed back to question the operation which constitutes metastable systems.

As usual, his strategy is to integrate into his philosophy models elaborated in fields of scientific research where the attack on substantialism seems stronger. In this case the notion of 'phase' serves the purpose: 'ontogenesis is the theory of the phases of being' (I 284).[3] The term 'phase' is widely used in scientific jargon: from the physics of waves to chemical-physics and astronomy. Simondon, as already seen, is often concerned with electronic wave examples, in which one can observe phenomena of constructive/destructive interference due to a phase-shift. However, it is mainly 'crystallisation' which represents in *Individuation* the notion of 'phase': in a saturated solution one can speak of different phases (solid and liquid, for instance) which can succeed one another, but can also coexist, and, under certain

[3] See especially I 321–32.

system conditions, it is the random presence of a seed crystal which can determine the passage (even partial) from one phase to another.

According to this model the concept of 'phase-shift' takes on a double meaning in *Individuation*. On the one hand it refers to a *succession* of states, a process which at times gives the idea of an 'evolution' from the physical to the biological until the psycho-social 'phases'. On the other hand, following Simondon's claim that 'the existence of phases of being cannot be understood as a simple sequence' (I 323), the term phase-shift refers to the simultaneous presence of multiple tendencies – not necessarily harmonised – which render the system metastable (Hottois 1993: Chap. 6).[4] In short, the physical notion of phase serves again to undermine a substantialist representation of the individual, conceiving it as developing and simultaneously crossed by different and divergent processes which are a part of it as well as its own phases. Thus each individuation can be considered the solution of problems posed by a previous phase-shift, and each 'solution' entails a change of scale, determining the emergence of a further process of individuation and a new individual rich in tensions, 'phase-shift' in relation to its pre-individual milieu. Thus Simondon can state that the individual, as a 'moment' in a process of individuation, is constituted by 'stages of stability jumping from one structure to another' (I 327).

In the paragraph *Topologie, chronologie et ordre de grandeur de l'individuation physique* [Topology, Chronology and Order of Magnitude of Physical Individuation] – the only part of the chapter on *Forme et Substance* [Form and Substance] to survive the editing process at PUF for the first edition of IGPB[5] – the quantum paradigm is still central. The pre-individual is defined there as an 'original reality' [*réalité première*], a source of both 'ontogenesis' and of 'operation':

> The oppositions between continuity and discontinuity, particle and energy, would therefore express not complementary aspects of the real, but rather the different dimensions emerging in the real when it individuates. The complementarity at the level of individuated reality would result from the fact that individuation appears on the one hand as *ontogenesis* and on the other hand as the *operation* of a pre-individual reality which not only produces the individual, the model of substance, but also the energy, or field associated with it. Only the couple individual-associated field can therefore explain the pre-individual (I 149, italics added)

By differentiating 'ontogenesis' and 'operation' Simondon presents here two ways of conceiving individuation: as a process the partial result of which is the structured individual ('individuated'), and as a process the complete result of which is the system individual-milieu. In fact, although it is 'pre-physical and pre-vital', nevertheless the pre-individual only reveals itself in a (partially) individuated system.

---

[4] 'As a phase we do not mean a temporal moment replaced by another [...] in a system of phases there is a relation of equilibrium and reciprocal tensions; the complete reality is the actual system of all phases, not each phase in itself' (MEOT 159). On the phrase 'phase-shift', see above Chap. 1, n. 11.

[5] It was the last part (out of 10) of Sect. 1.3 of the original thesis: in fact the only part of that chapter to be published (as Sect. 2.4) in IGPB (now in I 148–53). In the first edition of IGPB in 1964, not only were the subsections devoted to psychic and collective individuation omitted, but also the section concerning physical individuation was considerably cut, and this chapter was the only one to survive the editing of the third part. See the *Brief Note* on the editorial vicissitudes of *Individuation* (Appendix to this volume).

In this sense Simondon is allowed to state that 'one can consider being as a mixed set of individuated reality and pre-individual reality' (I 317), the operating regime of which varies according to the 'degree' of individuation. The physical field is characterised by an high degree of individuation and therefore of determinacy (for this reason physical and chemical sciences present considerable results in the study of the cause-effect relation). Biological individuation emerges 'interrupting' physical individuation, it is – more precisely – the 'slowing down' [*ralentissement*] of physical individuation (I 152)[6] which makes of living beings individuals whose functioning presents a higher level of indeterminacy. Finally, psychic individuation is a further 'slowing down of the individuation of the living being' (I 165) correlated to a complementary improvement of indeterminacy, which opens a path for collective individuation.

One can notice that, at each level, the more the individual is linked to its (pre-individual) milieu, the more it remains 'open' to a non-deterministic functioning. That is because 'after individuation being *has a past* and the pre-individual becomes a phase' (I 320) and nevertheless it remains 'pre-individual', i.e. the 'non-structured' phase of a system individual-milieu. In fact, the pre-individual is named 'phase' here in a quite different way compared to the other 'phases' of the individual, since the term rather indicates the individual's relation to his milieu, the energetic source for further phase-shift. Thus the individual 'reflects' 'the development, the regime and finally the modalities' of the operation of individuation from which it derives (I 24), and its pre-individual 'phase', as an 'associated milieu', maintaining the same regime and modality of functioning which characterised its ontogenesis.

Again, as far as he rejects the reductionist hypothesis which entirely consigns to fundamental physics the keys of ontology *tout court*,[7] Simondon must solve the problem of the ontological status of the kind of operation which the pre-individual would be 'in itself'. In fact, in its 'purest' sense the pre-individual is '*the being in which there are no phases*' (I 25), or it can be said to be a phase *sui generis*, which structures itself through phase-shift, i.e. ceasing to be itself. Now, since it has no structure, no theory of the phases of being can ever produce an adequate (structured) knowledge of it. Thus, if 'the pre-individual being is more-than-one [*plus qu'une unité*]' and the principles of 'identity' and the 'excluded third' do not apply to it (I 25), how can Simondon think it is possible to have adequate knowledge of it? Once more Simondon looks for a possible solution in quantum physics, where, according to him, field theory and corpuscular theory, although still partially dualist, '*are moving towards a theory of the pre-individual*':

---

[6] See also I 319, n. 4. To represent this schema of development Simondon refers to the notion of 'neoteny', which he seems to consider only metaphorically (see I 152 e I 324 for the expression *néoténiser*). In biology 'neoteny' or 'juvenilization' refers to the retention by adults of morphological and physiological features typical of previous developing phases. For an interesting attempt to link the concepts of neoteny and metastability, highlighting Simondon's Canguilhemian heritage, see Morizot (2011).

[7] 'Physics does not display the existence of a pre-individual reality, but it shows the existence of different individualised geneses starting from determinate state conditions' (I 327).

> [The new theories of quanta and of wave mechanics] might be conceived as *two ways of expressing the pre-individual* [...] Below continuity and discontinuity, there is the quantic and the complementary metastable, the more-than-one, which is the true pre-individual. (I 27)

In my opinion, the analysis of the concept of the pre-individual from the point of view of quantum physics reveals a genuinely Kantian epistemological problem haunting Simondon's approach. In fact, he defines the 'true pre-individual' towards which physics would *converge* as a 'unity doubling itself in aspects which *appear to us* complementary while *in themselves are* coupled in the continual and transductive unity of the intermediate being which we name here internal resonance' (I 151, italics added). The pre-individual is a proper object of knowledge only through its 'manifestations', since we perceive 'the dimension of the real rather than the real itself', i.e. we can grasp 'its chronology and the topology of individuation without being able to grasp the pre-individual real which underlies such a transformation' (I 151).

Precisely because he had assumed the radical impossibility of a 'pure' science of operations (against the Bergsonian possibility of direct intuition),[8] in the conclusions to *Individuation*, Simondon is compelled to go back again to his 'hypothesis of a pre-individual state of being' in order to justify it. What he tries to do there is to differentiate two levels, of 'latent and real potentials' and of 'structural and functioning actuality', both traversed by the *one and only pre-individual*, a 'pure omnipresent potential' existing before and after individuation (I 318). In short, Simondon concludes his main work describing the pre-individual on the one hand as 'being without phases' and on the other hand as 'monophased', thus posing the problem of a primordial [*originaire*] state of being through the logic of the *après coup* (I 320).

This obsessive compulsion – at each level of the analysis of individuation – to question the primordial state of being, is the mark of a heritage which appears in Simondon's 'philosophy of nature' through the veil of the undeniably aporetical concept of the pre-individual: the phenomenological legacy.

3.2 The Phenomenological Legacy: Nature and Sense in Merleau-Ponty

Considering Merleau-Ponty's *La structure du comportement* [The Structure of Behaviour] (1942) and the *Phénoménologie de la perception* [Phenomenology of Perception] (1945) – both published during the years of Simondon's intellectual growth – Simondon's thesis might appear to be the possible continuation of his master's trajectory towards questioning the subject and dissolving the phenomenological centrality of consciousness.

---

[8] As I will show, when Simondon admits the possibility of an intuitive knowledge, intuition does not exclude the concept (Sect. 4.3).

In *La structure du comportement* Merleau-Ponty recognises that the notion of form has succeeded in imposing itself only through experimental evidence against the substantialist ontology of classical physics, since it 'denies individuality' in the sense that classical physics affirmed (Merleau-Ponty 1942: 148). But *Gestalttheorie*, although intending to go beyond the antinomies of substantialism, eventually relapses, precisely due to an inadequate analysis of the notion of form, to which Merleau-Ponty prefers that of 'structure'. Structure would be the 'philosophical truth' both of naturalism and of realism. With this move he intends to overcome the duality of structure and sense – i.e. to unify the objective and subjective fields – in order to bring them back to the true focal point of philosophical reflexion: perception.[9] A few years later, in the *Preface* to the *Phénoménologie de la Perception* he defines phenomenology as an 'exact science' concerning essences, connected to what Husserl in his *Meditations* called 'genetic phenomenology' or 'constructive phenomenology' (Merleau-Ponty 1945: I). This science he definitively founds on 'perception', the common origin of the act of knowledge and its object. Perception is defined by a mixed status, it is a background preceding both subject and object as their condition of possibility, both consciousness *and* nature: a 'more fundamental' logos (Merleau-Ponty 1945: 419) in which the sky 'thinks itself within me' (Merleau-Ponty 1945: 248).

The concept will go through a series of metamorphoses during the 1950s, one of which will be *Le concept de nature* [The Concept of Nature].[10] In fact, the 1956–57 course was for Merleau-Ponty an analytical progression through the sciences of matter, life and culture with the aim of establishing 'the philosophical significance of the concept of nature'. Taking into account contemporary biology, and in particular Von Uexküll's ethology, he intended to revive the anti-anthropocentric power implicit in the ancient concept of *physis*. The problem was thus posed:

> Can we validly assume the notion of nature? Is not it something other than the product of a history [...] Nature is the primordial – that is, the non-constructed, the non-instituted [...] Nature is an enigmatic object, an object that is not an object at all; it is not really set out in front of us. It is our soil [*sol*] – not what is in front of us, facing us, but rather, that which carries us. (Merleau-Ponty 1956–60: 19–20)

But how should one conceive Nature's functioning, which is so far from the mechanistic picture we inherited from seventeenth century? One must notice, first

[9] *La structure du comportement*, in fact, already questioned the concept of perception. See in particular the third chapter, divided into three parts (concerning the physical, biological and human fields) converging towards perception.

[10] This is the title given to the text, the notes and the summaries [*résumées*] of three courses held at *Collège de France*: *Le concept de nature* (1956-57); *Le concept de nature. L'animalité, le corps humain, passage à la culture* (1957-58); *Le concept de nature. Nature et Logos: le corps humain* (1959-60). Since I am focusing on Simondon's problematic relationship with his master's philosophy before completing *Individuation* (1958), I will not refer to Merleau-Ponty's writings subsequent to the courses on nature of 1956-57 and 1957-58, and last of all to his posthumous *Le visible et l'invisible*. However, if a word is to be said, it is worth stressing that the concept of 'nature' can be inscribed in a long history of revisiting the same problem, the last modulation of which will be the concept of *chair* (Mancini 1987: 299).

of all, how Merleau-Ponty characterises organisms and animal societies one year later, in his second course on *Le concept de nature. L'animalité, le corps humain, passage à la culture* [The Concept of Nature: Animality, The Human body and the Passage to Culture], a course he held while Simondon was still completing *Individuation*:

> Organisms and animal societies do not depend on 'all or nothing' laws, but rather on unstable dynamical equilibriums. (Merleau-Ponty 1952–60: 136–37)

The consonance with Simondon's concepts of transduction, metastability and the pre-individual is quite recognisable. Merleau-Ponty and Simondon seem to converge: they give philosophy the task of questioning Nature as a 'primordial being' which is not yet a subject nor an object, thus giving voice to 'complete reality'. In this sense in *Individuation* Simondon will show his fidelity to the project and formulas of his master's '*entre-deux*' philosophy:

> The true first philosophy is neither a philosophy of the subject nor a philosophy of the object. Neither is it the philosophy of a God nor of a Nature investigated through the principles of transcendence or of immanence. It is, on the contrary, the philosophy of a real that precedes individuation, a real which can be found neither in the objectivised object nor in the subjectivised subject, but rather in the fine line between the individual and what is outside it, according to a suspended mediation between immanence and transcendence. (I 269–70)

The hypothesis could be advanced that in *Individuation* Simondon extended Merleau-Ponty's biological paradigm to matter itself, following the path opened by the Greek verb *phyo-*, 'which alludes to the vegetative' (Merleau-Ponty 1956–60: 19). But this approach would hide the true divergence between the two thinkers, which one could resume as follows: Simondon's fidelity towards Merleau-Ponty's project cannot evade a radical criticism of 'the subjectivity implicit in all conceptions of the individual, physical or biological, in the current doctrines' (I 321).[11] His constant reluctance to attribute to the term 'perception' the status of a philosophical principle, results quite evidently precisely in the *Course sur la perception* [Course on Perception] (1964–65), where Simondon reduces the alleged philosophical priority of the concept to a historical-genealogical matrix:

> Among the different forms of knowledge and belief, perception has actually gained a privileged position since the birth of occidental philosophy [...] in a way, the dawn of Greek philosophy coincides with the unreserved choice of perception as the unique source of knowledge [...] this choice is neither spontaneous, nor naive or primitive; it has been made possible by the transcultural characterisation of the Ionic poleis. (CSP 6)

Refusing the name of perception to what is 'primordial', Simondon is looking for a new perspective, independent to the phenomenological one.[12] A clear mark of this process of differentiation is the treatment of the problem of the emergence of 'sense'. Let me clarify this.

---

[11] The statement is clearly directed against phenomenology.

[12] What I am claiming is in contrast with Barbaras' assumption that Simondon's course is part of a phenomenological analysis of perception as 'the source and norm of the different modalities of our relations with the world, however complex and far from perception they be' (Barbaras 2005: XVI).

When Merleau-Ponty refers to perception, he aims to found the 'I think' on the 'I perceive', a 'living experience' from which the speaking subject is necessarily excluded:

> It is true that we should never talk about anything if we were limited to talking about those experiences with which we coincide, since speech is already a separation [...] however, the primary meaning of discourse is to be found in that text of experience which it is trying to communicate. (Merleau-Ponty 1945: 388)

In his intent this should be an exit from idealism, but it is clear that, when a 'text of experience' is given, it makes no difference whether its principle is a subjective consciousness or a more primordial relation preceding both subject and object. The alternative simply presents two different *forms* of idealism: the ingenuous one, according to which the subject would integrally write the 'text of experience' by himself, and the absolute one, according to which 'being' (or whatever else) writes – through the subject – the same 'text'. Both assumptions in fact presuppose the identity of being and sense: whether a product or a 'producer' of the subject, being would *coincide* here with sense (Descombes 1979: 83–86 and n. 21). In this light Merleau-Ponty's courses on the concept of nature are not a departure from that path. There he not only directly connects *physis* to sense ('There is nature wherever there is a life that has meaning, but where, however, there is not thought: hence the kinship with the vegetative'), but rather he identifies nature with the emergence of sense itself: 'Nature is what has a meaning [*sense*] without this meaning being posited by thought: it is the self-production of a meaning' (Merleau-Ponty 1956–60: 19).

Since the 1940s Merleau-Ponty's philosophy had the merit of shifting the phenomenological theme of the transcendental horizon from the issue of the subject-consciousness to that of the subject-world relation, conceived not in terms of representation but in terms of perception, i.e. an activity situated – so to speak – on the borders between interiority and exteriority. Also the subdivision of the third chapter of Merleau-Ponty's *La structure du comportement* into the three parts (*Ordre physique, ordre vital, ordre social*) clearly corresponds to the structure of *Individuation*. Despite this, from Simondon's perspective even the more advanced of Merleau-Ponty's attempts cannot avoid the presupposition of 'sense'. While Merleau-Ponty's analysis is based on the renewal of the *Gestaltic* concept of form in order to 'understand matter, life and spirit as three orders of signification' (Merleau-Ponty 1942: 147), Simondon, on the contrary, renews the concept of information in order to relegate perception, signification and sense to the domain of psychic and collective individuation, making them *depend* on ontogenesis itself and thus abandoning the phenomenological hierarchy still implicit in the choice of the term 'perception'. It is not by chance that in *Individuation* the word chosen by Simondon to name 'being' before the reciprocal institution of object and subject, world and consciousness, is in fact 'pre-individual'. As aforementioned, the term 'pre-individual' is neither equivalent, nor does it refer to perception: it entails no reference to 'sense', experience or anything else, although it could be possibly compared to Merleau-Ponty's late concept of 'raw being'.

Fortunately enough, Merleau-Ponty himself provided a clear differentiation of his philosophy from Simondon's. In some working notes, which are the one and only reference to Simondon in his entire work, Merleau-Ponty writes:

> Simondon's point of view is trans-perceptive: perception is for him on the order of the inter-individual, unable to account for the true collective – There is something true here [...] We do not constantly perceive, perception is not coextensive with our life – Nevertheless, one no longer knows what one is talking about if one *places oneself* in the meta-perceptual [...] For my part, the philosophy of brute (or perceptive) being takes us out of the Cartesian *cogito*, of Sartrean intersubjectivity [...] but for it, the nexus [*foyer*] remains the perceptive field, insofar as it contains everything: nature and history. Simply, instead of saying: to be perceived and perception, I should rather say: brute or wild being and "foundation" (*Stiftung*). (Merleau-Ponty 1959: 42)

Simondon, in fact, will try to differentiate the 'text of experience' the subject contributes to write (form, sense), from what operates without being a text (information), being rather its – non transcendental – condition of possibility. That is why 'perception', in *Individuation* (I 233 ff.) and everywhere else, is treated by Simondon as a psychological issue, and the subject can never be awarded a privileged relation with the 'active centre' or the 'central operative zone' of a system, as it happens, on the contrary, with the 'central sector' of which Merleau-Ponty speaks (Merleau-Ponty 1942).

In short, posing the 'phenomenological' problem of the pre-individual in terms of information allows Simondon to make a double move. Through the concept of the 'pre-individual' he detaches the question on the origin of sense from the perspective of the transcendental horizon, even from the one identifying sense and *physis*.[13] With the reformulation of the concept of information he tries to explain the emergence of sense through processes in which neither any human subject nor any 'consciousness' need be necessarily involved.

3.3 The Debate at the *Société Française de Philosophie*

On February 27th 1960 Gilbert Simondon, young Professor at the university of Poitiers, was invited to lecture at the prestigious *Société française de philosophie* – a common ritual in the Parisian philosophical scene. The lecture *Forme, Information et Potentiels* [Form, Information and Potentials] was intended to summarise his philosophical research concerning the axiomatisation of the social sciences.[14] The audience counted among its ranks excellent names, including J. Wahl, J. Hyppolite, P. Ricœur,

[13] About Simondon bending the Ionian concept of *physis* towards the notion of pre-individual, see also MEOT 203 and HNI 339–40.

[14] Initially published in the *Bulletin* of the *Société*, posthumously added as the second part of the introduction to IPC in 1989, the text can be found also in the Millon edition of *Individuation*, unfortunately still deprived of the debate we are going to analyse here. I shall therefore refer to the paper as FIP, giving the page numbers of *Individuation*, while the subsequent discussion will be referred to as FIPD, giving the page numbers of the *Bulletin* pdf version (unfortunately with different page numbering), which is now available in the official site of the *Société française de philosophie*.

G. Marcel and the president G. Berger. At that time the 'philosophical trend' had not yet moved from phenomenology (and existentialism) towards the emergent structuralism, and in the debate following his speech Simondon found himself defending a position perceived as almost heretical within the phenomenological entourage. His statements were quite concise, evidencing what in his writings sometimes remained in the background, i.e. his views on the problems of 'sense', 'language' and 'subject'. The exchanges with Ricœur and Hyppolite were quite emblematic of that, as was the way Berger addressed his criticisms to Simondon's questioning of the problem of consciousness.[15] Let us unpack the contours of the argument.

Ricœur immediately underlines that Simondon's proposal for an axiomatisation of social sciences starts from a domain – Nature – in which only apparently resides the original reciprocity of the 'Man+Nature' relation.[16] Trying to 'construct the universe of discourse from the region of nature which is itself something included within discourse', Simondon's proposal would therefore experience irremediable paralogisms. Ricœur's question well exemplifies the hypothesis of the 'hermeneutic circle', placing the birth of sense (or 'precategorical signification') within the constraint of the 'universe of discourse'. From this perspective Simondon incarnates the 'risk of objectivism', i.e. the 'assumption that consciousness is part of a total field and the speaker's significations are merely a part of the set of all things'.

Surprisingly enough, Simondon's answer is a crushing remark: 'how could one admit that nature is part of the discourse? This is the postulate underlying your argument, and this is what I shall definitely refuse'. Still more surprisingly, he seems to *accept* the 'objectivistic picture' of his philosophy Ricœur has just drawn claiming it is 'pre-critical' (or at least external to the dominant trend derived from the linguistic turn in twentieth century philosophy).[17] But Simondon cannot accept Ricœur's portrait without denouncing the limits of his interlocutor's position: 'he starts from a non transductive conception of signification'. In his view Ricœur's argument is grounded on the assumption of an alleged unaccountability of the emergence of 'signification' out of (human) discourse. If signification is reduced to discourse, which 'implies the word and the laws of signification', any signification is eventually rooted into the horizon of human sense. And if sense is the horizon in which the (linguistic) reality of nature itself is given, one does not abandon the perspective of a transcendental subject. The final point for Simondon is: 'there is no *Word* [la *Parole*], but there are *words* [les *paroles*], there are multiple typologies of words; there is signification, yes, but not the Word'.[18]

---

[15] Everything happens within a few minutes, which fill a dozen of pages. I will focus on three of Simondon's 'sub-discussions': with Ricœur (FIPD 758–60), Hyppolite (760–63) and Berger (764–65). I will avoid quoting the page numbers to prevent the overburdening of the text.

[16] 'In my view, what precedes the social sciences is not nature, but the totality Man+Nature; is it possible, starting from a structure of thought borrowed from nature, to provide an axiomatisation of the totality Man+Nature?' (FIPD 758).

[17] The theme of language is surprisingly marginal in Simondon's production (Van Caneghem 1989: 816). The reasons will emerge during the exposition.

[18] In the transcription of the debate, a capital initial for 'Signification' is presented, but I find this editing choice wrong as far as it suggests the idea of a 'totality' which is quite far from Simondon's final view: 'there is no universe of discourse, neither there is a signification of all significations' (FIPD 759).

In other words, Simondon refuses to assume language as a transcendental horizon, and in this perspective we can read his 'theory of nature' as an attempt to force the phenomenological postulate of 'sense' as the original warranty for any kind of discourse and *therefore* for nature itself. Through an adequate theory of information, the radical disjunction of the issues of signification and language in a way postpones the problem of sense to the ontogenetic one: 'There is a theory of nature in what I tried to present, which could not admit such a theory of signification as included in the word'.

Hyppolite's subsequent intervention goes straight to the problem of situating language and sense in what seems to be a philosophy of nature. Hyppolite, in order to bring the discussion back to the concept of information, refers to *Individuation*. He separates the achievements of the theory of information from the problems it creates, which he then summarises: it 'presupposes a sense which it cannot provide'; a 'sense' which must be discovered in the irreducibility of 'natural language' to information. In short, Hyppolite grounds his argument (in which he still *identifies* word and signification) on the hypothesis that the transmission of a message presupposes sense, and he concludes that Simondon is wrong in trying 'to solve the problem of sense through a philosophy of nature'.[19]

According to Simondon, his theory is not in principle incompatible with a theory of language and, to support that, he eventually claims it can explain the genesis of sense through the concepts of metastable equilibrium and structural germ. As already stated, for Simondon language is only made of signals, and signals are not actual information, they just *become* information under some structural *and* aleatory conditions of possibility:

> In order to understand a language, there must be a proper tension in the receiver. Thus, for instance, a language which does not interest, does not bring about a message concerning an actual problem, is a dead language [...] it is useless, it gives no information, since it is not a seed which falls upon us as it would fall on an as yet unstructured metastable soil – finally structuring it. (FIPD 762)

Also a 'word' can serve as a 'structural germ', then, but exclusively if it works at a level in which what matters is not its linguistic nature but its function of signification. And, for Simondon, this is not the strict pertinence of a theory of language: 'the origin of structural germs is a very delicate problem, but I do not think a theory of language can actually solve it'.[20]

---

[19] Hyppolite, *directeur de thèse* of Simondon for *Individuation* and a close friend of Merleau-Ponty's, was the only one among the 'public' trying to mediate with Simondon's position. Although no parts of *Individuation* had yet been published, Hyppolite implicitly referred to it when addressing Simondon as follows: 'you omitted the theory of information you had nevertheless begun to develop in your thesis'. He also conceded that the theory of information could possibly explain the genesis of sense, by explicating 'the difference between sense and message'. In his replies Simondon does not defend the cybernetic concept of information, and he rather rapidly presents his criticism of it. His choice is probably motivated by the fact that he admitted the problem of sense could not be resolved by a *cybernetic* theory of information, because of its incapability to differentiate significant and non significant randomness (Sect. 2.3).

[20] The issue of 'structural germs' is related to Simondon's conception of archetype and cultural legacy, as I will explain in Sect. 12.2.

It is clear that a reciprocal misunderstanding concerning the nature of language haunts the whole discussion, during which Simondon strenuously defends the conceptual apparatus of his philosophy of individuation from the charge of displaying mere 'metaphors'.[21] On the one hand we have the postulate of an original interconnection of sense and being, subject and object, where no 'theory of nature' is allowed if not inscribed within the horizon of a theory of language. On the other hand Simondon's 'theory of nature' challenges the problem of the emergence of 'sense' and consciousness, starting from the study of their conditions of possibility, and trying to elaborate the conceptual tools which can allow one to speak of reality *before* the subject. Simondon's project passed through a reformation of the concept of information which tried to explain the emergence of significations *before* the emergence of sense and, most notably, of language. Aware or not of the distance separating his project from the shared postulate of his interlocutors at the *Société* – Simondon could not cover the distance without a direct attack on transcendental subjectivity.

Berger's conclusive intervention – in its almost naïve clarity – points out the insurmountable gap separating the discussants:

> I would like to pose the question. Where are you situating consciousness? Is it to be presupposed since the beginning? […] Making consciousness intervene one could possibly clear up the difficulties presented by Mr Hyppolite and Mr Ricœur. (FIPD 764)

Berger's discourse carries on the equivocal identification of information, signification and sense, merging them all into the subject's consciousness: 'when you say that information is transmitted […] I translate your assertion in terms valid for the subject', 'information, i.e. consciousness of something [sic!]' 'does not appear until consciousness receives the message and can give a signification to it'.

In short, from a point of view that postulates a common origin for information and consciousness, thus identifying being and sense, Simondon's philosophy necessarily appears an 'objectivism in which a more complex form than others would emerge, a new reality called consciousness'. It is worth quoting Simondon's conclusive statement, where he tries *in extremis* to defend his philosophy from the charge of 'objectivism', with formulas recalling those of *Individuation*:

> It is not an objectivism; this system would rather be a trans-objectivism […] In fact the true real is not 'objective'; it has to be grasped beyond this reductive notion. Before any opposition between subject and object, a mode of being can exist prior to the subject-mode and the object-mode. This is the mode of being of the operation of formation […] Therefore I think the dualism which opposes subject and object cannot be maintained, on the contrary it must be considered the result of a process of formation which, in this case, is the process of individuation. The word *ontogenesis* summarizes the question. (FIPD 765)

The conclusion is quite clear and perfectly consistent with Simondon's claim in *Individuation* that 'ontogenesis precedes critique and ontology' (I 284), and therefore the true philosophical question concerns 'complete reality,

---

[21] First Ricœur: 'hence the metaphoric essence of your transposition from the level of nature to the level of human significations' (FIPD 759); then Hyppolite: 'you are not going further than me, since you do not generate sense. You just have imagined it with potentials and tensions' (FIPD 762); and eventually Berger: 'I am using metaphors as well' (FIPD 764).

preceding the individuation from which the subject of critical thought and of ontology emerges' (I 269). A philosophy of individuation shall necessarily require an ontogenesis of the subject. But is this enough to secure an exit from phenomenology? And would this not lead to a simple, although cautious, scientism?

3.4 Between Phenomenology and Positivism?

The concept of the 'pre-individual' is rather the mark of a problem than its solution: an hypothesis still too closely connected to Simondon's phenomenological legacy to have definitive consequences. The surprising absence of the term 'pre-individual' from his lecture at the *Société*, the almost total disappearance of it from Simondon's later writings, and the quite rare and prudent use he made of the notion of 'sense' in *Individuation*, demand further interpretation. At the *Société* Simondon confronted the genuinely phenomenological question on the origin of sense (or of 'sense' as the origin), which structuralism was in that period going to abandon as a false question, by ascribing the emergence of sense to the operating of the signifier. A question that made him apparently endorse a stance quite close to positivism during the discussion at the *Société*. In this sense it is worth understanding if in the development of Simondon's philosophy we are witnessing the disappearance of a problem or rather its reformulation in different terms.

As previously stated, the term 'pre-individual' will completely disappear in Simondon's writings during the 1960s. But the problem of the original 'subject-world relation', typical of phenomenology, could only disappear *if* phenomenology *as a problem* disappeared, which in fact never completely happened for Simondon, at least during that period. That the notion of information could actually serve the purpose of explaining 'sense', or whatever precedes it as its condition, is something Simondon's philosophy of individuation implicitly aimed at: in a philosophy of individuation, what was to be considered 'original' was not sense, but the emergence of metastable systems, i.e. the operation which, ultimately, the pre-individual consisted of.

This does not mean that Simondon in *Individuation* – although he actually never gave a fixed meaning to the term 'sense'[22] – avoided posing the typical problems of

[22] One cannot deny the existence, inside *Individuation*, of some passages in which the term 'sense' appears. Nevertheless, it happens in contexts 'naturally' connected to phenomenology (e.g. I 213–14), without entailing Simondon's unconditional adherence to that perspective. Thus the 'sense of the situation' is essentially the polarisation of the world for the perceiving subject, but the subject does not *precede*, as its condition, the moment in which 'information acquires a predominantly intensive meaning': it rather emerges *with* the world from a single operation of coupling [*couplage*] (I 242). And again, in the conclusion, the question returns as still more complicated when, discussing the possibility of making of individuation a theory of being, Simondon claims that 'information must have a sense in order to exist' (I 328). Here the term 'sense' refers to a peculiar structure of the signal, which renders it compatible with the receiving system making of it a piece of information.

phenomenology, as he claimed that 'a theory of individuation must implement a theory of sensation, of perception, of affects, of emotion' (I 321). But he shifted such problems from the field of a supposed subject-consciousness to that of the pre-individual phase of being, which should have given an explanation for the emergence of the subject *without* going through the notion of sense. In *Individuation* this path traversed the reformation of the cybernetic concept of information, a model for the understanding of the emergence of order *before* the alternative between sense and non-sense (which could actually appear only within the horizon of an already structured subject). Thus it is clear that the problem of the subject became central for Simondon precisely because it was the central problem for phenomenology, and it could not be abandoned *as a problem* without running the risk of assuming individuation as a process actually concerning the subject. On the contrary – said Simondon – one must at all costs avoid the spontaneous phenomenology according to which 'the individual is always in a certain sense *conceived* as a *subject*' (I 321).

The exigency of avoiding the relapse into an idealistic conception of sense as produced or 'instituted' within the consciousness of the subject,[23] pushed him far from his phenomenological legacy and quite close to positivism. In fact Simondon's theory of nature was suspended between the refusal of what he considered a philosophy of the subject (phenomenology) and the denunciation of the reductionism entailed by what he considered a philosophy of the object (positivism). He tried to think beyond this simplistic opposition starting from a phenomenological legacy, but without abandoning the critical fecundity of natural philosophy Merleau-Ponty had widely recognised. Therefore ontogenesis cannot be considered a phenomenology: it is on the contrary a clear attempt to abandon such a philosophical path, and in particular the presupposition of a subject-consciousness, but *without* choosing the alternative path of abandoning the question on the origin of sense, as structuralism was going to do at the time.[24] Following Merleau-Ponty, Simondon rather assumed the new perspectives opened by the study of the organism within the

[23] Thus J. Wahl in the discussion at the *Société*: 'There are some aspects of your thought I am inclined to approve and admire. All that puts your lecture beyond the classical attitudes of the idealistic theory of knowledge arouses my instinctive consent' (FIPD 755).

[24] According to Guchet 'by confronting cybernetics and the social sciences Simondon aims to provide a serious alternative to structuralism' (Guchet 2005: 203). This perspective would explain Simondon's attempt to 're-inscribe the transcendental into subjectivity, although without abandoning the acquisitions of the philosophies of concept', thus preluding Deleuze's transcendental empiricism (Guchet 2003: 140–41). Pursuing a different line of research, Barthélémy makes a massive use of the phenomenological notion of 'sense', pushing toward the hypothesis of a 'self-transcendence' of sense. From this perspective *Individuation* would have developed only the 'regional ontology' of a more generally inclusive relativisation [*relativisation englobante*] (Barthélémy 2005a: 48–59; Barthélémy 2005b: 231–86). Garelli even puts forward the project of a phenomenological analysis of the genesis of sense in Saussure by means of Simondon's concept of metastability (Garelli 2003: 109, n. 68). I believe that Simondon's philosophy can be situated *between* phenomenology and structuralism, without conceding to the first the privileged primacy of consciousness or subject, nor to the second its actual cancellation as a problem. And the key for the full understanding of his approach will be the reference to Canguilhem's philosophy of life (see in particular Sect. 9.4).

phenomenological tradition, and he adopted the cybernetic concept of information as a model for a non-anthropomorphic understanding of the operation of signification and therefore of the *emergence* of sense.

The determination with which Simondon tried to explain the emergence of the 'subject' through the epistemological apparatus provided by his philosophy of individuation was a true turning point against his phenomenological legacy. Starting from the concept of the pre-individual, which he built using schemas derived from physics and biology, he tried to avoid reducing the subject both to the organism and to consciousness. In order to explain how this was considered possible, in the next chapter I shall follow Simondon's conversion of the concept of information into 'signification' when, in *Individuation*, he enters the psychic and collective domain. There the conditions of possibility for the ontogenesis of the subject progressively emerge in what he calls 'transindividual individuation', and one will eventually also find there the 'place' of consciousness, the absence of which from Simondon's 'system' particularly troubled Berger:

> From this perspective, consciousness should not be considered through an 'all or nothing' schema, opposing subject or object, but rather starting from a more primitive transconsciousness. (FIPD 765)

If the concept of the pre-individual – due to its phenomenological matrix – compelled Simondon to delve into questioning the subject's ontogenesis, in the same way the reformulated notion of information eventually brought him to a 'transindividual' threshold beyond which the (inherently collective) subject of philosophy of individuation finally emerged.

References[25]

Barbaras, R. 2005. Préface. In Simondon, G. 1964-65. *Cours sur la perception (1964-65)*, IX-XVI. Chatou: La Transparence, 2005.

Barthélémy, J.-H. 2005a. *Penser l'individuation: Simondon et la philosophie de la nature*. Paris: L'Harmattan.

Barthélémy, J.-H. 2005b. *Penser la connaissance et la technique après Simondon*. Paris: L'Harmattan.

Descombes, V. 1979. *Le même et l'autre. Quarante-cinq ans de philosophie française (1933–1978)*. Paris: Minuit.

Garelli, J. 2003. Perplexité de Saussure. *Archives de Philosophie* 1: 98–117.

Guchet, X. 2003. Pensée technique et philosophie transcendantale. *Archives de Philosophie* 1: 119–144.

Guchet, X. 2005. Simondon, la cybernétique et les sciences humaines. Genèse de l'ontologie simondonienne dans deux manuscrits sur la cybernétique. *Chiasmi International 7: Merleau-Ponty. Vie et Individuation*: 187–206.

Hottois, G. 1993. *Simondon et la philosophie de la "culture technique"*. Bruxelles: De Boeck.

Mancini, S. 1987. *Sempre di nuovo. Merleau-Ponty e la dialettica dell'espressione*. Milano: Mimesis, 2001.

[25] *Simondon's complete bibliography and a list of abbreviations are provided in the* Appendix.

Merleau-Ponty, M. 1942. *La structure du comportement*. Paris: PUF (English trans. *The Structure of Behaviour*. London: Beacon Press, 1967).

Merleau-Ponty, M. 1945. *Phénoménologie de la perception*. Paris: Gallimard (English trans. *Phenomenology of Perception*. London/New York: Routledge).

Merleau-Ponty, M. 1952–1960. *Résumés de cours (Collège de France, 1952–1960)*. Paris: Gallimard, 1968.

Merleau-Ponty, M. 1956–60. *La Nature. Notes. Cours du Collège de France*. Paris: Seuil, 1995 (English trans. *Nature: Course Notes from the Collège de France*. Illinois: Northwestern UP, 2003).

Merleau-Ponty, M. 1959. Notes de travail inédites. Cours 1959. *Chiasmi International 7: Merleau-Ponty. Vie et Individuation*: 39–42.

Morizot, B. 2011. La néoténie dans la pensée de Simondon. Ontogenèse d'une hypothèse. *Cahiers Simondon* 3: 109–129.

Van Caneghem, D. 1989. Hommage à Gilbert Simondon. *Bulletin de Psychologie* 42(392): 815–836.

Chapter 4
Subject and Method of a Philosophy of Individuation

The methodological foundations of Simondon's philosophy of individuation can be disclosed by an analysis of the problem of the subject. Simondon questions the subject of knowledge as a living being *and* as a social product through the original concept of the 'transindividual', focusing on the psychic and collective processes from which shared significations emerge. Thus he maintains the ontological emergence of the subject and the epistemological meaning of its ontogenesis together, conceiving knowledge as the 'individuation of the subject's knowledge'. This allows Simondon to connect science and philosophy as different kinds of strictly interrelated practices, going back to Bergson's concept of intuition, and making it into a theory of invention. Thus, once the question on the origins of thought (and being) has disappeared, thought itself becomes action, praxis, and a risky enterprise necessarily connected to the functioning of social systems.

4.1 Ontogenesis of the Subject: Transindividual and Signification

In the paragraph entitled *Sujet et individu* [Subject and Individual] Simondon outlines the explicit differentiation of the two concepts, in order to present the subject as a system composed of different 'phases'. The name 'individual', according to Simondon, 'is abusively attributed to a more complex reality, that of the subject'. In fact, the subject is composed by an individuated and visible element plus its associated milieu, a non-individuated 'natural' energetic charge which determines its inclination towards further individuations, i.e. the pre-individual. Thus the distinction *subject/individual* must be explicitly theorised precisely because it is usually hidden under a false identification, a direct substantialist reduction of the subject to the individual (I 310).

A. Bardin, *Epistemology and Political Philosophy in Gilbert Simondon*,
Philosophy of Engineering and Technology 19, DOI 10.1007/978-94-017-9831-0_4

In order to treat the subject out of the substantialist paradigm which reduces it to an individual, Simondon refers to his theory of the phases of being: the subject will be thus constituted of different phases, i.e. structurally phase-shift into different regimes of individuation. It will not be a homeostatic system then, but a metastable one (with some 'associated pre-individual'). In fact, once crossed the quantum threshold of the chrono-topological structure we call life, the living being remains partially phase-shift, full of energetic potentials, and therefore capable of further individuation within the psycho-social domain where the collective arises. This particular kind of individuation Simondon often calls 'individualisation'.[1]

A patent hierarchy relates individuation and individualisation, but not exactly a succession, because, even if 'individualisation continues on from individuation', they are both phases which, as such, simultaneously constitute the subject. Biological individuation is the permanent condition of the possibility of individualisation. Once biological individuation has started, a quantum leap is possible into a regime of psychic and collective individuation, where different processes of individualisation indefinitely follow one another. This can be considered in fact a 'permanent individuation' which saturates the open 'assiomatics' of vital problems, thanks to an 'indefinite sequence' of individuations which 'absorb more and more pre-individual reality integrating it into the relation with the milieu' (I 29). Each psychic and collective 'individualisation', each thought, each conceptual discovery, and each emotion, repeats and develops as 'a partial but faithful' repetition the schema of that first 'absolute individuation' (I 264). Perceptions are the unification of a disparate series of sensations, emotions of a disparate series of affects, significations of a disparate series of signals, and so on: all of them are operations of 'individualisation' through which subjects invent new forms of metastable coherence ('coupling' or 'compatibilisation').

These coordinates necessitate an initial questioning of Simondon's problematic anthropology through the analysis of the concept of 'signification' [*signification*], a concept he especially develops in the last part of Individuation, entitled *Le collectif comme condition de signification* [The Collective as a Condition of Signification]. In order to understand the meaning of Simondon's problematic anthropology, it is worthwhile beginning by challenging the difficult and largely debated theme of the 'transindividual':

> The two individuations, psychic and collective, are reciprocal. They define the category of the transindividual as the systematic unity of internal (psychic) and external (collective) individuation. (I 29)

The individual/pre-individual phase-shift characterising the individuated organism is overcome by transindividual individuation. But this 'resolution' is different from the vital one: Simondon clarifies that the transindividual is not the 'synthesis'

---

[1] In the part concerning *L'individuation des êtres vivants*, the term 'individualisation' is also widely referred to the formation of organisms. Nevertheless, at the level of psychic and collective individuation, 'individualisation' clearly marks a conceptual difference from 'individuation'.

of the phase-shift, but rather its 'signification'. The transindividual production of significations, he says, 'envelops' the phase-shift of the subject without definitively resolving its tensions, because in the course of transindividual individuation the system-subject still remains phase-shift, i.e. 'more than individual, individual and nature [...], and at the same time both phases of being' (I 307). In this sense, the subject-being [*l'être sujet*] 'can be conceived as a more or less consistent system of three subsequent phases of being: pre-individual, individuated and transindividual' (I 310).[2] The subject, by definition biological *and* psychic-collective, is in this sense engaged in a transindividual structure of significations which allows it to endure its transformation process, going through subsequent individualisations.

Now, how can the transindividual emergence of significations be conceived? How can it give consistence to the subject's phase-shift? Signification, Simondon says, depends on the 'consistence of two orders of reality, individuation and individualisation' (I 267), a 'metastable' consistency which is not at all a synthesis. In order to make his point Simondon still refers here to the operation of 'coupling', through which the 'psychic living-being' maintains 'a plurality of signals' together:

> A being is never completely individualised. In order to exist it needs to continue to individualise itself solving the problems posed by its own surrounding milieu. (I 263–64)

Although here one might possibly understand signification as the product of the adaptive process of an organism, nevertheless it necessarily must be referred to the subject, since each coupling of signals becomes signification *exclusively* at the transindividual level, where the subject emerges. On the other hand, it would be wrong to suppose Simondon maintained the hypothesis that the incidence of *language* on the organism is what determines the emergence of the human being: transindividual *significations* contribute to constitute subjects independently of their species specific biological individuation. Thus one must reject both the hypothesis of a subject arbitrarily producing and/or utilising significations, and the hypothesis of significations determining subjects (in the form of discourses producing subjectivation). Simondon rather conceives a sort of paradoxical simultaneity of the two-sided process in which the 'individuation of significations' consists. Each 'individualisation' is in fact an 'individuation of significations' under the condition of the existence of individuated subjects: a process in which the subjects involved could be said – but this would still entail a substantialisation – to be simultaneously active and passive.

The problem must be challenged by assuming that the paradoxical status of the subject is in fact the effect of an unwarranted projection of the wrong classical

[2] In *Individuation* neither the level on which the emergence of the subject should be placed, nor the terminological distinction between what is psychic, collective and transindividual are clear. According to Barthélémy one can speak of 'subject' at the level of psychic individuation, while transindividual individuation would constitute what Simondon calls 'personality' (Barthélémy 2005a: 206 ff.). On the contrary, I assume that only starting from the transindividual regime of individuation can one understand the subject as the actual simultaneity of the three phases. I will clarify my own interpretation in Sect. 5.4.

substantialising approach, conceiving psychic functions *both* as a set of processes through which the organism tries to cope with reality (giving sense to it) *and* as the effect of an exchange of transindividual significations capturing the organism in a horizon of sense entirely determined by the collective. Both these assumptions reveal the same fundamental incompatibility with a philosophy of the *processes* of individuation. In both pictures the subject – involved in a never-ending sequence of operations of individualisation (i.e. of signification) – is clearly *identified* and thus mistaken for the *individual*: on the one hand for an individual organism, on the other for a collective individual which is the social group itself. This is precisely what Simondon often indicates as the mistakes of both psychology and sociology: what each miss in both cases, is the 'transductive' identity of the process, which Simondon names 'subject'.

It should be clear by now that neither the subject corresponds to the organism, nor can the object be conceived of as the supposed 'content' of a subject's knowledge. Philosophy of individuation is rather concerned with the system of potentials in which the operation of knowledge emerges as a subject-object relation. In short, the pre-individual milieu (the condition of existence for any individual), the individual organism (the absolute condition of any individu*alis*ation) and the transindividual (psycho-collective individuation) are processes of phase-shift which simultaneously and together are the conditions for the emergence of the subject-object relation, and therefore of thought and knowledge themselves.[3] Thus one must avoid both, on the one hand, considering the object through 'the poor and negative idea of what *is not* the subject, the remains of the subject's knowledge' (FIPD 765), and, on the other, substantialising the subject by following the spontaneous tendency of thought to self-identify with 'its own condition of existence' (I 321).

I will return to the ontogenesis of the subject when dealing with the problem of the transindividual in Chap. 5. I shall first take up the question from an epistemological point of view, by reformulating the two questions 'Who knows?' and 'What does knowledge mean?' in the following terms: how does the system of significations we call knowledge emerge *along with the individuation of what Simondon calls 'subject'*?

[3] The phenomenological question of 'origin' resurfaces here, because any attempt to understand the formation of the subject is forcedly concerned with the pre-individual being preceding it as its condition of possibility. On the other hand, since this condition of possibility is a process, it cannot be described without regarding the way in which subject and object emerge from it. But subject and object do not emerge *from* thought, rather they are its conditions of possibility: 'thought is a particular *mode of the secondary individuation*, occurring after *the fundamental individuation which constitutes the subject*' (I 321). To sum up: pre-individual being is the condition of the subject-object differentiation which, in turn, is the condition for cognitive operations. But it is worth highlighting that, among the conditions of the emergence of a subject (and therefore of thought) there is also a transindividual individuation as well: thought 'is secondary in relation to the condition of existence of the subject, but this condition of existence is neither isolated nor unique, because the subject is not an isolated, self-constituted term' (I 321). On Simondon's approach to the question of the transcendental, see Bardin (2008).

4.2 Individuation of the Subject's Knowledge: A Question of Method

It is now time to expose the contradiction which is progressively emerging through my analysis. Simondon's philosophy might seem on the one hand a plain, dogmatic naturalism in which knowledge entirely depends on the a-priori given structure of the organism, and on the other a kind of postmodern relativism suggesting the possible existence of as many 'individuations of knowledge' as there are processes of individualisation. Thus Simondon's philosophy would be no more than a naturalism of complexity or an interminable hermeneutic.

In order to find a way through this aporia, it is worth returning to the concept of transduction. In fact, transduction extends over all the domains of individuation, knowledge included, and 'is therefore both a metaphysical and a logical notion. *It applies to ontogenesis and is ontogenesis itself*' (I 33). In the presentation at the *Société française de philosophie*, Simondon explains the reciprocal implications between his 'analogical method' and ontological transduction:

> There is a kind of identity between the method I am using, which is analogical, and the ontology I presuppose, which is an ontology of the transductive operation within the process of formation. If the transductive operation of structuring does not exist, analogy is an invalid logic; this is a postulate. The postulate is simultaneously ontological and methodological. (FIPD 757)

Precisely because thought proceeds transductively, it cannot be consistently formalised through the classical principles of identity and the excluded third (I 324). More crucially, a logic in the classical sense is impossible. The problem can be posed in the following terms: if the 'transductive operation' is singular – i.e. its origin, course and results cannot be subsumed under any universal concept – then the foundation of *a single* logic of transduction will be impossible by definition. In fact a 'theory of being preceding any logic' according to which there are 'multiple kinds of individuation, should rather produce 'multiple logics, each corresponding to a definite kind of individuation' (I 36), thus resulting in the practice of a 'pluralism of individuation' (IPC 217).[4]

This kind of logic of singularity would incorporate and repeat the aleatory factors of the transductive process itself, and nevertheless the elevation of transduction to the role of a methodological paradigm does not entail that any *formal logic* can be simply deduced from the *ontological* assumption of transduction. That is how Simondon defends the validity of analogical thought: *grounded* on ontological transduction[5] but not *guaranteed* nor *established* by any possible logical formalisation. The attempt to define the conditions of ontogenesis as the knowledge of (transductive) individuation therefore compels Simondon to pose the problem of its

[4] This expression, added in IPC, was not yet present in the original thesis, where one could read 'pluralism of phases' (I 318).

[5] 'The possibility of using an analogical transduction for thinking a domain of reality, indicates that such a domain is the actual place of a transductive structuration' (I 33).

epistemological status and scientific value, since it concerns processes which are not reproducible, and therefore neither universalisable nor predictable through what early-modern science conceived as 'laws', at least within the horizon of classical mechanical physics.

Now, what does a *grounded* but non-*guaranteed* knowledge mean? The problem can be reformulated in Simondon's terms: if any act of thinking (an individuation of significations) is intrinsically transductive, what differentiates a thought limiting itself to an individuated object and a thought capable of grasping the transductive process of ontogenesis? Simondon's quite enigmatic response arises at the very end of his introduction to *Individuation*:

> Therefore it is neither an immediate knowledge, nor a mediate one, that we can have of individuation, but rather an operation of knowledge which is parallel to the known operation; we cannot, in the common use of the term – *know individuation*; we can only individuate, individuate ourselves, and individuate in ourselves [...] Beings can be known through the subject's knowledge, but individuation of beings cannot be grasped out of the individuation of the subject's knowledge. (I 36)[6]

Reading this passage against the background of my previous arguments, what one can understand is that: (1) according to Simondon there are two different modes of knowledge; (2) they are both operations of individualisation depending on the precondition of the transindividual individuation of a subject; (3) they are two distinctly different kinds of operations, corresponding to two different ways of conceiving being, as a set of structures or as a set of processes. Unfortunately, Simondon gives no more suggestions in his introduction. Nevertheless, by looking inside *Individuation* one can discover several clues to the double paradoxical function of knowledge as a subject-object relation. The first is a function of 'stabilisation' through the significations collectively elaborated and already *given* within the different fields of knowledge: 'true knowledge is the one which corresponds to the highest possible stability within the given conditions of the *subject-object relation*' (I 83). The second is rather a function of 'meta-stabilisation', where significations are kept in a state of motion in order to dispose them to further individuations of subject-object relations, according to the actual becoming of transductive processes: 'true knowledge is a relation, not a simple formal relationship' (I 83).

In the light of this structural shifting in Simondon's conception of knowledge, one can try to translate the opposition he establishes between 'the subject's knowledge' and 'the individuation of the subject's knowledge' in terms of two opposed functions within knowledge itself. On the one hand the 'stabilisation' of an accomplished and structured system, the individuated terms of which are subject and object, on the other hand the participation in transindividual processes within a metastable system of significations in which a (new) subject and a (new) object can emerge.

---

[6] The first to adequately highlight the crucial importance of this philosophical distinction was undisputedly Barthélémy 2005b: 242 ff.

From this perspective the 'individuation of the subject's knowledge' cannot be a kind of knowledge among others, it is in fact the peculiar case of the phase-shift of the pre-individual assuming the form of a relation between an object (of knowledge) and a subject (of knowledge), neither of them existing prior to the process (FIPD 765).[7] And yet, while simultaneous, the emergence of subject and object is entirely non-symmetrical, because both the 'subject's knowledge' and the 'individuation of the subject's knowledge', in fact, emerge *within* a subject. Why then should only an 'individuation of the subject's knowledge' grasp ontogenesis? Because it is precisely in the 'individuation of the subject's knowledge' that the process of knowledge can be grasped in its transductive and singular becoming, i.e. as a part of the 'real' process structuring both the subject and the object, and not simply as a process 'internal' to the subject. The subject-object relation, emerging in *any* act of the subject's knowledge, is part here of a peculiar operation of knowledge which, one could say, grasps insofar as it accomplishes the double individuation of a subject and an object in the transindividual domain of significations:

> This approach consists in *following being in its genesis*, accomplishing the genesis of thought together with the accomplishment of the object. (I 34)

In other words, this means to 'invent':

> To invent means to make one's thinking function [...] according to an experienced [*vécu*] dynamism, grasped insofar as produced, accompanied in its genesis. (MEOT 138)

On these bases, it is possible to understand the precise meaning and extension assumed by the theme of 'invention' in Simondon's work.

4.3 Invention, Analogy, Intuition and Bergson

The theme of invention, already present as a pedagogical issue in Simondon's *Réflexions préalables à une refonte de l'enseignement* [Preliminary Reflections on a Reform of Teaching] (1954), emerges especially during the 1960s as a technical issue and as a mode of existence of the 'psychic living being'. When the term invention defines the way transduction appears in the field of knowledge, it clearly does not concern its normal functioning, but a 'rare and often aleatory process' (IT 332). Just as any vital problem can have different solutions starting from a singular element which exceeds the given elements of the organism-milieu relation, each problem posed to thought can have solutions exceeding all given elements in the fixed relation subject-object. This exceeding operation of knowledge Simondon

[7] From pre-individual processes derives the becoming-object of an object and the becoming-subject of a subject in their reciprocal singularity. In FIP Simondon's argument moves around the notion of 'field' (FIP 540). See also I 270.

calls invention: it is precisely the transductive process as it presents itself at the level of thought (i.e. of transindividual significations).

The debt towards Bergson's *L'évolution créatrice* [Creative Evolution] is quite evident here, and it is probably worth considering the implicit reference to Canguilhem, which would seem to shift the problem of invention to the level of biological individuation.[8] But for Simondon invention is neither a category of biology nor, of course, a metaphysical category, linked to any absolute difference between human being and nature, or life and matter. Rather it describes a precise regime of transduction, the discontinuous process characterising any individuation which, at the transindividual level, proceeds through the collective institution of significations. This permits a response to a few previously asked questions: What is knowledge as an operation or process? What is a knowledge of the processes of individuation (grounded-on but not *guaranteed*-by them)? Knowledge in its 'core' is a process of transindividual invention, neither individual nor inter-subjective, but properly 'subjective' in the following sense:

> It is not the individual who invents, it is the subject. Wider and richer than the individual. the subject entails, in addition to the individuality of the individuated being, a certain charge of nature, of non individual being. (MEOT 248)[9]

In this sense invention is certainly an act of thinking, grounded nevertheless on its actual conditions of possibility: on the one hand the pre-individual milieu with which it composes a metastable system, on the other the aleatory 'encounter' of the system with a singular structure (idea, need, image, etc.) which triggers a transductive reaction.

The link now becomes clear with what in his former programmatic texts Simondon called the 'analogical act': the method of transferring a thinking process from one structure to another. Simondon contrasted there 'analogical transduction' to induction and deduction, thus justifying the paradigmatic transference of schemas inherited from empirical sciences to philosophy in order to gain some heuristic efficacy.[10] This does not exclude, it rather presupposes a coherent thinking, since structural formalisation is the precondition of any analogical transduction. It is not an analogy of structures, Simondon explains, which would produce only 'an association of ideas', but an analogy of operations relying on determinate conditions of state:

---

[8] See for instance MEOT, where Simondon refers to invention the anticipatory attitude of the 'schemas of creative imagination' (MEOT 58). In MEOT invention defines the transductive process through which an organism imagines and builds a 'compatibility system' analogous to its own functioning: 'it is precisely because it is an individual related to its associated milieu, that the living being can invent' (MEOT 58). But in what follows Simondon explains that invention can be placed in the peculiar recurrent causality which only takes place between life and thought, i.e. what in *Individuation* he includes in the notion of the transindividual when dealing with 'affectivo-emotivity' (see Sect. 5.2).

[9] In other words, 'significations integrated into the collective', function at an higher scale than the '*hic et nunc* of the individual within the subject' (I 311).

[10] What at the *Société* was interpreted as a 'metaphorical usage of concepts' (see above, Chap. 3, n. 21). See for instance *Individuation*: 'invention is neither inductive nor deductive, but transductive [...] it is the analogical operation' (I 33).

> For a schema to be effectively utilised as a paradigm, there must be a functional and operatory analogy between the original domain and the new domain of application of the paradigm. (I 319)[11]

This possibility depends on the ontological decision Simondon previously confronted in *Allagmatique*. There he asked to choose between three different hypotheses: (1) 'beings are to be defined through their operations and not through their structures'; (2) 'a being can be defined both through its structure and through its operations'; (3) 'the structure, not the operation, is primitive' (A 564). According to Simondon, only the first hypothesis founds and justifies the analogical method. In this sense he can speak of the pre-individual as 'nature', 'real' or 'energy of a metastable system' (I 313). In it knowledge as a 'real' process, a relation between two operations, can emerge, and analogical method 'can be unreservedly applied' (A 563–564). It is precisely this hypothesis which Simondon experiments throughout *Individuation*, traversing the fields of the natural and social sciences while his philosophy of individuation progressively takes shape, arising from paradigms analogically transposed onto the different regimes of individuation, thanks to an act of thinking typical of philosophical enquiry.

Now, since concepts are fit for individuals and not for individuation (I 27), a purely Bergsonian solution to the epistemological problem seems to re-surface. However, there is no textual evidence of such a result. Simondon always rejects a sharp distinction between sciences of structures and philosophy of operations and, furthermore, he explicitly criticizes two 'symmetrical' errors: the 'phenomenal objectivism' of Kant and Comte and the 'dynamic intuitionism' of Bergson, which relegates matter to a mere 'degradation of vital dynamism' (A 564).[12] In short, in Simondon's philosophy of individuation one witnesses an effort to delve into the conceptual gaps of scientific knowledge without assigning to them the aim of grasping the phantasmatic whole of intuitive knowledge. Thus, when in MEOT he defines intuition as the proper form of philosophical knowledge, Simondon is still differentiating himself from Bergson:

> Intuition is neither sensitive nor intellectual; it is the analogy between the becoming of what is known and the becoming of the subject, the coincidence of the two becomings [...], it is the peculiar knowledge of genetic processes. Bergson made intuition the proper method of the knowledge of becoming, but one can generalize his method without excluding intuition

[11] In MEOT 189 Simondon explicitly refers the notion of 'analogy' to Père De Solages. In a *Dialogue sur l'analogie* (1946), Bruno De Solages, rector of the *Institut Catholique de Toulouse*, provides the 'free transcription' of a series of seminars held at the beginning of 1943 at the *Société toulousaine de philosophie* on the theme of analogy which, – he notes – appears with surprising frequency in the writings of Luis De Broglie (De Solages 1946: 13). De Solages concludes the *Dialogue* with a claim apparently inspired to scholastic Aristotelianism: 'knowledge is an analogy. This analogy of knowledge presupposes the analogy of being' (153). In fact, it is rather Monsieur Cazals who, during the discussion, provides a definition of analogy close to the one adopted by Simondon: 'what makes the originality of analogy [...] is the fact that it is rather a similitude of relations than a relation of similitude' (15).

[12] Simondon's philosophy refuses all 'implicit spiritualism' that – incapable of understanding the organisation of matter – would hierarchise matter and life (I 159).

> from the domain of matter [...], in fact intuition can be applied to all domains in which a genesis takes place, since it follows the genesis of beings at their level of unification [...] For intuition the level of unity is neither the whole, as for knowledge through ideas, nor the element, as for knowledge through concepts. (MEOT 236)[13]

As usual, Simondon does not hesitate to adopt a terminology derived from the philosophy he his actually challenging. Slowly shifting its meaning into one suitable to his perspective, in *Individuation* he comes to refer the term intuition to a 'not only mental' transduction, but one also very close to the meaning of invention (I 34). In fact, intuition becomes an operation which is neither determinate as a concept nor synthetical as an idea, submitted to threshold conditions and tensions which constitute the field of transindividual individuation, as invention itself does.

The concept of 'intuition' therefore makes sense exclusively within the analogical operation of 'individuation of the subject's knowledge', the only one capable of grasping 'the individuation of beings' (I 36). And the 'analogical method' does not entail any 'immediate' kind of intuition, since it is possible only on the basis of some structured knowledge issued from a former individuation. In fact, only from the analysis of the given structures can one move towards a proper understanding of 'operation':

> The analogical method requires the possibility of *defining structures through the operations which dynamise them* instead of *defining operations through the structures between which they take place*. (A 562)

Again, in the 'individuation of the subject's knowledge' the subject is able to go back on its own conditions of possibility and identifies them as the causes both of individuation in general and of its individualisation in particular. This is the

---

[13] Simondon clearly states both the anti-mechanistic relevance and the epistemological limits of Bergsonian intuitionism: 'in Bergson the intuition of the *mouvant* became an essential category of philosophy, powerful enough to authorize the criticism of a whole intellectual system based on the primacy of forms (mechanism). This reform is parallel to the development of the life sciences, with the notions of evolution and transformation. But one can cast some doubts on the dycothomic hypothesis according to which movement can be grasped only by intuition while forms can be grasped only by concept; in fact they are two different ways of perception, both real' (CSP 201). But the intention of marking his detachment from Bergson is particularly evident in the way Simondon treats time as a modality of the structure subsequent individuation, thus not related to intuition in the Bergsonian sense: '*allagmatic theory* [...] does not grasp being beyond space and time, but before the division in spatial system and temporal schema' (A 565) (on 'intuition as a method' in Bergson, see Deleuze 1966. This essay, published in the same year of Deleuze's review to IGPB is possibly linked to some Simondonian fascination). In short, playing on Bergson's philosophy in order to explore the anti-mechanistic implications of quantum physics, Simondon not only is following Canguilhem (1952), but he also pays his debt towards De Broglie, who – in *Les conceptions de la physique contemporaine et les idées de Bergson sur le temps et sur le mouvement* – wrote: 'the main question of this paper [is]: does any analogy exist between the Bergsonian criticism to the idea of movement and contemporary quantum theories? It seems the answer should be yes' (in De Broglie 1947: 199). De Broglie precisely refers to a note of Bergson (1934: 61) to suggest that 'living beings would necessarily have a "mechanistic" perception only because in the macroscopic world an apparent determinism reigns, which allows them to act on things' (De Broglie 1947: 210–11).

operation in which the subject simultaneously finds and produces, i.e. invents, the universal grounding for knowledge:

> If knowledge can trace back the lines which allow for the interpretation of the world according to stable laws, it is not because in the subject some a priori forms of sensibility exist, the consistence of which with raw data derived from sensation would be inexplicable; this happens because being as a subject and being as an object come from the same primitive reality, and thought, which now seems to institute an inexplicable relation between the object and the subject, in effect continues the initial individuation. The *conditions of the possibility* of knowledge are the actual *causes of existence* of the individuated being. [...] It is because individuation is the universal ground of the relationship between the object and the subject, that knowledge can be universal. (I 264)

In this sense the analogical method is the subjective continuation of an actual transduction, which allows for the risk of a singular solution. Evidently, the success of such a reflexive operation cannot be guaranteed by any methodological formalisation. The analogical method is in fact the method of invention, an operation both theoretical and practical, which lacks any guarantees, as Simondon himself reveals when defining his own philosophy as 'a dramatic [*dramatique*] theory of the becoming of being' (FIPD 755).

4.4 Unfolding Sciences: The Emergence of a Philosophy of Individuation

In the conclusion to *Individuation*, after again questioning the possibility of a science of the pre-individual, Simondon confirms the conformity of the concept with his aim: although running the risk of indefiniteness, the concept of the pre-individual can avoid the implicit creationism of any philosophy which 'concentrates all becoming in its origins' (I 327). In front of a neat predominance of the *philosophical* concept of 'pre-individual' within the whole text of *Individuation* (1958), the massive appearance of the *scientific* concepts of 'order of magnitude', 'scales of reality' and 'stratifications' in twelve of the thirteen notes to the introduction is strikingly revealing of the growing importance Simondon attributed to quantum physics in order to support the philosophical hypothesis of the pre-individual.[14]

The practicing of modern science – or at least of its jargon – probably pushes one today to consider this approach better-chosen than the one relying on the notion of the pre-individual. Furthermore the 'multiscale model' can fit the interpretation both of *Individuation* and of other texts of Simondon, and seems to be suitable for export

[14]The theme is, of course, central in the concluding chapter of the section concerning physical individuation, in particular in I 148–49 (for a close analysis of this chapter see above, Sect. 1.4). In effect, reading the whole series of notes of the introduction in sequence suggests that Simondon had possibly decided to insert them *after* the conclusion of his work: most of the theoretical development of the 'multiscale' problems in *Individuation* is concentrated there.

to many fields of contemporary scientific research.[15] Thus the 'internal resonance' of a quantum system, the exchange between its different scales, allows explanation of both ontological and epistemological issues. At the ontological level it expands the concept of relation: 'relation exists physically, biologically, psychologically, collectively as the internal resonance of the individuated being' (I 313). At the epistemological level it justifies analogical knowledge, explaining why 'thought *is not necessarily capable of thinking being in its totality*' (I 321), because 'totality' is always relative to the scale of a relation in which thought emerges as a process of individuation itself, i.e. as a specific modality of the internal resonance of a system, the institution of a subject-object metastability.[16]

This 'multiscale model' can solve many problems within *Individuation* by translating them into terms which are more appreciable today, and nevertheless it is impossible to assume it actually *was* a solution according to Simondon. In his work in progress multiple hypotheses converged, superposed, dissolved and generated new hypotheses. Thus one cannot separate the point at issue in Simondon's notes to the introduction and what I previously discussed about the phenomenological matrix of the concept of the pre-individual. In the conceptual apparatus of *Individuation* the individual can be conceived as a mediator between different orders of magnitude precisely *because* the hypothesis of the pre-individual allows the understanding of the individual as the partial result, always incomplete, of a process of individuation which both *precedes* it and *continues* in it. The individual itself is thus a system made of phases and thresholds which can put different systems in relation.[17]

The terminological confusion which tends sometimes to overlap the individual and the process of individuation traverses the whole of Simondon's main work, and derives from the same metonymic attempt to define the relation between a system

---

[15] Simondon's philosophy has been and still is thoroughly explored in this direction by Vincent Bontems in his *atelier* at the Paris ENS. See for instance Bontems (2008), and Barthélémy-Bontems (2001) who refer to the astrophysical research of Laurent Nottale.

[16] No wonder the most adequate notion to cope with the problem of pre-individual is still information, since the pre-individual reveals itself in an individuated system as 'active communication' which forms a net of 'internal resonance' between different orders of magnitude. Thus 'information, conceived as the occurrence of a singularity creating a communication between different orders of reality' (I 151–52) also solves the problem of the different nature of modulation and crystallisation we treated in Chap. 1 (see in particular I 328–30).

[17] In the conclusion, the strategic function of the pre-individual is clearly aimed at shifting the focus from substantial being to the becoming of systems, *without downplaying the role of the individual*, since individuation necessitates the transductive function of the individual as the outcome of a process of individuation and the trigger for a further one. In short, in Simondon's theory of individuation the hypothesis of the pre-individual necessarily entails what he calls the 'amplifying' role of the individual. In this way individuality is not entirely reduced to the operating of systems: although the ontogenesis of the individual 'is inscribed within the becoming of systems' and its existence is relative to the scale of the system in which it appears ('the emergence of an individual corresponds to a certain state of the system' I 328), nevertheless – precisely because systems are multiscale systems – ' the individual is not a being but an act […] it is the transductive relation of an activity' (I 191).

and the processes which constitute it (structure and operation), that we will find again at the level of the social system in the concept of the transindividual. None of these conceptual couples can be dissolved into one of the two terms, because there is no term before the process which the couples emerge from: the individual neither precedes the process of individuation, of course, nor does the process of individuation exist before a partial individuation, i.e. the emergence of a structured individual. Although Simondon is well aware of many of the problems emerging within the conceptual constellation he is building in *Individuation*, he does not always reflect on the deeply rooted dualistic tendency which – even in his harsh criticism of hylomorphism – still partially persists. In fact, his entire philosophy *assumes* conceptual couples in the attempt to dismantle any *fixed* relation between them. This is why his philosophy of individuation often assumes also a pre-critical framework in order to grasp the concept of relation:

> Before any exercise of critical thought over the conditions of judgement and of knowledge, one should answer the following question: what is relation? (I 320)

It is nevertheless clear that 'relation' cannot be understood independently of a system of already instituted significations, i.e. the collective which is the result of a transindividual individuation in which information is transformed into signification (I 307). This should suffice to dismiss any charges of pre-Kantianism[18] against Simondon, who, from an anti-Kantian perspective, poses the epistemological problem too:

> In order to understand *how being can be conceived*, it is necessary to understand how it individuates, because this individuation supports the validity of all the logical operations related to it. (I 321)

If being is always ontologically derived (a being is always the *result* of an onto-genesis) such as the subject is, looking for a 'first philosophy' makes no sense, either in the form of a critical philosophy establishing the a-priori conditions of knowledge,[19] or in the form of a plain ontology.[20] But that is not all.

---

[18] Simondon's friend Mikel Dufrenne (see MEOT 7) was testing a phenomenological resolution of the Kantian a-priori problem which he called, before Deleuze, an 'empiricism of the transcendental' (Dufrenne 1959: 284. About the influence of Simondon's thought on Deleuze's 'transcendental empiricism', see Sauvagnargues 2012). In what follows Dufrenne patently contrasts Simondon's 'ontogenetic' solution, of which he declares he contests the results while sharing the inspiration: 'is it necessary then to go back to a philosophy of nature, a pre-critical ontology? [...] this ontology takes the problem of time seriously, the time of genesis [...] Nevertheless the project of a pre-critical ontology, although widely legitimate, seems to us impracticable: the subject as a transcendental cannot be generated starting from the world' (Dufrenne 1959: 284). For an attempt to connect Merleau-Ponty's and Simondon's ontologies from a phenomenological perspective, see De Bestegui (2005).

[19] Here is a clear statement where Simondon writes against subject and object as the 'terms' of *Kantian criticism*: 'it seems, in effect, that a certain conception of individuation is entailed by the notion of "term". When reflection, before any ontology, aims to define the conditions of validity of judgement, it refers [...] to the subject and the object as terms' (I 320).

[20] 'As for the reproach that I do not commence from the study of being, I hold this to be impossible' (FIPD 756).

Thanks to the hypothesis of the pre-individual, according to Simondon one can conceive individuation itself as different from the 'origin', subsequently refusing the name of 'first philosophy' to ontogenesis: 'individuation is *the event of a moment of being* which is not primordial' (I 320). Philosophical thought is therefore always 'second': phenomenologically second to common sense and the sciences concerning structures, ontogenetically second to pre-individual processes. Of the pre-individual, philosophy can only grasp – but this is already everything – the 'remains' of that which individuation is derived from: 'not only it is not primordial, it also carries on a partial *remainder of the pre-individual phase*' (I 320). This is like saying that philosophy must abandon the imaginary (or worse, ideological) horizon implied by any question about the 'origin' which, instead of disclosing new paths for research, confines it within the idealistic presupposition of a given sense.

And yet science offers certain knowledge of structures and of their functioning, and philosophy is always concerned with the risk of an operation founded on scientific achievements but not reducible to them. In the terms Simondon used in his programmatic writings, one can conclude that, despite there not being science of individuation, a philosophy of individuation is only possible on the basis of the sciences of structures. The hypothesis would explain both the absence of the expression 'science of individuation' and the presence of an ethical exhortation at the heart of *Individuation*'s conclusions,[21] where Simondon theorises an ethics capable of 'grasping' [*saisir*] and 'accompanying' [*accompagner*] the individuation of being (I 331), i.e. an ethics of invention.

This should be an ethics capable of keeping the twofold function of metastability, structural stabilisation and processes triggering, therefore preventing the accomplishment, the end of individuation. Thus philosophy as 'individuation of the subject's knowledge' is grasped by operating against the most probable result,[22] against – one could say – death due to an excess of perfection:

> Only death will be the resolution of all tensions; and death is neither the resolution of all tensions nor the solution to any problem. The decisive individuation is the one that maintains tensions in a metastable equilibrium rather than exhausting them in a stable equilibrium. (I 206)

Once it is assumed that the study of individuation, i.e. of the ontogenesis of the individual from pre-individual tensions, cannot be axiomatised according to the typical modalities of the sciences, the question still remains open as to an alternate path of axiomatisation. The brief programmatic writings and the paper presented at the *Société française de philosophie* evidence the persistence of Simondon's project. The encyclopaedic extension of *Individuation* gathers all the materials for a unified theory of the sciences; the wide breadth of this study progressively transfers the ontological assumption of the quantum paradigm into the epistemological

[21] Simondon just provides a sketch of it in I 330–35.

[22] 'This state of non-functioning is stable, and it is also the most probable. In all domains, the most stable state is a state of death; it is a degraded state in which transformations are no longer possible' (FIP 541).

creation of a 'non-deterministic' theory of information, in order to provide an axiomatic for the social sciences.[23] In fact, with his frontal attack on substantialism on the basis of a non-deterministic conception of *all* processes, Simondon aims to reconstruct the field of social sciences providing a dissolution of classical dualisms (body/mind, immanence/transcendence, necessity/liberty) within ontogenesis. This epistemological production of compatibility is precisely what he usually calls 'axiomatisation'.

According to this view, *Individuation* looks like a huge encyclopaedic work which, starting from the 'facts' and models of sciences, *experiments* with the philosophical possibility of unfolding their ontological and epistemological presuppositions. But Simondon's undertaking is twofold: on the one hand it emerges as the repetition of this philosophical attempt next to any term, concept and image he draws from each field of scientific research, and on the other it repeatedly collapses into a systematic, totalising project. For this reason the problem of a theory of ontogenesis in *Individuation* both recurs under the imaginary shape of a dream of axiomatisation, and appears as its genuinely philosophical face. The philosophical force of Simondon's oeuvre, of which he is not always aware, emerges in the obstinate repetition of the same operation of structuring a subject-object relation (i.e. of knowledge) at the exact scale of each of the systems concerned. It is an irremediably singular operation about which he at least once explicitly assumes the impossibility of providing a definitive formalisation:

> It might be that ontogenesis cannot be axiomatised. This would explain the existence of philosophical thought as perpetually marginal in relation to all the other studies. Philosophy would be the kind of thought set in motion by the implicit or explicit research of ontogenesis in all orders of reality. (I 229)

Only in this sense is ontogenesis for philosophical thought the true mode of grasping actual becoming which, in structuring being, constitutes itself (also) as thought. Thus, once the ideological question on the origins of thought (and being) has disappeared, thought becomes action, praxis,[24] and the question becomes the following: What is thought capable of? What are its preconditions and its possible

---

[23] This is without doubt one of the most plausible interpretations of Simondon's research, at least as long as he is pursuing the axiomatisation of that 'theory of operations' he calls allagmatics (See Barthélémy 2008). According to Guchet, Simondon's program of an axiomatic of the social sciences must be equally distant from a 'scientific positivism' and a phenomenology of the immediate access to proto-experience. This program would link Simondon and Merleau-Ponty to the aim of 'reaching the concrete human being, starting from positive knowledge (psychology, sociology, history)' (Guchet 2001: 103). Although this interesting analysis tends in a way to explain Simondon against the backdrop of Merleau-Ponty's work, it still has the merit of showing some important points of contact between the two which I have tried to take into account.

[24] In fact, Simondon's problem of ontogenesis is 'as directly epistemological as it [is] ontological', it brings us beyond the apparent alternative between a naive 'nature philosophy' and cultural constructivism (Massumi 2009: 37–38). As I will explain in Sects. 12.1 and 12.2 thought itself is in this sense a risky enterprise necessarily connected to the functioning of social systems, i.e. it can be political.

effects? It is only over the threshold of the transindividual that some kind of answer can be given to these questions, in the practices through which living beings actually build the material and symbolic instruments from which collective life continuously emerges.

References[25]

Bardin, A. 2008. Simondon: transcendantal et individuation. In *Les métamorphoses du transcendantal. Parcours multiples de Kant à Deleuze*, ed. G. Rametta, 189–215. Hildesheim/Zurich/New York: Georg Olms Verlag.

Barthélémy, J.-H. 2005a. *Penser l'individuation: Simondon et la philosophie de la nature*. Paris: L'Harmattan.

Barthélémy, J.-H. 2005b. *Penser la connaissance et la technique après Simondon*. Paris: L'Harmattan.

Barthélémy, J.-H. 2008. *Simondon ou l'encyclopédisme génétique*. Paris: PUF.

Barthélémy, J.-H., and V. Bontems. 2001. Relativité e Réalité. Nottale, Simondon et le réalisme des relations. *Revue de Synthèse* 2001(1): 27–54.

Bergson, H. 1934. *La pensée et le mouvant*. Paris: Alcan.

Bontems, V. 2008. Quelques éléments pour une épistémologie des relations d'échelle chez Gilbert Simondon. Individuation, Technique, et Histoire. *Appareils* 2. http://appareil.revues.org/595. Accessed 30 Dec 2014.

Canguilhem, G. 1952. *La connaissance de la vie*. Paris: Vrin (English trans. *Knowledge of life*. New York: Ordham University Press, 2008).

De Bestegui, M. 2005. Science and ontology: From Merleau-Ponty's 'reduction' to Simondon's 'transduction'. *Angelaki* 10(2): 109–122.

De Broglie, L. 1947. *Physique et microphysique*. Paris: Albin Michel.

De Solages, B. 1946. *Dialogue sur l'analogie à la société toulousaine de philosophie*. Paris: Aubier.

Deleuze, G. 1966. *Le bergsonisme*. Paris: PUF.

Dufrenne, M. 1959. *La notion d'«a priori»*. Paris: PUF.

Guchet, X. 2001. Merleau-Ponty, Simondon et le problème d'une «axiomatique des sciences humaines». L'exemple de l'histoire et de la sociologie. *Chiasmi International 3: Merleau-Ponty. Non-philosophie et philosophie*: 103–129.

Massumi, B. 2009. 'Technical mentality' revisited: Brian massumi on Gilbert Simondon. *Parrhesia* 7: 36–45.

Sauvagnargues, A. 2012. Simondon, Deleuze, and the Construction of Transcendental Empiricism. *The Warwick Journal of Philosophy* Special Volume. Deleuze and Simondon: 1–21.

[25] *Simondon's complete bibliography and a list of abbreviations are provided in the* Appendix.

Part II
Organism and Society

> *Donnons donc au mot biologie le sens très compréhensif qu'il devrait avoir, qu'il prendra peut-être un jour, et disons pour conclure que toute morale, pression ou aspiration, est d'essence biologique.*
>
> (Bergson, *Les deux sources de la morale e de la religion*)
>
> *Henri Bergson, prenant un point de vue différent, a clairement défini dans* Les deux sources de la morale et de la religion, *un* état statique, *dans lequel les groupes humains tourneraient en spirale, changeant de génération en génération un nombre restreint de concepts, de prescriptions progressivement compliquées, et un* état dynamique *où les groupes prendraient en ligne droite le sens réel de leurs tendances. Nous serons portés à reprendre, en l'adaptant au point de vue qui nous préoccupe, cette vue extrêmement féconde.*
>
> (Leroi-Gourhan, *Milieu et techniques*)

After displaying the epistemological and ontological framework of Simondon's philosophy – with its basic reliance on quantum physics, its critical reference to concepts derived from *Gestalttheorie* and Cybernetics, and its problematic encounter with both the Bergsonian and phenomenological heritages – it is time to start enquiring into the social and technological implications of a philosophy of individuation. The thresholds between physical, biological and social fields are an important philosophical problem for Simondon. He proposes a method which, instead of delimitating different orders of beings, describes different kinds of processes in mixed systems. Hence I will show how he explains the emergence of human social systems, thematising the peculiar role played by 'affectivity' in this process.

A detailed analysis of *Individuation*, of MEOT and of the posthumously published *Note complémentaire* [Complementary Note] (1958), will clarify the meaning of the concept of the 'transindividual'.[1] My method consists in utilising the

[1] Despite its appeal, the term 'transindividual' is not widespread in Simondon's work. In fact, after appearing both in *Individuation* and in the conclusions of MEOT, it completely vanished. The *Note* was written in the period of the two theses but first published in the 1989 edition of IPC, which Simondon was preparing during the last days of his life. It is a very important *trait-d'union* between *Individuation* and MEOT: in effect, it explicitly treats the relationship between the individuation of technical objects and collective individuation.

bibliography of *Individuation* to unveil the sources that shaped Simondon's peculiar approach[2] and to show his debt to Georges Canguilhem.[3] This will allow me to display the impact of biological concepts on Simondon's theorisation of the genesis and functioning of social systems. Simondon's main philosophical references in this field are Bergson's biological and social theories, Canguilhem's philosophy of life sciences and techniques, Wiener's cybernetics of society and Leroi-Gourhan's palaeoanthropology. This background will introduce a thorough analysis of the crucial role played by technical normativity in Simondon's theory of the social system.

[2] As already pointed out, the bibliography of *Individuation* counts just 20 references, 5 of which concern books on human psyche studied from a biological point of view: Gesell (1946), Goldstein (1934), Kubie (1949), Lewin (1946), Rabaud (1951).

[3] Georges Canguilhem (1904–1995) during the 1950s became professor of philosophy at Sorbonne, succeeding Gaston Bachelard as the director of the *Institut d'histoire des sciences*. A former hero of French antifascism, after the second World War he was a key figure in the French educational system and he inspired plenty of young philosophers educated in the period. See Roudinesco 2005: 15–69. Canguilhem was Simondon's *directeur de thèse* for MEOT.

Chapter 5
From Life to Signification

According to Simondon quantum theory will introduce models of discontinuity in the understanding of organisms. Referring on the one hand to the 'polarisation' of matter[1] and on the other hand to the 'quantic' nature of life,[2] Simondon aims to circumscribe a not yet structured intermediate domain – 'neither continuous, nor purely discontinuous' – in which it would be possible to study the relationship between physical and biological individuation in terms of the theory of information. This perspective of radical immanence 'presupposes a concatenation of physical reality up to superior biological forms, without establishing any distinctions between classes and genres' (I 158).

On this basis, when closing the part of *Individuation* devoted to living beings, Simondon refuses both Bergson's sharp differentiation of continuous life and discrete matter, and Goldstein's 'Parmenidean ontology', since both would contribute to the cancellation of any possible *relation* between the study of biological and

[1] Simondon's hopes for a theory of the polarisation of matter, which should prove 'the relation between what is called living matter (or organised matter) and inert or inorganic matter' (I 203) reminds one of Canguilhem, who often refers to the 'dynamic polarisation of life' starting from which many forms of the organisation of the inorganic might prelude the functional organisation of organisms (e.g. Canguilhem 1943: 155). In this sense, within the bibliography of *Individuation* a text on the *La polarisation de la matière* (1949) stands out, containing the proceedings of a *Colloque International* concerning 'the complementary outcome of the magnetic and electric polarisation of molecules' (1949: 3). About the continuity between physical and biological structures and processes, see also I 151, I 324, I 320. Needles to say, the progresses made during the last 50 years in molecular biology makes this idea of a 'theory of the polarisation of matter' a completely out-dated one (see, for instance, I 203 and IMIN 38 about the function of polarisation in growth processes).

[2] 'Although in an organism everything is linked with everything else, from a physiological point of view different regimes of causality can be isolated, thanks to laws of quantic functioning' (I 204). Also the discontinuities between species 'seem to be connected to the quantic issues which appear in physics' (I 158). On the hypothesis of a 'quantic biology, the reply to modern physics', see Tétry 1948: 322, where the author examines the joined functions of technical instruments and biological organs (the text is part of the bibliography of MEOT).

A. Bardin, *Epistemology and Political Philosophy in Gilbert Simondon*,
Philosophy of Engineering and Technology 19, DOI 10.1007/978-94-017-9831-0_5

physical processes (I 228–29). Simondon does not intend to ground his project on any definition of what the living is, in order to classify a *structure* as organic or non-organic. He rather tries to find a criterion for classifying the different *processes*, consequently understanding them as possibly coexisting within the same individual conceived as a phase-shift system (I 204–5). It is thus the relative independence of the living from its milieu that will offer the model for understanding life as a process of mixed 'temporal series' – what he calls 'vital transduction' (I 164). As is Simondon's common practice, he tries here to analyse the chosen domain through a study of the processes of individuation which would provide ontogenetic 'schemas' one could apply to different fields. These schemas actually become the *paradigms* which Simondon refers to when analysing single processes, not in order to classify them, but to establish their threshold conditions.

With this very strategy in mind Simondon also treats the distinction between vital and psychic life when, in the paragraph *Les niveaux successifs d'individuation: vital, psychique, transindividuel* [The Subsequent Levels of Individuation: Vital, Psychic, Transindividual], he asks himself: 'how can the psychic and the vital be distinguished?'. He declares he will avoid any classification according to genres, and he looks for processes and threshold conditions in order to provide a 'psycho-biological' analysis of the living being (I 127). He thus conceives the relation between organic and psychic life as a relation between two different kinds of individuation: 'psychic individuation is a dilatation, a premature expansion of vital individuation' (I 166). Again, it is a 'slowing down' [*ralentissement*], a 'neotenic amplification' occurring when the organism faces new problems which make its 'affectivity' unable to exercise the normal regulatory action, thus changing the terms of the organism-milieu relationship:

> Actual psychism emerges when the vital functions cannot solve the problems posed to the living being, i.e. when the triadic structure of perceptive, active and affective functions is no longer effective. (I 166)

The quantic nature of the thresholds which separate vital and psychic individuations prevents the conception of psychic life as incorporating ('psychic life is not [...] a superior rearrangement of vital functions' I 165) or 'solving' vital problems (I 166). In short, there is no synthetic *Aufhebung* of life in the thinking individual as its internal phase: on the one hand vital processes continue within psychic individuation as a persisting set of problems (I 166) and, on the other hand, psychic processes do not unify, but rather 'upset' [*dérèglent*] the normal functioning of organic processes. No prevalence of a pole on the other, then: just a phase-shift individual in which neither an organic determinism of psychic life, nor a conciliation of vital forces in a superior psychic unity can make any sense. This stance explains Simondon's conception of a certain 'intermittence' between life and thought in animals and man:

> This means that there are not on the one hand only-living beings and on the other hand living and thinking beings: animals probably just find themselves less frequently than humans in a 'psychic situation'. The human, having available more extended psychic possibilities, in particular due to the resources of symbolism, more frequently calls on psyche; it is the vital situation that is exceptional in the human, and thus humans feel more destitute in it. But it is not a matter of a nature, an essence, serving to found an anthropology;

> it is simply that a threshold is crossed. Animals are better endowed for living than for thinking, human beings better for thinking than for living. Both of them live and think, normally or exceptionally. (I 165)

If it is quite evident that psychic processes cannot be understood in terms of perception, action and affectivity (the 'triadic structure' Simondon refers to *organic* processes), it should also be clear that biological processes take place in phase-shift individuals, i.e. they imply a systemic lack of regulation. From here one can follow the way Simondon deduces the model, the 'schema', of the processes which stand as the preconditions of psychic life and *therefore* of collective life from the study of the living being.

5.1 The System Organism-Milieu: Beyond Homeostasis

For his critical re-elaboration of the notion of organism, Simondon draws on some of Canguilhem's characteristic references: Claude Bernard's 'internal milieu', Walter Cannon's concept of 'homeostasis', the Cybernetic conception of a 'feed-back' machine.[3] All these notions are intertwined for the explanation of complex self-regulatory processes.

A first explicit reference to the concept of 'internal milieu' appears in *Individuation* when Simondon discusses the limits of physical individuation. When he calls Bernard into question, he is trying to differentiate the physical individual from the biological one. Simondon denies that the latter is characterised by an interiority (its 'internal milieu') which would be more 'substantial' than the physical one: the 'being' of all individuals resides in fact in the relationship between the internal and the external milieus. This relationship is not what 'expresses' being, but rather what in general 'constitutes' it (I 128). Also in the case of organisms, the view of a supposed 'interiority' is in fact an anthropomorphism analogous to other attributions of 'interiority' at the basic physical level.[4] On the contrary, Bernard's 'internal milieu' is read by Simondon in the terms of his relational ontology:

> The notion of internal milieu built by Claude Bernard for the purpose of biological research shows – because of the mediation it institutes between the living being and the external milieu – that the substantiality of being cannot be confused with its interiority, even in the case of the biological individual. (I 127)

---

[3] Claude Bernard (1813–1878) and Walter Cannon (1871–1945) were respectively a French and an American physiologist, whose concepts were quite disputed in Paris during the 1950s, where Canguilhem – as the President of the *jury d'agrégation* in philosophy – dictated a good part of the philosophical agenda.

[4] Such an anthropomorphism would depend on an 'immediate belief in interiority' grounded on the perception of one's 'own body': 'the conception of a physical interiority of the elementary particle is a subtle and rooted biologism, which can be found also in the most rigorous theoretical mechanicism of ancient atomists' (I 127). Only the appearance of the theory of relativity allows such a 'biologism' to give way to a 'more rigorously physical conception of individuation' (I 127). The equivoque depends on considering the organism at the scale of its macroscopic structures rather than of the processes it is made of.

When Simondon eventually returns to the notion of internal milieu, he is using the concept of homeostasis to differentiate physical and biological processes. Homeostasis is a system of regulations which grants the stability of the individual's internal milieu thanks to constant interaction with the eternal milieu: while in the 'purely physical' being the relation with the external milieu is limited to the points of direct contact distributed on its surface, in the living being 'thanks to the nervous system and the environment, interiority is everywhere in contact with a relative exteriority' (I 161). In this sense Simondon can represent both the internal and the external relationships into the unique concept of 'associated pre-individual milieu': 'in the living being interiority and exteriority are distributed everywhere' (I 161).[5] Referring to an individual and its associated milieu means for Simondon to figure out an individual-milieu system defined by a relational activity simultaneously involving processes of differentiation of the internal milieu *and* processes of integration of the external milieu into the internal one:

> The action of the individual on itself is of the same kind as its action on exteriority: the individual grows by constituting, in itself, a colony of reciprocally intertwined subsets. (I 209, n. 17)

According to this schema, Simondon reads the structurally twofold activity of all organisms, from the elementary to the most complex, by focusing on the different locations of the processes of integration: the mediation between the inside and the outside. At the elementary level the processes of integration depend on the individual. In the organism, which only lives *as* a colony (coelenterates), the mediating function of integration between the external and internal milieus is carried out by the group. But at both levels the process of differentiation is – according to Simondon – entirely dependent on species characteristics and thus independent of the processes of integration: only in complex organisms is it the vital activity of the individual itself to conjugate the processes of integration with its own differentiation. What is true in all cases, the activity of mediation between internal and external never arrives to merge the two kinds of processes into a unique static structure because, whether concerning the group or the individual, vital activity can only make the two processes (and the corresponding 'physical structures') compatible in a metastable system.[6]

What finally differentiates the activity of structuration we call 'life' from the one which operates for instance in the physical process of crystallisation is the fact that the whole internal milieu of the living being is 'topologically in contact with the

[5] 'In fact, a sufficiently profound psycho-biological analysis would reveal that in a living being the relation to the external milieu is not distributed on its external surface' (I 127). Thus Canguilhem: 'The individuality of the living being does not stop at its ectodermic borders any more than it begins at the cell. The biological relationship between the being and its milieu is a functional relationship, and thereby a mobile one; its terms successively exchange their functions' (Canguilhem 1952: 144).

[6] When the living being appears, 'its equilibrium is the one entailing metastability: it is the case of a dynamical equilibrium which presupposes a series of subsequent new structurations without which the metastable equilibrium would not be maintained' (I 237).

content of the external one' and therefore it tends to support the 'continuation of individuation', while, on the contrary, the internal milieu of crystals is 'in general non homeostatic in relation to exteriority' (I 227). This not only complicates any attempt to define living beings on the basis of the structural differentiation of an internal and an external milieu, it also criticises the concept of homeostasis, strictly linked to the 'stability of the internal milieu'.[7] In fact – as Simondon points out – 'homeostasis is not the whole of vital stability' (I 161). Once again Simondon refers to a profound 'triodicity' of the living being, in which two complementary activities are linked to a third responsible for 'actualising their integration' (I 162).[8]

Now, what is an activity that provides the compatibility of integration and differentiation without being a function of the organism itself? Why does the concept of homeostasis not suffice to describe the dynamical equilibrium of a living system, whether an individual or a group? The point is the following: the complexity of the living being cannot be reduced to the simple homeostasis of the 'internal *milieu*', because it depends on the structural phase-shift of the system composed by the individual *and* its associated milieu, which can only be stabilised through a series of 'subsequent assembling of structures and functions' (I 205). Thus, it is finally possible to understand why in Simondon's perspective one cannot properly speak of *an* homeostasis of the living being. It is evident that multiple homeostatic processes cohere in a system of which the living being is the (always partial) metastabilisation through a transductive process of invention. And this serial repetition of an always-singular invention of new compatibilities between organism and milieu, and not the organism alone, is called life:

> The living being relies on homeostases for developing and becoming, instead of perpetually remaining in the same state [...] there is a power of absolute event, which, although resting on homeostases, uses and exceeds them. (MEOT 151)

In the light of this conception of life as a process concerning the system individual-milieu in its entirety, a whole series of categories related to the common representation of the relationship between the individual, the species and the milieu ought to be reformulated. And first of all, the notion of 'adaptation'.

Simondon's criticism of the biological notion of adaptation in *Individuation* (I 209–14) counterpoints the *Insuffisance de la notion d'adaptation pour expliquer l'individuation psychique* [Insufficiency of the Notion of Adaptation for the Explanation of Psychic Individuation] (I 273–76). In both sections the heuristic efficacy of the distinction between the normal and the pathological is in question, a theme inherited from Canguilhem and burdened with a considerable sociological

[7] The expression comes from Claude Bernard. On the hypothesis of a straight connection between the concepts of 'stability of the internal milieu' and homeostasis, see Sinding (1991).

[8] In the course *Initiation à la psychologie moderne* (1966–67) one can also find a 'triodic theory' of the organism (which is supposed to be also the 'basic schema of the course'), in which the input and output systems are linked by motivation, conceived as 'potential energy' (IPM 290). The technical model of the triode reappears in *Le relais amplificateur* (1976): it is a 'device through which some weak energy, usually carrying information, governs and doses some strong energy [...] thus allowing the actualization of the latter as work' (MEC 135). On the triode see also MEOT 28–29.

relevance, which it will be worth retracing later in order to extend the analysis of normativity from biological systems to social systems.[9] According to Simondon any explanation of the constitution, development and behaviour of an individual in terms of theory of adaptation is ruined by an 'implicit sociology' which disregards the process of individuation. At each level the individual in question – electron, organism or subject – is conceived as structured and oriented, caught into a field of forces to which it adds its own force, entering the relationship with the forces exerted by other individuals, eventually adapting its 'activity' and thus contributing to the final shape of the field concerned. All theories based on a conception of the individual as a stable system (a position which – according to Simondon - Darwinian and Lamarckian Evolutionism, Goldstein's gestaltism, Lewin's theory of 'hodological space' and Dr. Kubie's 'cybernetic psychiatry' equally share)[10] end up assuming the category of adaptation and *therefore* setting the individual and the milieu above the relation from which they emerge. This stance is implicitly (and unwittingly) sociological: it theorises a system of *inter*-individual relations, i.e. relations external to the individuals as such.

The notion of a metastable system, instead, allows Simondon to conceive a multiplicity of layers, or 'phases', which can account for the relative identity of the individual without dissolving it into the system, as the concept of 'normality' precisely does:

> The physical individual cannot be understood starting from laws derived from the study of interindividual relations. In fact, if the individual exists, it is precisely because at its level what becomes preponderant is the action of laws that at the interindividual level are not observable. If a unique kind of relationship existed, the individual would not be isolated from the whole into which it is integrated. Similarly, in psychology, it is not possible to define the normality of the individual through a law which would express the consistency of the human world. If such a law was the only valid one, there would be no individual reality and no problem concerning normality could be posed. (I 275)

---

[9] A full chapter of Kurt Goldstein's *The Organism* (Goldstein 1934) focuses *On Norm, Health, and Disease. On Anomaly, Heredity and Breeding*. The full title of Canguilhem's doctoral thesis of 1943 is *Essai sur quelques problèmes concernant le normal et le pathologique* [An Essay on Problems Concerning the Normal and the Pathological]. A further essay also titled *Le normal et le pathologique* was published in *La connaissance de la vie* (1952), which consists in a brief summary of the former. I will not consider here the *Nouvelles réflexions concernant le normal et le pathologique* Canguilhem added to the 1966 edition of *Le normal et le pathologique*, because they are written after the works Simondon examined here. Also Durkheim, in the *Les règles de la méthode sociologique* (1895), had dedicated a whole chapter to the *Règles relatives à la distinction du normal et du pathologique*.

[10] 'What links the three notions of adaptation, good form and hodological space is the condition of stable equilibrium' (I 213). Simondon attacks both Darwin and Lamarck for sharing the same 'objective conception of the milieu' (I 212; on Lamarck's mechanistic conception of the milieu, see Canguilhem 1952: 131–32). Furthermore, Simondon critically recalls the 'theory of hodological space' of Kurt Lewin, first formulated in Lewin (1935) (I 210–13). And finally, on the same grounds he criticises Dr. Lawrence S. Kubie's intervention at the 6th *Macy* conference on cybernetics in 1949 (I 274–75).

Life is the process of production of a world ('an hodological space') *starting from* tensions and polarisations among sets and processes. 'Normativity' neither pertains to the milieu nor to the individual: it pertains to the process of the individuation of the living system, which is life, i.e. a transductive series generating individuals and milieus. The accent Simondon poses on the theme of relations is so strong because 'the notion of milieu itself becomes deceiving', since it risks to support the hypothesis of a 'given' world, already structured *in front of* the individual, as it happens in the 'type of relation prospected by the theory of adaptation', which takes a result of life processes (the 'unified world') for its precondition (I 211–12).[11] This is the criticism Simondon addresses to Lewin: adaptation is *one* of the modes of life, such as interindividual 'hodological space' is *one* of the modes of group relations, therefore retroactively projecting it as the original condition of all relations, the 'field'. According to Simondon, Lewin's stance can be assimilated to Goldstein's, which he repeatedly attacks in *Individuation* as paradigmatic of the holistic presupposition of a totality (I 214).[12]

Although recognising the organism's power to produce norms, according to Simondon Goldstein is compelled to reduce this normativity to adaptation because of his conception of the milieu as an independent structure functioning as a principle of order which directs the organism's adaptive processes both at the biological and at the psychic level. In *The Organism* Goldstein defines the 'fundamental biological law' as the 'maintenance of a relative constancy, distinctive to each organism, only possible when there is a definite configuration of the stimuli, that is, [a] milieu' (Goldstein 1934: 105). As he clarifies in chapter X, only in relation to a sufficiently constant milieu can the organism maintain its 'individual norm', i.e. an 'adequate' physiological functioning. Not only is this norm different for each individual, but it also changes within the same individual according to the variation of its health conditions. Recovering does not consist in the restoration of the original functioning, but rather in the establishment of a new relationship between organism and environment, whether through a partial restructuration of the individual functioning or, in the worst hypothesis, through a 'restriction of the milieu' (Goldstein 1934: 339).

Notwithstanding the extreme variability of norms and the active stance Goldstein attributes to the organism, his vision is dominated by the reference to the *milieu* as an independent order on which any regulatory process of the organism must depend in order to adopt a 'normal', i.e. adapted, functioning. Similarly, in the brief essay titled *The Concept of Health, Disease and Therapy. Basic Ideas for an Organismic Psychotherapy* (1954), Goldstein defines disease as a 'disordered behaviour' and

[11] In Simondon's words, the world is always phase-shift and therefore 'non coincident with itself', while the milieu as a 'space-one' [*espace un*] is always a partial result of a process of individuation (I 211–13). Here the crucial reference is Uexküll's concept of *Umwelt* (Sects. 9.4 and 10.1).

[12] On Goldstein, see I 213–14, 229, 289. Simondon repeatedly highlights that Goldstein's Parmenidean ontology not only 'prevents the correlation study of the living and the study of the inertial' (I 229), it also tends to entirely absorb the psychic into the organic (I 289).

health as an 'ordered functioning', thus confirming that his approach is based on an adaptive conception of order:

> All treatment of a condition in which a full restitution cannot be achieved, consists in a transformation of the individual [...] The more the patient will *accept this role without resentment*, the more he will be able to realize himself, the more happy (or less unhappy) he will be, the more "healthy" – even in spite of irreparable defects. (Goldstein 1954: 763–64)

In his masterpiece *Le normal et le pathologique* [The Normal and the Pathological] (1943) Canguilhem grounded his teaching on the refusal of the concepts of normality and adaptation as they were currently assumed in medicine, thus placing the study of normativity at the centre of the knowledge of life. The condition for such a use of the concept of 'norm' was its radical detachment from any moral assumption and its dependence on 'the dynamic polarity of life'. The whole of his reading derives from a conception of life as 'normative invention' grounded on the 'propulsive value' of 'physiological constants' (Canguilhem 1943: 155–57). According to Canguilhem, if confronted with the invariant normativity of inorganic matter, organisms show an evident exception, as he clearly reaffirms in another brief essay titled again *Le normal et le pathologique*:

> We can therefore conclude that the term normal has no properly absolute or essential meaning. In an earlier work, we proposed that neither the living being nor the milieu can be called 'normal' if we consider them separately. Only by considering them in relation can we maintain the guiding thread without which we would necessarily have to treat as abnormal (that is to say, we believe, pathological) every anomalous individual, every carrier of anomalies – every individual aberrant in relation to a specific, statistically defined type. Insofar as the anomalous living being ultimately reveals itself to have been a mutant at first tolerated and then invasive, the exception becomes the rule in the statistical sense of the word. But even as biological invention appears to be an exception to the current statistical norm, this invention must be normal in a different, though unknown sense. Otherwise, one would arrive at the biological contradiction that the pathological could engender the normal through reproduction. (Canguilhem 1952: 161–62)

Although Canguilhem explicitly recognises his debt towards Goldstein in theorising the active role of organisms, he seems to take some distance when he disjoins the concepts of normality and health, by asserting the non-contradictory relation between the concepts of 'normal' and 'pathological' and eventually ascribing normality to disease itself: 'life in the pathological state is not the absence of norms but the presence of other norms' (Canguilhem 1952: 166).[13] Life is therefore characterised by the intertwining of different normativity processes the normal/pathological distinction cannot describe, since it rather demonstrates the necessarily 'anthropological' – and in the last instance 'moral' – status of sciences such as human biology and medicine.[14]

---

[13] Canguilhem refers here to the Bergson of *Les deux sources de la morale et de la religion* (1932). As I will show, this reference will be crucial to the study of social systems dynamics both for Canguilhem and for Simondon.

[14] 'In conclusion, we hold that human biology and medicine are, and always have been, necessary parts of an "anthropology". But we also hold that there is no anthropology that does not presuppose

Radicalising Canguilhem's stance, Simondon refuses any scientific value to the normal/pathological distinction, since it avoids neither adaptationism nor anthropocentrism. Furthermore, he not only refers normativity to 'living systems' but to matter itself (not 'inert' anymore), thus building an ontological paradigm which can cover other fields of research. In Simondon, Canguilhem's heritage extends to all the regimes of individuation,[15] from the physical to the social: the different regimes of individuation always cross in metastable relationships since the different normativities produce systemic effects. The dynamism of each phase is thus treated by Simondon in the terms he used to frame the physical paradigm in the first part of *Individuation*: also the organism, '*such as the physical individual*, is made of the consistence of a domain of transduction' (I 276, italics added). Therefore, just as the physical individual is unstable so too is the organism, since the purely homeostatic relationship between the internal and external milieus (i.e. an adaptation relationship) is only *one* of the disparate normativities the system is composed of.

The exceeding of vital normativity on homeostatic processes compels Simondon to a similar reform of the notion of evolution. He uses the term 'evolution' according to the model depicted by the American psychologist and paediatrician Arnold Gesell in order to uniformly describe the process of growth, from embryology to the somatic-psychic development of the child during the first 2 years.[16] In *Individuation* this model is first of all paradigmatic for biological individuation: growth, conceived as the progressive integration of the relationship with the external milieu in the individual through an internal differentiation of the latter, is for Simondon 'the model of *any* vital process' (I 209, italics added). If processes of growth already take place in the physical individual, the 'internal resonance' becomes in the living being a 'rhythmical activity' (I 195): growth, reproduction, learning, are nothing but different aspects of the same 'transductive amplification' of information, which simultaneously produces the individual and its associated *milieu* (I 191). In short, the general functioning of a biological system – at any scale – can be read as a dynamical equilibrium between processes of integration and differentiation 'coupling' divergent processes not necessarily functional to life's *conservation*. There the individual always functions by coupling the processes of integration and differentiation as a mediator, a 'transductor' between internal and external milieus: 'it is the equilibrium between integration and differentiation [that] characterise life' (I 161).

---

a morality, such that the concept of the "normal", when considered within the human order, always remains a normative concept of properly philosophical interest' (Canguilhem 1952: 169).

[15] Canguilhem will carry on his own analysis of biological normativity in the *Nouvelles réflexions concernant le normal et le pathologique* (1966), by critically testing the extension of the paradigm on the social field. I shall consider the essay on *Le problème des régulations dans l'organisme et dans la société* (1955), presumably known by Simondon when he was writing *Individuation*.

[16] 'The description offered by Gesell of human ontogenesis and of the principles through which he interprets it, would prolong the results of general embryology; according to him, these principles are not only metaphorical and descriptive, they also traduce a general aspect of life' (I 207). Simondon's references Gesell (1946) here.

A correct understanding of such a function necessitates the adoption of the terms metastability and information:

> In order to describe the activity of the living, it is necessary to substitute the notion of stable equilibrium with the notion of metastable equilibrium, and the notion of good form with the notion of information; the system in which the living being exists is a metastable universe [...] it is the living being which, through its activity, maintains such a metastable equilibrium, transposes, prolongs and supports it. (I 213)

The same model describes for Simondon the evolution of the species. The schema does not change, because the whole system is still made up of three elements (species, environment and relational tension, I 235) the intertwining of which also at this scale repeats the 'triadicity' of life, i.e. the exceeding of relational activity on the two related terms, in this case species and environment.

Evidently enough, the notion of 'evolution' is clearly differentiated from the one of adaptation: either when it explains the species-milieu*s* relationship, or when it defines the development of the single organism or individual, 'evolution dis-adapts as much as it adapts. The realisation of adaptations is only one aspect of life; homeostases are just partial functions' (MEOT 105). In this sense, the dominance of the relation of adaptation exclusively defines pathological states, in which the 'restricted' normativity linked to the individual-milieu relationship dominates. Pathological means here, in a kind of Bergsonian or rather Nietzschean attitude, regressive. In effect Simondon seems to imply that the return of life on its own steps can be considered pathological, while its continuation corresponds to the essential feature of life itself.

Life is in fact the repeated activation of a metastable system of relations which constitute the organism from the beginning, and is not reducible to adaptation: it is a 'transductive' excess. At the level of the biological colony it is through reproduction that the individual endorses this transductive function:

> The individual is therefore the system of compatibility between two antagonistic functions which correspond, respectively, to integration in a vital community and to the amplification through which the individual transports life out of it. (I 173)

This works for both coelenterate colonies and for groups of more complex species (I 331, n. 12) where the 'concrete vital unity' is a single individual being (I 157). But only in the second case – thanks to the emergence of sexed reproduction – the transfer entails the creation of a community.

5.2 From Affectivity to Emotion

Whether functional to the group (as it happens within insects, where 'the group integrates', I 157) or partially independent of it, the individual is the necessary part of the transductive process called life. But the role it plays varies according to the system: with the appearance of sexuality a threshold is crossed, and the transductive activity is integrated within the individual. Here a true social group emerges, because independent biological individuals appear, with a singular story, which through reproduction do not merely repeat themselves. When firstly describing an 'elementary collective' Simondon refers to the exemplary nature of sexuality: 'sexuality [is] a mix of nature and individuation; it is a suspended individuation, arrested in the asymmetrical determination of an elementary collectivity, the unified duality of the couple' (I 308). The basic couple of sexed individuals is already a step towards community: it is precisely individual reproduction that marks the exit from the status of the colony in which individuality is instead still 'diffused'. Here we witness the true emergence of a social system:

> When the individual, instead of founding a new colony, reproduces itself as an individual, the vital functions of continuity (nutrition, growth, differentiation and motion) must be fulfilled with a new stratification of individual behaviours, the social ones [...] germinal functions are reserved to the same individuals that express somatic functions. There is no more colony then, but community or society. (I 174)

At this level psychic life emerges too: sexuality, as a relation between individuals with completely differentiated 'somatic and germinal functions', is the basic structuring of a new 'field' (both psychic and social) for *all* organisms in which sexual desire is a critical point of intersection between the biological and other levels of complexity.[17]

Simondon places his criticism of Freudian psychoanalysis here. According to him, Freud's doctrine would not succeed in differentiating two irreducible and divergent kind of processes, 'instincts' and 'tendencies', and would reduce both to the concept of 'drive'. Simondon's stance corresponds to his critical identification of Freudianism with a reductionist 'pure organicism' (NC 504)[18] guilty of a double fault: firstly, it would reduce the entire psychic activity to a biological mechanism; secondly, it would conceive that mechanism as intrinsically homeostatic. As we already know, in Simondon's view the tension is instead – both at the biological and at the psychic level – *between* the complex functioning of a system *and* an exceeding process which renders it metastable. In order to explain sexuality Freud had thus to substantialise two principles instead of conceiving the 'field' of tensions which constitutes the psycho-social regime of individuation: for Simondon *Eros* and *Thanatos* are indeed discontinuity itself as opposed to the homogeneous continuity of tendencies that 'can be socially integrated'.[19] As a result, Freudianism would be ineluctably marked by its implicit 'normalising' biologism:

> Freud's doctrine does not sufficiently differentiate instincts and tendencies. It seems to consider the individual univocally and – although structurally and dynamically distinguishing a certain number of zones in it – it allows the idea that the individual could reach a complete integration thanks to the construction of the super-ego; as if being could ever discover a condition of absolute unity. (I 170)

[17] It is worth underlining that in his analysis of the genesis of the group Simondon is not referring exclusively to homo sapiens or, more in general, to mammals, but to the colonies of coelenterates (I 167–71); and even sexuality as a 'complication' is not the monopoly of humans (I 177).

[18] This is what he states in the *Note complémentaire* criticising both Freud and Marx for their reduction of culture to a simple superstructural 'expression' of – respectively – the biological and the economical (NC 504).

[19] Death drive is 'the dynamic limit of the exercise [of life drive], not a different one'; between the two there is therefore a 'functional homogeneity' (I 171).

Freud's error, in short, can be assimilated to the typical *Gestaltic* one: to have considered the possibility of a virtuous resolution of a system of tensions thanks to the notion of 'stable equilibrium' (I 205). In this sense Simondon can maintain that 'Janet's idea of a doubling of personality is perhaps closer to reality than the idea of the unconscious' (I 286).[20] The limits of psychoanalysis reside for him in its inability to consider consciousness other than as an epiphenomenon of the individual organism: an error mirroring the rationalist one of conceiving it as one, clear and distinct. Consciousness functions according to a causal regime which excludes the concept of stable equilibrium, and therefore of individual identity: 'if one supposes that the individuality of the states of consciousness is quantic [...] then a regime of intermediate causality appears' (I 247). In this way all reflexive activities can be understood as effects of the circular or 'cumulative' causality eminently characterising biological and psycho-social systems. But the core of psychic systems is rather the 'intermediate' activity which relates an obscure unconscious causality with clear conscious scopes (I 248). This subconscious 'layer' [*couche*] of transductive relations Simondon calls 'affectivo-emotivity'.

Affectivity and emotions are the 'transductive form of psychic life *par excellence*' because, given the quantic characterisation of consciousness, they are both relational activities which contribute to build the transductive identity of the individual: 'psychism is neither pure interiority nor pure exteriority, but permanent differentiation and integration, according to a regime of associated causality and finality which we will call transduction' (I 247).[21] So, how and why does Simondon, at this point, institute a further distinction between emotion and affectivity? The gestaltic model still constitutes the framework here, since the function of emotion is compared to that of perception. Perception and emotion are operations of unification, respectively of sensations and affects: they are 'two psychic individuations which prolong the individuation of the living', the first by 'discovering a unity of the

[20] Although this does not necessarily imply a fundamental Janetism in Simondon (as advanced by Stiegler 2007: XIV), it probably explains his choice of Jung as a recurrent reference. In Simondon's view, the theory of a subconscious so to say 'distributed' between the individual and the collectivity seems perhaps more suitable to cover the field neglected by the too 'vitalist' and 'hylomorphic' Freudism (I 170), on the one hand, and by the structuralist attempts to provide a linguistic-symbolic inscription of the unconscious, on the other. He probably includes both in the following prise of distance: 'the thesis we'll present will differentiate from the doctrine generally named Psychoanalysis. Psychoanalysis correctly remarked that an unconscious exists in the individual, but it considered it as a complete psychism, according to the model of the observable conscious' (I 248). On the relationship between Simondon's thought and psychoanalysis, a suggestion can be found in Aspe (2002), Chabot 2003: 107–23, Garelli (2005). Simondon only partially inclines towards the phenomenological understanding of the unconscious as 'ambiguous perception' (Merleau-Ponty 1960: 291 ff., see also the critical remarks of Descombes 1979: 87). Nevertheless, Simondon is not at all interested in Lacan's 'work in progress' on the concept of death drive, which was possibly pointing to a similar direction, in particular in seminars from VII to X, held from 1959 to 1963. In the end Simondon's theory of *Imagination et Invention* (1965–66) avoids the use of the notion of the unconscious, although sparse references to Lacanian psychoanalysis appear there (Sect. 9.2). To Simondon's Jungian references and to his peculiar adoption of the notion of archetype I will devote Sect. 12.2.

[21] In this sense 'present' is for Simondon the peculiar time dimension of the psychic being, 'transduction between the field of future and the reticulated points of the past' (I 288).

world', the second by 'discovering a unity of the living being'. In short, perception and emotion are complementary processes through which the living being 'discovers' external and internal consistence: perception is 'not yet constituted emotion' and emotion is 'a kind of insular temporality with its own structure' (I 260).

And nevertheless the two processes differ because of the different level of functioning: perception can entirely take place between an organism and its milieu, while emotion, in order to coordinate 'in the subject' the different affective dimensions, requires the mediation of the collective (I 258). Although emotion and perception may appear to hold a symmetrical role in producing the collective, the point is that while perception structures only the relation between the organism and the external milieu, emotion structures the relationship of the organism with itself *through* the relation with the external milieu, i.e. also with other organisms as *subjects*. Even if a few pages later Simondon seems to take for granted that emotion can be the activity of an organism 'alone', it is patent that a threshold must be crossed before a further mediation *between* perceptions and emotions can successfully take place, a threshold marked by the emergence of 'the domain of the collective *or* transindividual':

> A mediation between perceptions and emotions is conditioned by the domain of the collective, or transindividual; for an individuated being the collective is the mixed and stable point of fusion in which emotions become perceptive points of view and, conversely, perceptive points of view are possible emotions [...] The collective is the stable space-time, the milieu of an exchange, principle of conversion between these two sides of the activity of the living being, perception and emotion. By itself, the living could not go beyond perception and emotion, i.e. of perceptive and emotive plurality. (I 261)

It is worth noting that the paragraph in which Simondon summarizes the entire part devoted to psychic and collective individuation is titled *La zone opérationnelle centrale du transindividuel*; *théorie de l'émotion* [The Central Operating Zone of the Transindividual: Theory of Emotion], where he shows how emotion is converted into signification. Collective systems are in effect characterised by their own regime of individuation, the ontogenesis of the transindividual. Through this process a peculiar metastable system emerges, characterised by inter-subjective (*not* inter-individual[22]) relations structuring the pre-individual potentials which still existed *between* individuated beings as a 'residue' of previous individuations. These potentials simultaneously become *emotion* within the individual and *signification* within the collective. Emotion is therefore equally as internal as external to the subject: for this reason, as an activity *in* the subject and *between* subjects, it 'prefigures the discovery of the collective' (I 314).[23]

[22] Simondon always uses the term 'interindividual' in differential opposition to 'transindividual', while the expression 'intersubjective' – which is of no systematic use – appease here as a synonym of transindividual. Simondon's reluctance to speak of 'intersubjectivity' depends on the clear phenomenological connotation of the term, entailing the reference to a sort of anteriority of the subject in relation to the social: as I will show, this is perhaps a major reason for Simondon's choice of the term 'transindividual' (see Chap. 9, in particular Sect. 9.4).

[23] It has to be remembered that, although the term 'discovery' might seem to refer to a given state of things to be assumed as such, all discovery of significations is for Simondon a paradoxical operation of invention.

5.3 Emotion and Transindividual Individuation (After Goldstein and Sartre)

In the essay *On Emotions: Considerations from the Organismic Point of View* (1951), Goldstein critically refers to Sartre's *Esquisse d'une théorie des émotions* [Ideas for a Theory of Emotions] (1939) where the latter endorsed Janet's theory on emotion as a 'lower level substituting behaviour' (Janet 1903). According to Sartre, emotion would occur in order to substitute the individual's failures – Sartre writes – as an attempt to 'magically transform' the world:

> The origin of emotion is an experienced degradation of consciousness vis-à-vis the world. What consciousness cannot tolerate, it tries to manage falling asleep, imitating the consciousness of sleep, dream, hysteria. (Sartre 1939: 187)

Explicitly contrasting Sartre, Goldstein underlines, on the contrary, the 'active' function of adaptation carried out by emotions and their direct link with the organism's action on the environment: emotions are in fact intertwined with the actions they activate and conduct, and therefore – as Goldstein claims – 'there is *no behavior without emotion*', and '*no action occurs without emotion*' (Goldstein 1951: 38, 47). As Goldstein underlines, this is true also for anxiety, although it is rather a limit-case of fear, so peculiar and undetermined to be at the borders of what we can legitimately call emotion: 'It might serve to avoid confusion if we would not label that condition emotion, but designate it as the *inner experience of catastrophe*!' (Goldstein 1951: 46).[24]

Simondon follows Goldstein in underlining the active functionality of emotions as the necessary de-structuration preluding any re-structuration of identity. In particular, in his analysis of anxiety Simondon seems to endorse Goldstein's perspective, according to which the study of emotions would have revealed 'an *essential character of man, his not being primarily concerned with security*' (Goldstein 1951: 45). It is nevertheless important to notice that, although concerned with distancing himself both from a positivistic and from a phenomenological approach, Simondon ends up posing the problem of emotion exactly in the terms Sartre advanced in his *Ésquisse*, i.e. as a problem concerning the relationship between emotion and affectivity, and the emergence of 'signification':

---

[24] Although the English term 'anxiety' can mean both '*anxiété*' and '*angoisse*', it is to be intended here as '*angoisse*'. In effect, Goldstein makes direct reference to Sartre, who connotes the term as existential, according to the Heideggerian matrix; and furthermore, when writing in German, Goldstein uses the term '*Angst*'. Also Sartre (1939) highlights a certain continuity between fear and anxiety when speaking of an 'indefinite anxiety [*angoisse*]', a 'limit-case' of a fear so intense as to result *almost* 'without object', although still determined: 'one is always afraid of determinate aspects of the night, of the world' (Sartre 1939: 172). On anxiety as the proof of the transindividual in Simondon, see Combes (1999). Although it is perhaps more precise to make reference to the cultural psychoanalysis of Karen Horney (1937), recalled by Simondon in *Psycho-sociologie de la technicité* (1960–61), Combes' book still undoubtedly remains the best introduction to the 'political' Simondon through the lens of the transindividual, and my interpretation of that concept derives from both assuming and challenging her inaugural reading.

> The study of emotions verified this very principle: an emotion refers to what it signifies. And what it signifies is the totality of the human relations to the world [...] the psychological theory of emotion presupposes a preliminary description of affectivity inasmuch human-reality [...] is in fact affective human-reality. (Sartre 1939: 198)

Against Goldstein's 'biological reductionism' Simondon would thus seem to confirm Sartre's attitude according to which 'an emotion refers to what it signifies', i.e. it entails a properly *human* symbolic function (I 289). But something deeper marks the distance between Simondon and Sartre: for Sartre emotion is properly inscribed in human nature, it is a 'mode' of consciousness within the unique horizon of a 'total human reality which *becomes* moved, watchful, percipient, determined, etc.' (Sartre 1939: 197–98, italics added).[25] On the contrary, for Simondon the 'central zone' of the transindividual is 'affectivo-emotivity' which is not *essentially* human, since it is just *a potential start* from which the transindividual can emerge. In fact, no sexualised living being is deprived of the quantic functioning of an 'affectivo-emotivity', which is the very condition of any emergence of 'collective grouping' (I 248–249). This tendency of living beings to become collective still remains inexplicable both for the psychological and the sociological approach,[26] while the transindividual tendency must be understood on the basis of a complex emotional reality which – although preceding the formation of the collective as its condition of possibility – only as 'signification' can become transindividual and, in this sense, *also* human.

In short, 'human reality' is the given horizon of sense in Sartre's research, while in Simondon's that horizon instead appears open, a domain featuring determinate threshold conditions which do not depend on any – either biologically, psychically or transcendentally – pre-defined, and therefore presupposed, human nature.

In *Individuation* one can appreciate a sort of sliding from the 'human-reality' of affectivity to the conception of 'affectivo-emotivity' as the pre-individual (and, of course, pre-human) pole of the emotive process, the potentials of which become collectively structured only through the production and exchange of significations. This shows that Simondon is still working within the phenomenological tradition in order to demount its framework, leveraging on paradigms borrowed from the natural sciences (here biology), and – one might say – playing one contrary reductionism against the other: Goldstein's biologism and Sartre's transcendental 'reduction'.

What about the individual organism as the ground for psychic-life then? The 'pre-individual associated to individuated living organisms' is not as limited as the organism, and therefore, starting from the pre-individual, a single living being can,

---

[25] 'Emotion is one of the modalities in which it [consciousness] grasps (in the Heideggerian sense of *Verstehen*) its own "being-in the world"'; emotion is therefore originally 'provided with sense, it means something for my psychic life' (Sartre 1939: 195).

[26] Concerning the criticism to the psychological approach, one must recall Merleau-Ponty's polemics against Sartre, who would make of the Freudian unconscious 'a case of bad faith, an hesitation of imaginative liberty', subsequently establishing the priority of consciousness (Merleau-Ponty 1952–60: 69, see also Merleau-Ponty 1954–55: 202–03).

through a 'new immersion into pre-individual reality', take part in a new individuation which 'exceeds its own limits and emerges into functions and structures which cannot take place within the limits of the individual living being'. In this sense, 'if one calls the living organism individual, psychic life emerges on a transindividual domain of reality' (I 165–66). In conclusion, the emergence of psychic life in organisms *immediately* entails the transindividual: 'the psychic is the birth of transindividual being [*est du transindividuel naissant*]' (I 165–66).

This is quite evident if one considers the topology of the transindividual. There are determinate conditions in which the living being is obliged to 'intervene [...] as a *subject*' (I 29), i.e. to calculate itself as a factor of the system-problem to be solved, and in order to do that it has to become – I would say – external to itself. This process in which a living being actually *becomes* a subject, determines the paradoxical topology of the transindividual, which 'is not exterior to the individual and nevertheless is partially detached from it' (I 281).[27] Although over the threshold of transindividual individuation psychic life can be defined only through a reference to the collective, this does not mean – as the whole third part[28] of *Individuation* clearly demonstrates – that the analysis of psychic life can be reduced to a phenomenology of social relations. At the level of transindividual individuation, the activity of the psychic living being and the emergence of the collective must not be understood as two different individuations, but rather as different phases of the same individuation. Transindividual individuation simultaneously takes place at different levels, it entails different structures and functions, and the space of communication it opens is a kind of exteriority situated within subjects which extends to their external milieus: it is the complex topological space characterising a system of production *and* exchange of significations.[29]

5.4 Signification and the Emergence of the Subject and the Collective

It is now the moment to recognise the exact function fulfilled by the notion of signification in Simondon's 'system'. As already explained, in a phenomenological milieu it is unavoidable to connect the question of language with the problem of the

[27] It is quite understandable then in which sense Simondon must necessarily criticise Bergson when he explains the relationship between the biological and psychic-collective individuation as irreducible to any of the two terms: 'the transindividual cannot be understood as élan vital, because it is not precisely in continuity with vital individuation' (I 303).

[28] See above Chap. 1, n. 28.

[29] The topology of the transindividual might be named 'extimate' [*extime*], a term which in Lacanian topology refers to the mathematical concept of a torus. When Simondon asserts that 'the transindividual, being non structured, crosses the individual; it is not in a topological relation with it', with the term 'topology' he is polemically referring to two specular ways of understanding the process of individuation starting from the individual: '*immanence or transcendence* can only be defined in relation to individuated reality' (I 304).

emergence of sense from the original relation between the subject and the world. Nevertheless, the concept of information allowed Simondon to conceive forms of organisation and 'polarisation' which *precede* the emergence of the subject *without* the mediation of the concept of 'sense', and therefore to explain communication as an activity not human in itself, but only relative to determinate threshold conditions.[30] According to Simondon's ontogenetic approach, the emergence of sense only takes place within transindividual individuation: 'the disparation existing between the two phases of being contained in the subject is enveloped [*enveloppée*] in signification by the emergence of the transindividual' (I 307). And this production of sense is just the way the subject prolongs, at the level of collective individuation, a process of information exchange which was already present at the biological level, and, far before, in matter itself. Precisely because 'the tension of information supposes a series of possibly open receptors' (FIP 544), whatever system can – by functioning as an 'amplifying *relay*' – receive a signal, modify and re-transmit it, is part of a transductive process. In the case of human species, this function does not depend on any particular organ or instrument, it depends on affectivity:

> Between the information input and the action output, human being lacks something capable of orienting these two extremes and making them communicate. The mediator is here still ill defined: it is affectivity. (HO 32)

It is, in short, the transindividual regime of individuation that, relieving the affective-emotive 'charge', produces and circulates (i.e. 'individuates') significations, thus compensating for the functional deficiency featured in humans. Thanks to the emergence of the collective, the 'tension of information' in the system thus crosses the threshold beyond which an 'amplified' emotion becomes 'signification'.

In this sense, it is again the concept of information that explains signification as a process shaping the field of the transindividual. As aforesaid, Simondon makes the concept of information into a paradigm for all domains of individuation. Although widespread across the whole text of *Individuation*, it is the subject of intense enquiry in the part devoted to the individuation of living beings.[31] At the end of this part, when moving from biological individuation to psychic and collective individuation, Simondon introduces and discusses his criticism of the cybernetic conception of information in a paragraph titled *De l'information à la signification* [From Information to Signification]. The transition from information to signification does not entail a change of paradigm; it prefigures the different declination the concept of information will assume in the new domain, becoming the key concept for the

---

[30] As demonstrated in Chap. 4, signification does not function as a mere 'linguistic instrument', but rather as a 'structural germ' and therefore it cannot be the object of a theory of language. Neither – as Hyppolite proposed during the above commented discussion at the *Société* (Sect. 3.3) – of a theory of 'natural' language. In fact, in Simondon's view, such a 'natural language' from which sense would emerge should be rather the object of a pre-linguistic theory of information: 'What would a natural language be? Is that still a language?' (FIPD 186).

[31] Divided in two chapters: *Information et ontogénèse: l'individuation vitale* and *Individuation et information*.

understanding of psychic and collective individuation. Although grounded on different structures, signification will work there as information does in the physical and biological domains: 'disparation does not give birth to a signal, but to a signification which only makes sense within a determinate functioning. A receptor is needed for disparation to take place; a system with structures and potentials [i.e. metastable]'(I 224).

Now, which kind of system 'with structures and potentials' is the one in which signification emerges? If it is true that 'the existence of the collective is necessary for an information to signify', a signification is born in a system where 'the original charge of nature [i.e. pre-individual] carried by individual beings' is structured and organised in a field of forces: this is the collective (I 307). The collective is therefore simultaneously the condition of possibility *and* the effect of the emergence of signification from an exchange of information: 'to receive information means, for the subject, to operate an individuation in itself; this creates a collective relationship with the being from which the signal is coming' (I 307). In fact, the emergence of the collective and of signification are the same operation of individua(lisa)tion, i.e. – in Simondon's terms – the emergence of a transindividual relation:

> There is no difference between discovering a signification and being in a collective relation with the being in relation to which a signification is discovered, because signification [...] is transindividual. (I 307)

And the final condition of possibility of this transindividual individuation is the existence of 'functioning receptors' which can grant the disparation of 'a system with structures and potentials': this is what Simondon calls a 'subject'.

Then, in which sense is the existence of the collective 'necessary for an information to signify'? The collective and the subjects do at least logically precede signification as its conditions, or are they simply its effects? It is worth recalling that any transductive process institutes both differential relations and their terms. At the level of psychic and collective individuation these 'terms' are 'subjects', which must not be understood as the 'terms' of a transcendental philosophy, but rather as the effects of a real relation.[32] In fact, Simondon names 'subject' both – in biological individuation – the system individual + pre-individual, and – in psychic and collective individuation – the system of the three phases individual + pre-individual + transindividual. Furthermore, in the same instance, after stating that the subject 'bears within itself, more than individuated reality, a non-individuated aspect, [which is] pre-individual', he claims that 'the subject-being [*l'être sujet*] can be conceived as a more or less consistent system of the three subsequent phases of being: pre-individual, individuated, transindividual' (I 310). Given such premises, it is clear that the collective can be neither the system in which subjects emerge (since as individual + pre-individual charge, they would *precede* the collective), nor the

[32] In the conclusion, Simondon states: 'it seems, in effect, that a certain conception of individuation is already contained, at least implicitly, in the notion of term. When, previously to any ontology, reflection tries to define the validity of the conditions of possibility of judgement, it recurs to a determinate conception of judgement and, correlatively, of the content of knowledge, of the object and of the subject as terms' (I 320).

effect of a relationship between pre-constituted subjects (since as individual + pre-individual + transindividual, they would *presuppose* the collective).

To solve this theoretical problem it is necessary to assume the paradoxical matrix of both 'signification' and the collective as Simondon conceives them in order to explain the emergence of transindividual individuation. In this light signification – i.e. the transindividual modality of information – will be what (re)structures the subject in two ways: 'a signification has two meanings: one depending on the structure, the other depending on a functional becoming' (I 264). This means that signification appears both as a structured meaning and as the process from which meaning emerges, and therefore it *must* be simultaneously understood both as structuring the psychic and collective subject and as emergent from the individuation of multiple biological subjects. Similarly, the collective is both a structured system of relations and a process. It is, on the one hand, a system derived from the exchange of significations *between* phase-shift subjects, living beings transforming their pre-individual charges in subsequent individualisations which assume the form of norms, beliefs, actions, words, concepts, etc. And it is, on the other hand, a transindividual operation of signification in which new subjects emerge as 'coherent systems of the three phases'. In short, the collective necessitates the previous individuation of different subjects in order to emerge *through* the production of subjects. The concept thus maintains the paradoxical chronology featuring each process of ontogenesis, as far as it aims to cancel the question of origin as false and deceptive.

In conclusion, Simondon transfers into the field of psychic and collective individuation the basic matrix of the individual/individuation relation which traverses the whole text of *Individuation*, determining the semantic shift I already highlighted in Sect. 1.1 between the alternate reference of the same notion to structure or to operation. The collective also undergoes this. Although the use of 'transindividual' to indicate the process, and 'collective' to indicate its structured result is prevalent, the explicit identification of the collective and transindividual is so frequent that the two terms can be in fact considered synonyms. What actually takes place are processes constituting subjects psychically and collectively individuated *and* relational activities between subjects producing structures: this paradoxical simultaneity defines the transindividual regime of individuation in which what is 'collective *or* transindividual' emerges (I 261, italics added).

References[33]

1949. *La polarisation de la matière*. Colloque international du CNRS (Paris, 4–9 April).

Aspe, B. 2002. La pathologie au lieu du transindividuel. In *Gilbert Simondon: une pensée opérative*, ed. J. Roux, 19–34. Saint-Etienne: Publications de l'Université de Saint-Etienne.

Bergson, H. 1932. *Les deux sources de la morale et de la religion*. Paris: Alcan (English trans. *The Two Sources of Morality and Religion*. London: Macmillan & Co., 1935).

[33] *Simondon's complete bibliography and a list of abbreviations are provided in the* Appendix.

Canguilhem, G. 1943. *Le normal et le pathologique*. Paris: PUF, 1966 (English trans. *The Normal and the Pathological*. New York: Zone Books, 1998; First edition: *Essai sur quelques problèmes concernant le normal et le pathologique*. Paris: Les Belles Lettres).

Canguilhem, G. 1952. *La connaissance de la vie*. Paris: Vrin (English trans. *Knowledge of Life*. New York: Ordham University Press, 2008)

Chabot, P. 2003. *La philosophie de Simondon*. Paris: Vrin (English trans. *The Philosophy of Simondon: Between Technology and Individuation*. London: Bloomsbury, 2013).

Combes, M. 1999. *Simondon. Individu et collectivité, pour une philosophie du transindividuel*. Paris: PUF. (English trans. *Gilbert Simondon and the Philosophy of the Transindividual*. Cambridge, MA: MIT Press, 2013).

Descombes, V. 1979. *Le même et l'autre. Quarante-cinq ans de philosophie française (1933–1978)*. Paris: Minuit.

Durkheim, É. 1895. *Les règles de la méthode sociologique*. Paris: Alcan (English trans. *The Rules of Sociological Method: And Selected Texts on Sociology and its Method*. London: Macmillan, 1982).

Garelli, J. 2005. La remise en cause de l'inconscient freudien par Merleau-Ponty et Simondon, selon deux notes inédites de Merleau-Ponty. *Chiasmi International 7: Merleau-Ponty. Vie et Individuation*: 75–87.

Gesell, A. 1946. The ontogenesis of infant behavior. In *Manual of child psychology*, chap. 6, ed. Carmichael, L., 295–331. New York/London: Wiley (French trans. *L'ontogénèse du comportement de l'enfant*. In *Manuel de psychologie de l'enfant*, Paris: PUF, 1952, vol. I, Chap. 6, 470–527).

Goldstein, K. 1934. *Der Aufbau des Organismus*. Haag: Nijhoff (English trans. *The Organism: A Holistic Approach to Biology Derived from Pathological Data in Man*. New York: Zone Books, 1995).

Goldstein, K. 1951. On emotions: considerations from the organismic point of view. *The Journal of Psychology* 31: 37–49.

Goldstein, K. 1954. The concept of health, disease and therapy: basic ideas for an organismic psychotherapy. Proceedings of the association for the advancement of psychotherapy. *American Journal of Psychotherapy* 8/1: 745–764.

Horney, K. 1937. *The neurotic personality of our time*. New York: Norton.

Janet, P. 1903. *Les obsessions et la psychasthénie*. Paris: Alcan.

Kubie, L. S. 1949. The Neurotic Potential et Human Adaptation. In *Transactions of the sixth Conference on Cybernetics (1949)*. New York, 1950.

Lewin, K. 1935. *Principles of topological psychology*. New York: McGraw-Hill.

Lewin, K. 1946. Chapter 16: Behavior and development as a function of the total situation. In *Manual of child psychology*, ed. L. Carmichael, 791–844. New York/London: Wiley.

Merleau-Ponty, M. 1952–60. *Résumés de cours (Collège de France, 1952–1960)*. Paris: Gallimard, 1968.

Merleau-Ponty, M. 1954–55. *L'institution. La passivité. Notes de cours au Collège de France (1954–55)*. Tours: Belin, 2003.

Rabaud, É. 1951. Sociétés humaines et sociétés animales. *Année Psychologique* 50: 263–272.

Roudinesco, E. 2005. *Philosophes dans la tourmente*. Paris: Fayard.

Sartre, J.P. 1939. *Esquisse d'une théorie des émotions*. Paris: Hermann.

Sinding, C. 1991. Du milieu intérieur à l'homéostasie: une généalogie contestée. In *La nécessité de Claude Bernard*, ed. J. Michel, 65–81. Paris: Klincksieck.

Stiegler, B. 2007. L'inquiétante étrangeté de la pensée et la métaphysique de Pénélope. Preface to Simondon, G. *L'individuation psychique et collective. À la lumière des notions de forme, information, potentiel et métastabilité*, I – XVI. Paris: Aubier.

Tétry, A. 1948. *Les outils chez les êtres vivants*. Paris: Gallimard.

Chapter 6
Genesis and Structure of the Collective: The Transindividual

Although, particularly in *Individuation* and in the *Note complémentaire*, Simondon tries to understand human 'society' by differentiating it from biological 'community', this does not mean that an anthropology aimed at defining the human domain as opposed to that of animals is at stake in his philosophy. And nevertheless Simondon openly adopts some concepts of social psychology as 'in group' and 'out group', namely from Kurt Lewin and Gordon Allport, that allow him to describe the fundamental processes shaping the domain of collective individuation, and to challenge Bergson's distinction between a 'closed' community and an 'open' society. Reconstructing Simondon's sources is necessary to understand how he tries to provide an analysis of the social system without presupposing a given anthropology, but rather exploring different perspectives on the human/nature threshold through the concept of transindividual.

In his study of the social system Simondon relies on a twofold approach. In *Individuation* he progressively gives conceptual shape to the processes he calls transindividual thanks to the ontogenetic analysis of the phenomena of belief, work and language. In all these domains, a 'basic community' emerges just over the threshold of the biological group, but it is clearly only through 'significations', 'technicity' and 'implicit belief' that the structural ambivalence of all collective processes as simultaneously closed and open becomes clear. Hence my analysis of the *Note complémentaire*, where Simondon adopts the distinction between a 'closed' community and an 'open' society in order to describe the field of tensions traversing a social system, will finally confirm Simondon's reliance on the Bergsonian closed/open paradigm, and frame his call for a theory of the regulation of social processes.

A. Bardin, *Epistemology and Political Philosophy in Gilbert Simondon*,
Philosophy of Engineering and Technology 19, DOI 10.1007/978-94-017-9831-0_6

6.1 The Terms of the Problem: Community and Society

When analysing adaptation processes in animal societies, Simondon rhetorically asks: 'is it necessary to state that sociality resides in the characteristics of the species?' (I 300). Resting on long dated examples, Simondon claims that 'sociality' can be localised in the characterisation of a species only when a morphologic and functional distinction genetically compels individuals to be constantly part of a group, as it happens for ants. In species with less differentiated individuals, as mammals, on the contrary 'the group can be intermittent': in this case individuals, due to their lack of biological specialisation, maintain a certain margin of independence from the group. Societies of insects form a much more cohesive whole than societies of mammals, and this is why only the latter are open to further individuations. The less a behaviour can be reduced to the expression of predetermined patterns, the more individual development is independent of the group: in this sense Simondon can characterise homo sapiens as the living being whose functional specialisation is peculiarly unachieved, and whose development is 'evolutive individual by individual' (I 301).

But it is only when in *Individuation* he is achieving the study of the individuation of living beings, that Simondon elaborates the concepts for the understanding of the complex social systems emerging along with psychic and collective individuation. In order to avoid reducing the collective to the product or the function of a biological species, his strategy consists in defining the different processes involved in the formation of groups, presupposing neither their identification nor their opposition to what would be 'specific', 'communitarian' or 'individual', i.e. without ever reducing group cohesion to a genetic outcome or to the virtuous result of the initiative of some individuals. This conceptual task is pursued along with a process of terminological reform that determines the oscillation of many of the terms Simondon often uses without univocally defining them. For this reason the restricted use of the term 'community' in the third part of *Individuation*, can be a good occasion to fine-tune the instruments for the analysis of the peculiar open-close dynamics characterising transindividual individuation that I will carry on in this chapter.[1]

In the *Note complémentaire* we can find a clear differentiation of 'community' from 'society' at the level of psychic and collective individuation: 'community is biological, while society is ethical' (NC 508). The indisputable anteriority of community seems to confirm here ethology as the privileged field for the understanding of social ontogenesis, since societies cannot exist without (biological) communities,

---

[1] In the whole part devoted to *Les fondements du transindividuel et l'individuation collective* the term appears just once, in the expression 'community of beliefs' (I 294). Otherwise it appears in the part concerning psychic individuation: firstly, in the expressions 'community of action', 'affective community' and 'community of yoke' (I 248–49); subsequently in strict connection to the notions of 'personality' and of 'interpersonal relation' (I 265); and finally as an adjective qualifying a 'deviation' which is said to be 'communitarian' (I 281–82). The concept is otherwise crucial – as I am going to explain – to the *Note complémentaire*, in particular in the section concerning *Individuation et invention*.

while – Simondon writes – 'it would be untrue to assume the contrary' (NC 508). The same structure of *Individuation* – where the use of the term 'community', central in the second part on living beings, tends to leave a place to the term 'society' in the third on psychic and collective individuation – would seem to confirm the hypothesis of a biological characterisation of community shared by human beings and animals: on this biological basis, the incidence of what one presumes properly human (language, work, or something else) would subsequently contribute to the emergence of a more complex social system.

But this is not the case. In fact the 'community' we are concerned with at the transindividual level *is not* the biological community. On the contrary, at that level the terms 'community' and 'society' denote, in Simondon's argument, different tensions which only make sense as simultaneous internal tendencies of the same system entailing different levels of cohesion. This is what he tries to represent with a simple image: magnetised metals – he says – have different degrees of resistance to demagnetisation, depending on whether they have been smelted below or above their Curie temperature.[2] He is speaking of a different scale cohesion producing different structural effects: on the one hand we have a simple 'group phenomenon', while, on the other hand, we have a 'magnetisation and orientation of each single molecule individually considered' (NC 508). What Simondon aims to suggest with this 'structural analogy'[3] is that the pervasive cohesion which keeps together social systems cannot be described in merely structural terms, since what actually appears as an identity of structures can hide different processes of individuation and, consequently, quite different performances. It is consequently impossible to maintain that, for Simondon, society is simply made of the biological nature of community *plus* something else, because both community and society actually coexist as different tendencies in the same social system.

The path towards understanding what is at stake in this apparently simple opposition of community and society as internal tendencies of the social system is quite winding. Indeed, Simondon's analysis of collective individuation is complicated by so many factors that one cannot hope to clearly understand his point without knowing the sources and the philosophical discussions from which his jargon concerning transindividual individuation originates. This is mainly forged in the paragraph *Groupes d'intériorité et groupes d'extériorité* [Groups of Interiority and Groups of Exteriority], where Simondon enacts a direct and explicit polemic against Bergson's distinction between 'closed' and 'open' society in *Les deux sources de la morale et de la religion* [The Two Sources of Morality and Religion] (1932). Simondon's claim that 'it is useless to follow Bergson, by separating open and closed groups',

[2] The 'curie point', is the temperature at which induced magnetism changes the original magnetism of a determinate material.

[3] This is – Simondon says – '*only* a structural analogy', an explanatory image deprived of any scientific or philosophical value. Simondon is evidently referring to the way Rabaud explains, according to a clearly Durkheimian scheme (mechanical/organic solidarity), the functioning of animal societies (Rabaud 1951: 269) in order to highlight that any attempt to study the origin, nature and meaning of social life as *exclusively* human, would be deprived of any foundation (263).

points to the fact that Bergson's conceptual distinction cannot unveil the operative 'central zone' of social systems precisely because it makes of the two internal coexisting tensions two opposed principles, while 'the social is at short distance open and at long distance closed' (I 294). As I am going to show in what follows, a basic open/close paradigm grounds also the functioning of 'community' and 'society' in Simondon's study of the dynamics of social systems.

6.2 In-Group, Out-Group and Group Personality

In order to adequately conceive society as a process, Simondon puts forward the expressions 'in-group' and 'out-group', which he declares to derive from some 'American researchers'.[4] The terms, which recall but do not parallel the Bergsonian 'dyad' – *open society* and *closed society* – are translated by Simondon himself as 'interiority group' (*group d'intériorité*) and 'exteriority group' (*group d'extériorité*). The distinction is further complicated by Simondon's reference to the concepts of 'personality' and 'group personality', which he probably derives from Abram Kardiner through his friend Michel Dufrenne.[5] I shall try to reassemble the puzzle before presenting the whole picture.

In Kardiner's psychodynamics of social organization, the 'basic personality structure' is in homo sapiens – the most 'plastic' of animals – the 'constellation' which shapes the primary form of collective identity. This individual *and* collective basis constitutes the 'sense of reality' of the individual and the support for each subsequent restructuration of personality aiming at adaptation (Kardiner 1939: 469). Although the effect of collective processes, this basic structure is assumed by Kardiner as an individual result of different 'in-group formations', which he mainly identifies with primary institutions, such as families or clans (21–22).

[4] In 1952 Simondon spent some time at the University of Minneapolis (Minnesota), where he attended a course of social psychology (Cuviller 1962: 157).

[5] The psychoanalyst Abram Kardiner was the author of *The Individual and its Society*: *The Psychodynamics of Primitive Social Organization* (1939). In 1953 Dufrenne wrote *La personnalité de base*, in which he proposed a phenomenological reading of Kardiner's book, on which C. Lefort had opened the discussion since 1951 with his *Notes critiques sur la méthode de Kardiner* (now in Lefort 1978: 113–130). According to Lefort, through the concept of 'basic personality structure' '[Kardiner] is interested in individuating in a culture the equivalent of a Subject or, to put it better, a general experience of the world which would be the matrix of any individual experience' (Lefort 1978: 55). In 1969 Lefort eventually edited the first French edition of Kardiner's book, thus continuing the discussion in his introduction: *Ambigüités de l'anthropologie culturelle*: *introduction à l'œuvre d'Abram Kardiner* (Lefort 1978: 131–87). But Simondon's debt can also be traced back to Durkheim: 'we say *our individuality* and not *our personality*. Despite this the two terms are often confused, it is necessary to explicitly distinguish them. Personality is essentially made of supra-individual elements' (Durkheim 1914: 215, n. 1). The term also appears referred to groups in Leroi-Gourhan (1943). On the possibility of reading the traces of this Durkheimian heritage in Simondon, see in particular Chap. 11.

Simondon welcomes and re-elaborates this concept of 'personality', more broadly linking it to the process of individuation, an operation Dufrenne already derived from Merleau-Ponty: 'we are not the group, we *become* it, and it is precisely in becoming that the reality of the group coheres; we become constituent as far as we become constituted' (Dufrenne 1953: 12).[6] For this reason, in order to avoid making terrible blunders, one cannot understand the way Simondon conceives 'personality' without referring to group individuation, and particularly focusing on the way he conceives the functioning of the in-group. And this cannot be satisfactorily analysed without directly referring to the 'American researchers' Simondon refers to in *Individuation* (I 294).

It is not easy to divine Simondon's precise sources for the expression 'in-group', which already appears in Kardiner's book (1939), and is used during the 1940s in Kurt Lewin's studies on group dynamics and the formation of stereotypes, and rapidly spreads in the technical literature eventually becoming of common use.[7] In the article *Conduct, Knowledge, and Acceptance of New Values* (1945), Lewin and Grabbe, assuming a pedagogical stance of *social engineering*, advance the hypothesis that 'the processes governing the acquisition of the normal and abnormal are fundamentally alike' and thus norms in general can be accepted or refused by individuals through the mediation of the group, the interactions with which in fact determine 'what exists as "reality" for the individual' (Grabbe and Lewin 1945: 57). According to the authors, the fact that the cognitive structure, valences and values depend on *different* laws, makes any intervention of 're-education' particularly difficult (59): for an effective and permanent change, the individual's entire system of values must be involved, i.e. what someone calls 'a change in the culture of the individual; by others, a change of his super-ego' (64). In the Lewin-Grabbe model the concept of *in-group* answers therefore the question 'How, then, can acceptance of the new values be established if not by an item-by-item change in conviction?' It is precisely through 'the establishment of what is called an in-group,

[6] In accordance with the phenomenological quest for 'invariants' in human nature – through the concepts of 'signification', 'intentionality' and 'intersubjective relationships', Dufrenne uses the concept of 'basic personality' to shape 'the human universal' (Dufrenne 1953: 34, 321). However, with the expression 'human nature' Dufrenne neither refers to the biology of homo sapiens, nor to the product of primary institutions, but to what is called 'human as a spring or a potential, rather than as something given' (Dufrenne 1953: 71), where 'potential' means something 'not too far from the idea of structure, as the one Merleau-Ponty opposes to the idea of substance' (Dufrenne 1953: 69, n. 2).

[7] Again, the bare bibliography of *Individuation* does not help here. There is no mention in it of Simondon's 'American' sources, apart from Charmichael (1946), in which the quoted article of Lewin can be found. Although all the themes linked to the concept of 'in-group' appear in it, the term does not. And nevertheless, the previously quoted book on Kardiner by Dufrenne (1953) presents all the 'overseas' references that Simondon will make use of in his subsequent works: K. Lewin, G. W. Allport, F. Moreno, K. Horney, R. Benedict. In this sense it seems probable to consider Dufrenne's work an hypothetical bibliography of what Simondon refers to as the 'American researchers'. A text well used by Dufrenne is Lewin (1948), a collection of essays containing Grabbe and Lewin (1945) where a fully shaped concept of 'in-group' appears, as far as I know, for the first time. On the importance of Dufrenne for Simondon, see Carrozzini 2011: 215 ff.

i.e., a group in which the members feel belongingness' (67) that a solution to this problem can be found. The emergence of such a feeling of 'belongingness' to the *in-group* is the only effective means for the members of the group to be turned towards new values.

Now, the technical use of the in-group/out-group conceptual couple is already well established in the period preceding Simondon's writing of *Individuation*, within G. W. Allport's *The Nature of Prejudice* (1954).[8] In the chapter *Formation of Groups Internal to the System* Allport defines in-groups as plural, as the groups the individual is involved in:

> It is difficult to define an in-group precisely. Perhaps the best that can be done is to say that members of an in-group all use the term *we* with the same essential significance. Members of all family do so, likewise schoolmates, members of a lodge, labor union, club, city, state, nation. In a vaguer way members of international bodies may do the same. Some we-organisations are transitory (e.g., an evening party), some are permanent (e.g., a family or clan). (Allport 1954: 31–32)

According to Allport the reference to in-groups is a vital necessity for the individual who, pushed by its innate desire for security, adopts a conformist attitude, thus absorbing values and forming her/his personality and prejudices altogether. Individuals do belong to some in-groups since the first years (e.g. family), while during the rest of their lives they strive to or they actually arrive at belonging to other groups. Actual 'belongingness' defines an 'in-group', while desire to belong points to a 'reference group':

> The concepts of in-group and reference group help us to distinguish two levels of belongingness. The former indicates the sheer fact of membership; the latter tells us whether the individual prizes that membership or whether he seeks to relate himself with another group. In many cases, as we have said, there is a virtual identity between in-groups and reference groups; but it is not always so. Some individuals, through necessity or by choice, continually compare themselves with groups which for them are not in-groups. (Allport 1954: 38)

Whether the two kinds of groups coincide or not, the multiplicity of groups explains on the one hand the influence of the collective and on the other hand the individual's relative liberty: not exactly the liberty to choose which group to belong to, but rather the liberty to choose different 'reference groups' and *try* to become part of them. On the other hand, the groups to which the individual neither actually belongs nor desires to belong, Allport calls 'out-groups'. Although the attitude of the individual towards out-groups can occasionally reinforce the in-group's sense of belongingness, it is not necessarily hostile in itself (Allport 1954: 37–46).

[8] Allport (1945) is also quoted by Grabbe and Lewin in the above article ('The individual accepts the new system of values and beliefs by accepting belongingness to a group. Allport formulates this point as a general principle of teaching people when he says, "It is an axiom that people cannot be taught who feel that they are at the same time being attacked".' Grabbe and Lewin 1945: 67). Furthermore, Simondon's recurrent references to the theory of prejudice makes him the main candidate among the 'American researchers'. The term 'in-group' was also used by Coser (1956), who derived his thesis concerning the relationship between social conflict and collective identity from Simmel's theory, in a criticism internal to Talcott Parsons' sociology; but Coser's text will gain a certain success only from the 1960s onward, and Simondon never refers to it even indirectly.

A concentric structure of groups results, in relation to which the personality of the individual is defined by belonging to internal groups, by not belonging to out-groups and by pointing to reference groups which can be both 'internal' or 'external', even though the individual tends (and tries) to make them all internal. It is therefore always possible that, on the grounds of a personality structured through one or several in-groups, the individual might choose an out-group as a reference group and thus eventually arrive to institute an actual 'belongingness' to a new in-group. From family to humanity, Allport draws the horizon of a cosmopolitan perspective, endorsing an explicit ecumenical project to which, as I will show, Simondon will not be completely indifferent.[9]

It is precisely within the emergence of group relationships *between* psychic and collective individuation that the concept of 'personality' appears in *Individuation*: personality is the relational activity that keeps an individuation linked to its subsequent individuations (I 267).[10] Or, better, *it is* the system of individuation-individualisations which, through subsequent de-structurations and re-structurations, achieves its discontinuous shape, due to the fact that its domain is quantic (I 268). This process through which personality is structured does not depend on the individual only: personality is 'of group' by definition, it emerges by 'syncrystallisation' from some 'interiority group' (I 298). The 'interiority group' is in fact the minimum quantum of personality that necessarily precedes – as its condition of possibility – the formation of a structured 'personality' which *does not* coincide with the individual:

> It is necessary that a community of given conditions of personality allows the formation of a single mediation, a single personality for two different individuations and individualisations. (I 265)

The single mediation between two processes of individualisation does not entirely cover the involved individuals, which are in fact part of several similar processes (or which – one might say – belong to different in-groups): 'the interpersonal relation just involves a certain zone of each personality' (I 266). It is therefore a true superposition of 'parts' that partially exonerates individuals from communication, since it rests on a partial identity of lived experience, a kind of 'communication of consciousnesses' to ground which subjective consciousness does not suffice (I 266).

In this sense Simondon can say that the in-group entails an 'interior coincidence' of past and future in different individuals, while 'out-group' is a given 'reticular structure' through which each individual must necessarily pass in the course of his personal individuation (I 294). This model gains a double perspective on the relationship between the individual and the collective: on the one hand individuals

---

[9] Allport even quotes pope Pio XII's encyclicals *Humani Generis Unitas*: 'The unity of people, he said, is a unity of attitude – of tolerance and love – not a unity of uniformity' (Allport 1954: 42–43). Simondon himself endorses, particularly in PST 329 ff., a quasi-ecumenical perspective (see Chap.11).

[10] An editing mistake might deceive the reader: Simondon inverts the meaning of the terms *individuation* and *individualisation* when he designates 'individuation' as the kind of process 'which requires the support of the individuated living being in order to take place' (I 267).

structure the collective by coinciding in the same in-group, on the other hand the collective is the already structured symbolic framework of the individual's psychic and collective individuation. Once again, the process must be grasped in its centre, in this case *between* the individual and its social milieu. That is where Simondon situates 'society': 'at the border between the *in-group* and the *out-group*' (I 294).

From the point of view of the life-story of the individual, in-groups represent the primary individuations through which the individual is structured and through which it relates to out-groups. The first function of the in-group is to ground simultaneously personal identity and collective identity[11] and therefore mediate the relation with out-groups, which represent individuations that are originally external to the individual, but which can force the individual to re-enact previous achieved individuations. This mediating role of the in-group is often overlooked by interpreters, even though Simondon makes it quite clear that 'the social is made up of the mediation between the individual being and the *out-group* through the *in-group*' (I 294).[12]

This finally explains why Simondon explicitly connects the in-group to the transindividual individuation which exceeds all biologically determined social formations:

> Beyond these biological, bio-social and interindividual relations, another level exists which can be called the level of the transindividual; *this level corresponds to interiority groups*, i.e. an actual group individuation. (I 302)

Both personality and signification contribute to the emergence of affectivo-emotivity within social systems, but on different levels. As previously mentioned, personality is essentially 'of the group', and it makes part of the individual processes of individuation converge.[13] It is not a bond, it is an identity *tout court* which, superposing only 'parts' of individuals creates the *illusion* of a link between structured personalities and projects it onto an imagined community: 'the particular consistence of each personality allows for the belief that the community exists for the whole set of the two personalities' (I 266). On the contrary, transindividual signification emerges from the crisis of group identities; it is collective and concerns the relationship between structured individuals in course of individuation on the condition that they *do not* coincide. On the basis of a shared living-experience forming a group personality, the collective emerges as a newly shared symbolic

[11] Such as the partial identifications of 'group personality' and community demonstrate (I 265–66 and I 299). On this topic see also MEOT: 'the human individual is not linked to the group through its basic functions, whether they be active or perceptive, it is linked to it through the self-regulation enacted by its personality' (MEOT 125).

[12] Lewin expresses the same concept: 'this linkage between acceptance of new facts or values and acceptance of certain groups or roles is very intimate and [...] the second frequently is a prerequisite for the first' (Grabbe and Lewin 1945: 68).

[13] And nevertheless, 'personality' as a structure is conceived differently from 'personality' as an operation: the former is only a 'moment' of the 'quantic process' of personalisation represented by the latter (I 268). I will limit here the use of the term 'personality' to its structural meaning, in order to avoid terminological misunderstanding. The problematic use of the concept of 'personality' in Simondon was first evidenced by Barthélémy 2005: 206 ff., who referred to the concept of 'personalisation' elaborated by Teilhard de Chardin (Barthélémy 2005: 45).

order, a new kind of unity in which individuals maintain different personal features: 'the collective is what makes sense of an individual action as a symbol for other individuals' (I 219).[14]

What about the in-group then? Is it the vehicle of conservative tendencies towards the building and conservation of collective identity, or is it the field of the transindividual processes of innovation exceeding the normative regulation of the consolidate automatisms of group life? The final answer entails an explanation of the use Simondon makes of the terms in-group and out-group in *Individuation*, where any simple attempt to institute a direct correspondence or even a parallelism with the terms community and society, as many interpreters have tried to do, would inevitably lead to a whole series of contradictions and misinterpretations.[15]

In *Individuation* the expression in-group indicates a process of individuation in which a group personality emerges, i.e. a collective identity made up of the 'superposition' of the personalities of different individuals in course of individuation.[16] But this kind of 'structural' identity does not exhaust the potentials of the phase-shift system made of the individual and its associated pre-individual. Therefore, within a structured collective system the different processes of in-group structuration continue to take place, thus *menacing* the homeostatic mechanisms of the system. This is precisely what prevents the system from being completely stable and allows the possibility of further individuations. In this sense the expression 'in-group' refers to a transindividual process grasped from within which forms *and* at the same time metastabilises the social system (while 'out-group' refers to the same process grasped from without, from the point of view – so to speak – of another in-group process entirely external to the former). Thus the term 'community' refers to a social system 'closing', i.e. producing a collective identity through

---

[14] At this level, Simondon also states that the action becomes 'presence', i.e. a 'category of the transindividual' (I 219). This requires further attention, because in *Individuation* Simondon intends 'presence' in two different ways. On the one hand the 'presence' of the individual is the identity of the group itself, it is the process of a collective individuation – made consistent in *one* 'group personality' – through which a 'community' emerges (I 299). On the other hand the individual, insofar as it is capable of further individuations, always exceeds this basic identity thus opening to further processes of collective individuation (I 294). Therefore the singularity of an individual does not reside in its personality, which makes of it an essentially 'group individual' (I 298), but rather in the singular relation between the pre-individual potentials associated to the individual and group personality.

[15] Most interpreters, following Hottois 1993: 88 ff., take the problematic identification of *in-group* and community for granted. This interpretation is misleading as far as it tends to reduce Simondon's choice for 'in-group' and 'out-group' to a merely terminological innovation which would be in some way redundant in relation to Bergson's concepts of 'closed society' and 'open society'. On the contrary, my aim is to show how Simondon's innovation was in fact conceptual.

[16] Simondon has never openly stated the reference to a multiplicity of in-group processes taking place simultaneously in collective individuation. Nevertheless, my reconstruction of his sources shows how this is implicit to his argument. It is worth noting that, although an actual 'multiplicity of personality' can only be pathological (I 286), 'mental pathology is at the level of the transindividual', since it derives from the incapability of the individual to discover a collective signification and therefore to re-structurate the group personality in itself (I 309). On Simondon's supposed 'janetism' see Chap. 5, n. 20.

in-group processes which fix the shared significations, while the term 'society' refers to a social system 'opening', i.e. involved in multiple in-group processes producing new significations. In short, while the first kind of in-group processes express a group personality building a closed community, the second kind of in-group processes exceed this stable identity since they simultaneously open to different communities.

As a result, one must assume that each social system hosts a multiplicity of opening and closing processes, which are *always* defined in Simondon's jargon by the same expression: 'in-group'. The ambivalence of the term expresses the fundamental ambivalence of the transindividual processes from which social systems emerge. Such processes produce collective identity and at the same time do not fix individuals to it because the system thus constituted (the collective) is metastable, thanks to the unexpressed potentials (the 'pre-individual phase') that continues to persist *between* individuals:

> Individual personalities constitute themselves together by superposition [...] the transindividual does not localise the individuals: it makes them coincide, communicate through significations [...] this coincidence of personalities does not entail reduction, because neither is it founded on the amputation of individual differences, nor does it aim at their functional differentiation (which would reduce the individual to its particularities). On the contrary, it is founded on a second structuration which starts from what the biological structuration producing living beings has left unresolved. (I 302)

Transindividual individuation, in short, is simultaneously the condition of possibility and the main risk for the actual existence of groups, it is their psychic and collective life, the non-deterministic core of social systems: 'there is something hyper-functional in groups, i.e. their interiority' (I 301). This internal becoming of social systems is what social sciences have to describe, and 'in-group' is one of the concepts Simondon puts at stake for this purpose. Armed with tools apt to describe this structural ambivalence, Simondon's study of the processes of psychic and collective individuation crosses different domains, where different themes are analysed in the light of the concept of the transindividual.

6.3 Belief, Work, Language

In the part dedicated to psychic and collective individuation, the transindividual shapes issues that traditionally dominate psychological and sociological research. The themes of belief, labour and language are a major example of the conceptual framework Simondon is trying to 'transduct' into the field of social sciences.

6.3.1 Belief

Simondon only occasionally treats the theme of 'belief' in *Individuation*, although it is central to the sociological debate concerning the formation of the social bond. Out of the main thesis, it remerges when the problem of the homeostatic normativity of groups is directly challenged under the 'label' of 'sacredness'[*sacralité*].[17] As far as *Individuation* is concerned, Simondon's conception of the transindividual allows us to focus on the fundamental ambiguity of 'belief' and of related institutions. The undisputable role of belief in the ontogenesis of the collective does not imply that it can be uncritically considered its ultimate core. Professed beliefs neither are the source of the social bond, nor what is to be preserved at all costs as far as the preservation of community itself is at stake. They are rather the visible sign of an internal rift in the group, the symptom of an emerging defence mechanism within a menaced community: 'belief is a phenomenon of association or alteration of groups, not the basis of their existence' (I 299).

Beliefs mark an actual crisis, by emerging when communitarian cohesion has been weakened and the group is becoming more and more static: ageing and decaying the group tends to mechanically repeat itself, and is closed to any possible normative innovation. The production of beliefs as 'myths and opinions' is the remedy – only effective in the short term – through which the collective tries to contrast its progressive decadence (I 305). Myth and opinion are respectively the collective and the individual forms assumed by belief, and they carry on the same fundamental function: they are structures that – the result of a partial sclerotisation of in-group processes – appear when such processes lose part of their dynamic and expanding power, thus withdrawing into a self-referential and static representation. They are strongly cohesive structures, vectors of collective identity, whose efficacy is closed to any future change. By crystallising social dynamics in static structures, community eventually arrives at constructing its relationship to the outside (other communities, the state, etc.) more and more difficult, and consequently its own survival more and more problematic. In biological terms, the rigid organisation of its internal milieu becomes more and more dependent on the invariance of the external milieu, and community appears as a fragile organism, defending itself from an alien milieu by which it is in fact intimately crossed, fed and solicited. The collective identity thus imagined assumes in this case the characteristic shape of an autoimmune response (Esposito 2004).

But to deny to myth and opinion any foundational function, it is necessary to get rid of the conceptual framework deriving from the individual-society distinction. According to Simondon it is the methodological exigency of social sciences themselves (the 'trap of psychological and sociological surveys', I 296) to attribute to belief - or, better, to the way it is expressed through myth and opinion – a causal primacy in relation to the phenomenon of group belonging (I 299). And again, against any reduction of the concept to its institutionalised manifestations, Simondon invites us to conceive of belief as a process.

In effect, as any phenomenon taking place in the transindividual domain, also belief is apparently contradictory for Simondon. On the one hand, as an 'implicit belief', it is a tendency carrying on the process through which the sense of belonging emerges: 'belongingness [...] in the form of belief [is a] non structured tendency' (I 295), which in effect coincides with the formation of the collective.

[17] This theme is largely developed in the third part of this book. The topic of belief in relation to closed communitarian identity has been previously treated in Bardin et al. (2009).

On the other hand, as an 'explicit belief' presented in the forms of myth and opinion, it is the effect of the withdrawing of the same process beyond defensive structures.[18] This twofold characterisation is the consequence of conceiving (group) personality as the ultimate ground of belief:

> Belief presupposes a foundation which is the personality produced by group individuation [...] a foundation which is not only interindividual, but actually groupal [*groupal*]. (I 299)

Belief is a tendency internal to a field of forces by definition collective and constellated of structured personalities: it is the 'latent set of references in relation to which significations can be discovered'. Its paradoxical nature derives from the fact that belief can actually produce identity only *if* it functions latently, i.e. – to be rigorous – *if* it 'does not exist as such'. In this very sense belief is 'collective individuation in course of existence', the very presence of energetic potentials in the group (I 299).[19] This potentials can result – through the mediation of individuals – into technical, linguistic, ethical and political inventions precisely because, at the same time, it prolongs the common inheritance that similar past activities provided:

> In effect, such a group can be characterised by a community of beliefs implicit and explicit in all the members of the group [...] the belonging to an interiority group can be defined as a non structured tendency, compared to the future of the individual, since it merges with the individual future, but it absorbs its past too, when the individual attributes to itself an origin in the interiority group – actual or mythical. (I 295)

The actual 'presence' of this twofold dynamic in the collective depends on a fragile equilibrium: it is fully possible only when groups are able to maintain the partial latency of belief, without fixing it in the paralysing contents of myth and opinion. In the course of its individuation, the collective is always suspended to the double risk either of not achieving a structuration or of suffocating its energetic components in a cage of rigidly structured shared beliefs which are, in the end, imagined.[20] This twofold process is what Simondon names transindividual individuation. On the contrary, the growth of fear marks the predominance of one face of the process: the one which, aiming at the apparent diminution of the risk, entails its failed endorsement and a consequent communitarian closure. In this case, such as in 'biological' communities individuals exclusively interact on the basis of their structural-functional differentiation, in human community the obsessive

---

[18] It would be particularly interesting to read the relationship between 'implicit belief' and 'explicit belief' through the lens of both the fabulatory attitude of 'static religion' and the mystical force of 'dynamic religion' theorised by Bergson respectively in the second and third chapters of the *Deux sources*.

[19] 'Belief is this collective individuation in course of existence [*en train d'exister*]; it is presence [...] it is as belief that personalities superpose one another. More precisely, what is called collective belief is in the personality the equivalent of what would be a belief in the individual; but this belief does not exist as such [*à titre de croyance*]' (I 299).

[20] My use of the term imagination follows here the well-known suggestion of Anderson (1983) rather than the text of Simondon. And nevertheless the latter's analyses in his course on *Imagination and Invention* does not contradict my lexical choice, as will be clarified in Chap. 9.

repetition of 'implicit beliefs' serves the aim of reinforcing a supposed identity, an origin - actual or mythical – of the social system which, by granting its provisional survival, in fact marks the beginning of its end.

6.3.2 Work

If the phenomenon of belief allows us to describe the ambivalent dynamics of the in-group, it is in the analysis of work that the theme of transindividual individuation touches within *Individuation* the central concern of Simondon's entire oeuvre: technics or, better, the tendency he calls 'technicity'. Although only in MEOT and in the *Note complémentaire* Simondon explicitly matches the themes of social relation and technicity, also in *Individuation* one can find the traces of the problem, treated there in the light of the open/close paradigm, in order to contrast what he considers Marx's conception of work.

Work is the way in which, in order to dominate nature, human beings gather in 'groups which correspond to a determinate kind of behaviour according to the milieu' (I 301). The human relationship to the world takes place through community (NC 512), i.e. through a minimum level of organisation and the division of labour in view of common goals. Work is therefore always referred to a task-oriented organised group: however limited in space and time, work grounds the group on the 'predominance of finality on causality' (MEOT 119). In this sense Simondon can define the community of labour [*communauté de travail*] in a Durkheimian mode, as a 'social group of functional solidarity' (MEOT 248) which leads back (both in MEOT and in *Individuation*) to the biological category of 'community of action' [*communauté d'action*].[21] When Durkheim (1893) refers to the passage from 'mechanical solidarity' to 'organic solidarity' as an evolution through differentiation, he presupposes the homogeneous functioning of the biological and human domains, in continuity with the 'essential properties of organised matter'.[22]

Also for Simondon the social function of work lies at the threshold between the 'biological community' and society. In effect, for Simondon work *is* the model of a sort of 'basic collective' in the domain of psychic and collective individuation, such as in the biological domain the sexed couple was the model of a 'basic community' (I 308) leading to group individuation. This is why qualifying a group on the basis of 'functional solidarity' means to implicitly assume that it belongs to the horizon

[21] '*Communauté de travail*', which I translate here as 'community of labour', is also used by Simondon as a synonym of 'community of action' or 'group of functional solidarity': all expressions refer to goal oriented groups, characterised by interindividual relationship.

[22] 'It is no longer a mere social institution whose roots lie in the intelligence and the will of men, but a general biological phenomenon, the conditions for which must seemingly be sought in the essential properties of organised matter. The division of labour in society appears no more than a special form of this general *process*. In conforming to this law societies apparently yield to a movement that arose long before they existed and which sweeps along in the same direction the whole of the living world' (Durkheim 1893: 3–4).

of biological association insofar as it *does not* reach the properly collective modality of individuation. Such a group in fact just fulfils 'interindividual concrete functions', such as all relations operating at the scale of individuated beings (I 268), and these rigidly structured 'interindividual' relations are different from those characterising a 'second individuation' which is inevitably defined by a different term, i.e. the transindividual: this is a relational modality 'beyond biological, biological-social and interindividual relations' (I 302). And in fact the community of labour itself is not 'pure', it always hosts the tendency towards an interiority group which could give rise to a 'second individuation': 'exploiting nature does not completely satisfy; in front of the world, the species is not an interiority group […] a second genesis is necessary' (I 301).

Thus the community of labour cannot be simply reduced to a kind of biological community through which human beings exploit nature basically thanks to their interindividual relations, the organisation of the 'specific group' being thus an adaptive response of the homo sapiens to its environment. Although methodologically useful, this definition of work on the basis of its adaptive function does not satisfy Simondon. The 'community of labour' is a group of 'functional solidarity' which emerges beyond the threshold of the 'community of action'; and, yet, it is not necessarily peculiar to human beings, as Simondon seems to imply with his hint to the oxen's *suzughia*.[23] The process that, starting from an internal tension, might lead from a 'community of action' to the emergence of 'spirituality' in collective individuation concerns any kind of biological community:

> Nothing proves that human groups are the only one to own the characteristics here defined. Animal groups might imply a certain coefficient which corresponds to what we designate as the basis of spirituality in human groups, just in a more subtle manner, less stable, less permanent. (I 301)

What surprises us here is that, even far beyond the adaptive function of the community of labour, neither the 'second individuation' can be considered exclusive of human beings as a species. Mammals and other animals not only share the same distribution of roles within their groups, but also something of the 'second individuation' Simondon calls 'spirituality'. Again, the human/animal difference appears as a difference of intensity rather than a substantial one.

This is where Simondon's criticism of Marx can be found: the latter would transform into a specific anthropological feature (i.e. the predominant role of work in the human species' adaptive relationship with nature) what is a historical fact typical of nineteenth century (I 302).[24] On the contrary, for Simondon

---

[23] 'In order to express the strong and silent relationship typical of experienced sympathy, the Greeks used the term, referring to human couples as well, *suzughia*, community of yoke' (I 249).

[24] It is worth recalling that the object of Simondon's polemic is Marx and Engels's conception of labour as the activity through which human beings 'begin to distinguish themselves from animals as soon as they begin to produce their means of subsistence'. Although 'the existence of living human individuals' is the 'first premise [*Voraussetzung*]' of all human history, the 'young' Marx and Engels conceive the 'production of material life' as the very 'fundamental condition of all history [*Grundbedingung aller Geschichte*], which today, as thousands of years ago, must daily

> A naturalistic definition of work is insufficient. Stating that work is the exploitation of nature by socialised humans, means to bring work back to a basic reaction elaborated by humanity as a species in front of nature in search of adaptation. (MEOT 241)

According to him work is a relational modality that cannot define human nature: 'it is difficult to find the criterion which would allow for the integration of this relation into an anthropology' (I 302). This explains in which sense the analysis of the community of labour is central in *Individuation* for the understanding of *La réalité sociale comme système de relations* [Social Reality as a System of Relations]. If the 'community of labour' is for Simondon the key for entering social analysis, this is not because it defines the threshold of the human collective (in fact it rather makes that threshold indiscernible), but rather because it entails an internal disparation. Its twofold tension recalls the couple interindividual/transindividual, thus situating the community of labour in an intermediate zone between the in-group and the out-group: 'the human relations characterising work and emerging through it [...] are at the frontier between the interiority group and the exteriority group' (I 296).

It is on this line of thought that in MEOT Simondon definitively challenges Marx, reinterpreting work in light of technical individuation. According to Simondon technical activity is alienated in the working process *not* because of the capitalistic production relations, but for the very nature of the 'group of functional solidarity'. One can in effect 'define a pre-capitalistic alienation essential to work as such' (MEOT 248), since production is alienation in itself in two ways at least; firstly, because it determines group association only at the interindividual level; secondly, because it produces detached objects in which human work is incorporated. Now, these two sides of alienation correspond to the double dynamic structuring the community of labour as a system of interindividual and transindividual activities respectively defined by work and 'technicity'.

Interindividual activity is merely functional to 'work conceived as productive', and it is typical of 'the social group of functional solidarity, [or] the community of labour' which 'only puts in relation individuated beings' (MEOT 248). But another kind of activity emerges in the community of labour through the relation to technical objects, which on the contrary involves a relationship between subjects. Technical objects are born from a kind of primary alienation, a process including the 'crystallisation' of human activity in an object subsequently detached from the producer. This detachment is double and risky; it is both a menace for psychosocial alienation *and* a condition of possible transduction. As a commodity overdetermined by property and market relations, the produced object is, accordingly with Marx, a condition of alienation.[25] And nevertheless, as 'crystallised

---

and hourly be fulfilled merely in order to sustain human life'. Labour would be, in short, the very 'first historical act' that made and still makes homo sapiens properly human, determining the emergence of a specifically human social relationship, i.e. the 'relations of production' (Marx and Engels 1845: 10, 17).

[25] According to Simondon in his epoch technical objects are *prevalently* overdetermined by the market relationship, while in other cultures different forms of separation of human being and technical objects do exist (PST 127). But for Simondon work does not become alienated *because of* the

human activity', the fruit of collective invention, technical objects express the actual social relationship and can thus be the germ of further processes of individuation. In this sense they have a different function, they circulate, triggering possible processes of individuation between groups, far beyond the static life of the closed community of labour.

The temporal dimension thus acquired by the 'technical effort' is peculiar: while work is accomplished and exhausted in its result, technicity 'remains present', crystallised in the technical being. This confers to the technical effort 'an autonomy which community does not allow to labour' (NC 512).

> The possibility of detaching itself from the initial human operator – artist or producer – is, for the produced object, the beginning of a free adventure which entails, throughout the years, as many possibilities of survival and transmission as dangers of reduction to slavery or – in fundamental ambivalence – of possible alienations for the human activity included and crystallised in human work and productions. (PST 127)

Here it is why Simondon can consider technical activity 'the model for collective relationship': since 'it pertains neither to the pure social domain, nor to the pure psychic domain', but to the domain where properly transindividual relationship emerges (MEOT 245).[26] The social bond is therefore generated *between* individuals and *within* the community of labour, as long as some kind of technical activity is in the course of invention, and eludes any exclusively psychological or sociological approach. Contrasting both approaches, in fact, Simondon neither ascribes alienation to working conditions under the domain of capital nor reduces it to a mere problem of adaptation of the individual to the working milieu, a reduction functional to the administrative exercise of power:

> The right way to reduce alienation can be found neither in the domain of the social (of the working community and of the working class), nor in the domain of interindividual relations social psychology is concerned with, but rather in the domain of the transindividual collective. (MEOT 249)

The entire *Conclusion* of MEOT, devoted to define the ontological nature of the technical object out of the 'paradigm of labour', aims to shape a pedagogic-political alternative both to the project of integration and normalisation often implicit in the

---

market economy. In production itself (and therefore at the level of the biological community) technicity – teleologically organised as work – is already partially alienated. Also according to Marx production can be 'degraded' to the simple function of adaptation of the species to the environment, but this only takes place in the case of alienated labour. In this sense Simondon can be said to be one of those readers of the earlier Marx according to whom – in Althusser's words – Marxism is in the end 'a *Weltanschauung* of nineteenth century' (Althusser 1963: min. 2.25–3.40). For a wider discussion of the theme of alienation in Marx and Simondon, see Bontems (2013). On Simondon's criticism to the essentialism implicit in what he calls 'Marxist communism', see Bardin (2013).

[26] The proper level of technical activity is in effect defined by 'the centre of group relationship and of interindividual relations' (MEOT 253) which are, in fact, extreme limit cases of the same original transindividual relationship.

psycho-sociological approach, and to the Marxist dream of revolutionary emancipation.[27]

In short, if 'work as such is a source of alienation' (MEOT 249), the exercise of technicity is instead directly connected to transindividual ontogenesis, whether related to techniques depending on the instinctive, initiatory, and in the end artisanal knowledge, or to formalised techniques, and therefore to a science that is universalisable. In both cases one witnesses the possible birth of an in-group. In the first case a 'secret' technicity provides the sense of sacred, which 'produces the structure of groups' following the mechanism of recurrent causality that dominates community-like social relationships. In the second case the same force that inspired the project of *Encyclopédie*, gathers together 'researchers, editors, correspondents, giving to this *team* made of collaborators a faith without being linked to any social or religious community' (MEOT 92–93). In both cases a process of formation of an interiority group is described, the true genesis of a transindividual relationship, originally conceived so to say *before* its possible communitarian or social results.[28]

6.3.3 Language

As explained in Sect. 3.3, Simondon refutes the widespread conceptions of his time concerning language, i.e. the hermeneutical and the structural, clearly opposing those according to whom language is what marks the nature/culture threshold. Countering different declinations of the *linguistic turn*, Simondon clearly does not make of language the centre of his philosophical research, nor of his theory of the social system. The reason is that he considers language no more than a set of signals. As already said, 'the signal' – in distinction to what is properly 'information' – 'does not constitute the relation' (I 224). And therefore language, i.e. a set of signals, is nothing more than an instrument for the propagation of information which is 'particularly developed when the parts of a system are far from each other, as is the case with a macro-organism or a society' (I 195, n. 2).

Thus information crystallised as a 'word' has the same function as other objects which cross the collective field:

> Passing through the word in order to go from one individual to another, information makes a detour through the social institution of language. (MEOT 98)

---

[27] Simondon's hypothesis, as I will show in section 11.2, is to produce a 'technical culture' functional to a program of liberation of technicity from the paradigm of labour, and matrix of transindividual individuation. In *Psycho-sociologie de la technicité*, when he challenges the 'profound reality of technicity', Simondon clarifies that 'the technical product liberated in the social universe poses different problems than those related to work and production' (PST 128). His project might be compared to Marcuse's in surpassing 'the separation between work and invention' see (Toscano 2007: 203–204); on this topic see also Feenberg (1991).

[28] Similarly, in *Psycho-sociologie de la technicité* Simondon will theorise the possible overcoming of the alienation typical of the industrial era through the reticulation of technicity: firstly, 'the openness of the handcrafted object', secondly, 'the closure of the industrial object' and, finally, 'industrial production as a condition of openness' (PST 232–36).

This detour can determine a restructuration of the social field, i.e. a transindividual individuation. And this is why for Simondon it is not possible to build a theory of the social system based on language: for such a purpose the transindividual configuration of information is to be taken into account, grounded in extra-linguistic processes producing 'significations': 'significations constitute individual being [*constituent de l'être individuel*], although they presuppose the existence of a partially individuated being' (I 263). And significations are not 'language' but 'real' 'relations of being' (I 83).

Thus the linguistic 'object', as the technical, the esthetical and the sacred ones are, in fact, some crystallised transindividual activity which can eventually be re-activated (as the tobacco mosaic virus) and play a role among other objects, such as the technical object does (PST 324).[29] In short, as any other 'crystallisation' of transindividual processes, language is double sided: it is the stable remain of exhausted processes of collective individuation and at the same time the possible vector of further transindividual individuations.

However, a theory of signification is not accomplished in *Individuation*, and for sure it is not the focus of Simondon's theory of psychic and collective individuation. No wonder he cannot retain it as a solution to the problem of language, thus compelling him to develop in *Imagination et invention* a more complex theory of the symbolic function capable of explaining how 'the information related to the pre-individual real' in a system can become 'the beginning of the transindividual' (I 220). Thus Simondon's conception of language must be understood as a pragmatic of communication working through the crystallisation of transindividual individuation into symbols.[30] Already existing in the individual as a 'not yet individuated reality', only through transindividual individuation information can actually become an action endowed with symbolic value, thus contributing to 'open' the collective (I 219).

6.4 The Closed/Open Paradigm

In all the considered domains, a 'basic community' is initially shaped just over the threshold of the 'biological' goal-oriented group. From this perspective it is easy to understand how the 'transindividual' regime of individuation takes place in social systems as the continuative re-emergence of new group processes of individuation crossing collectively established structures. Transindividual individuation is

[29] This extension of the duality of the technical object to other objects is directly developed by Simondon in an interview with *Yves Deforge*, where he highlights the transductivity of the 'object' in general: 'in general one can designate by "object" what can be lost, abandoned, rediscovered. In short, what has a certain autonomy and an individual destiny' (ET 33; see also MEOT 10 and IMIN 178–79). This is also what Michel Simondon, following his father, calls the enigmatic 'ambiguity of the technical object' (Simondon 1994: 98).

[30] According to Montebello 'the question of social individuation cannot avoid confronting a reflexion concerning the pragmatic of language' (Montebello 1992: 85–86). I will give my interpretation of Simondon's understanding of the symbolic function in Chap. 9.

eminently marked by relational activities which complicate the organic nature of group relationships, by instituting new relations of belonging and thus new tensions making the group metastable. Through belief, emotion becomes signification, produces and circulates symbols that augment social cohesion. The true collective, the field of transindividual tensions, is thus instituted: signification is 'exteriorised' in symbols.

Along with technical activity, belief is the necessary base for the constitution of the social bond and *simultaneously* the main menace to its existence, since it tends to crystallise in myths and opinions, assuming the form of a closed community. In the same way one can discriminate the two different functions of language: the purely homeostatic function of its circulation in established forms and the unexhausted potentialities it carries on within those forms. Therefore emotion-signification (not language), technicity (not work) and implicit belief (not myths), actually express (and therefore allow us to define) what is 'primordial', i.e. the transindividual processes of information exchange that produce individuation (I 302). This shapes the different forms of relationship which can take place as work, belief and language, and thus the themes of belief, work and language pose to Simondon a problem of consistence analogous to the one derived from his use of the concept of in-group, the function of which *cannot* be defined in relation to one term of the conceptual couple community-society. Enquiring into the relation between the terms of this conceptual couple requires therefore the whole series of precautions I have advanced above. Furthermore, a final clarification of the issue necessitates the re-interrogation of the epistemological assumptions that ground Simondon's philosophical project.

Simondon establishes the distinction between in-group and out-group in *Individuation*, in the middle of the analysis concerning the *process* of transindividual individuation, while he establishes the distinction between community and society in the *Note complémentaire*, where he aims to unfold the internal tensions of the *structure* of the normative system of the collective, as it results from the former process. In the first case we have a true ontogenesis of the collective and of subjects, while in the second case we have a structural analysis of the social system in the light of the relations between norms and innovation. The two levels of analysis cannot be superposed, or – worse – confused, because the latter only concerns a partial aspect of the former, its provisional epiphenomenon.

Consequently, a structural analysis of social normativity necessarily requires an ontogenetic analysis of the social system, because this is needed in order to grasp the tendencies which still determine the potentials present in the analysed structure, i.e. the actual processes which are themselves part of a metastable structure. The complex epistemological relation between these concepts is further complicated by the spontaneous tendency to substantialise the terms chosen to indicate them. It is therefore necessary to reaffirm that, when Simondon speaks of in-group (and of out-group), he is in fact analysing the peculiar *processes* involved in the emergence of the collective.

Instead – at the level of the already structured collective – 'community' and 'society' must be read as internal and divergent tendencies which, in a way continuing the former processes, keep the social system in metastable tension. In conclusion, if in *Individuation* the analysis of the relationship between processes of individuation and the conceptual couple in-group/out-group is – so to speak – part of an ontogenetic deduction of society, in the *Note complémentaire* the analysis of the relation between different tensions within the collective rather puts the basis for a science of society that starts from a phenomenology of the social system in order to grasp within its metastable configuration the internal processes underway.

The whole question can be definitively unpacked by stressing how Simondon builds his entire theory of the relation between in-group and out-group starting from a criticism of the Bergsonian concepts of 'closed society' and 'open society'. The question posed by a Bergsonian approach to the problem of psychic and collective individuation concerns the difficulty of conceiving the individual at the right scale, neither out of community, nor entirely absorbed in it. A problem Bergson solves by attributing a twofold tendency to life itself:

> Everywhere the tendency to individualise is opposed and at the same time completed by an antagonistic and complementary tendency to associate, as if the manifold unity of life, drawn in the direction of multiplicity, made so much the more effort to withdraw itself into itself. A part is no sooner detached than it tends to reunite itself, if not to all the rest, at least to what is nearest to it. Hence, throughout the whole realm of life, there is a balancing between the individuation and association. Individuals join together into a society; but the society, as soon as formed, tends to melt the associated individuals into a new organism, so as to become itself an individual, able in its turn to be part and parcel of a new association. (Bergson 1907: 212)

While distancing himself from Bergson and consequently turning to the concepts of in-group and out-group in *Individuation*, Simondon still refers to him in the *Note complémentaire*, when he adopts the latter's distinction between 'closed society' and 'open society' by redefining it in terms of the opposition between community and society:

> Bergson's distinction of closed society and open society is valid beyond any doubts, once it is assumed that the open society corresponds to an influence of the individuals on their reciprocal relations, while community is the institutional [*statutaire*] form of the same relations [...] a society the sense of which is lost because its action becomes impossible becomes a community, and it consequently closes itself, creates stereotypes; a society is an expanding community, while a community is a society become static. (NC 509)[31]

Simondon confirms here his original adherence to the Bergsonian closed/open paradigm, by converting it into the privileged instrument to explain in general the metastability of social systems, permanently crossed and structured by processes,

[31] Although the two sections constituting the *Note* were published only posthumously in 1989 as an appendix to IPC, the contemporaneity of the *Note* and the two theses is not only evident by the style and themes there debated, it is also confirmed by the private correspondence of Simondon's son Michel, who attended to his father's work until he died when he was editing his last book. However there might be a change of mind from one text to another, I am trying to show in which theoretical sense an oscillation in Simondon's attitude towards Bergson is inherent in his philosophy. For further references to the complex publishing process of Simondon's works, see the Appendix to this volume.

relations and tensions, that this paradigm allows to conceptualise without improperly substantialising them, when read in the light of the twofold concept of in-group.

In this sense it is possible to conclude that 'open' and 'closed' are two different modalities of the same process of individuation, which Simondon calls transindividual and, in the third part of *Individuation*, identifies with the concept of the in-group he derives from psycho-social research. Thus translated in terms of in-group, the process of transindividual individuation presents two possible directions and final configurations. On the one hand, when opposing one or more out-groups, communities contrast one another by closing their members within a reactive and static identity aimed at restricted goals, which then brings about a rigid relational stability. On the other hand, the internal tension towards an identification with different out-groups, opens the community to a possible straight communication between individuals belonging to different communities, in a process of socialisation irreducible to simple mechanisms of homeostatic regulation. Simondon's basic tool for grounding his philosophical project of axiomatisation of the human sciences is, in short, the Bergsonian 'dyad'.[32] By introducing this metaphysical 'seed' in his epistemology of social systems, Simondon is allowed to keep a distance from the positivistic illusion of the disappearance of politics and the accomplishment of ethical life through social progress:

> It is a retroactive illusion to believe that historical progress steadily opens ethics by replacing closed moralities with open moralities: each new state of civilisation brings about new opening and closing processes starting from a unique centre: opening and closing are the dimensions of an indefinite, mono-dimensional and bipolar dyad. (I 333)

References[33]

Allport, G.W. 1945. Psychology of participation. *Psychological Review* 53: 117–132.

Allport, G.W. 1954. *The nature of prejudice*. Cambridge, MA: Addison-Wesley Publishing Company, 1955.

Althusser, L. 1963. Marx et la science; la coupure marxiste. (R.T.F. 16 février 1963). In *Anthologie sonore de la pensée française*. Vincennes: Frémeaux & associés.

Anderson, B. 1983. *Imagined communities: Reflections on the origin and spread of nationalisms*. New York/London: Verso.

Bardin, A. 2013. De l'homme à la matière, pour une *ontologie difficile*. Marx avec Simondon. *Cahiers Simondon* 5: 25–43.

Bardin, A., et al. 2009. Creencia y fundación de la identidad comunitaria: Simondon, Nancy y Lacan. *Kath'autón* 3: 38–45. https://www.academia.edu/1270420/Kathauthon_3. Accessed 30 Dec 2014.

Barthélémy, J.-H. 2005. *Penser l'individuation: Simondon et la philosophie de la nature*. Paris: L'Harmattan.

Bergson, H. 1907. *L'évolution créatrice*. Paris: Alcan (English trans. *Creative Evolution*. London: Macmillan & Co., 1954).

[32] In IMIN Simondon refers to a tradition, going from Plato to Bergson and Teilhard de Chardin, in which philosophy would be the 'knowledge of the mixed, of the indefinite dyad' (IMIN 62).

[33] *Simondon's complete bibliography and a list of abbreviations are provided in the* Appendix.

Bontems, V. 2013. Ésclaves et machines, même combat! L'aliénation selon Marx et Simondon. *Cahiers Simondon* 5: 9–24.

Carmichael, L. (ed.). 1946. *Manual of child psychology*. New York/London: Wiley.

Carrozzini, G. 2011. *Gilbert Simondon filosofo della Mentalité technique*. Milano: Mimesis.

Coser, L. 1956. *The functions of social conflict*. New York: The Free Press.

Cuviller, A. 1962. *Anthologie des philosophes français contemporains*. Paris: PUF.

Dufrenne, M. 1953. *La personnalité de base*. Paris: PUF, 1972.

Durkheim, É. 1893. *De la division du travail social*. Paris: Alcan (English trans. *The Division of Labour in Society*. London: Macmillan, 1984).

Durkheim, É. 1914. Le dualisme de la nature humaine et ses conditions sociales. *Scientia* XV: 206–221.

Esposito, R. 2004. *Bíos. Biopolitica e filosofia*. Torino: Einaudi.

Feenberg, A. 1991. *Critical theory of technology*. Oxford: Oxford University Press.

Grabbe, P., and Lewin, K. 1945. Conduct, knowledge, and acceptance of new values. In K. Lewin 1948, 56–68 (First edition: *The Journal of Social Issues* 1/3).

Hottois, G. 1993. *Simondon et la philosophie de la "culture technique"*. Bruxelles: De Boeck.

Kardiner, A. 1939. *The Individual and its Society. The Psychodynamics of Primitive Social Organization*, 1961. New York/London: Columbia University Press.

Lefort, C. 1978. *Les formes de l'historie*. Paris: Gallimard.

Leroi-Gourhan, A. 1943. *Evolution et techniques I. L'homme et la matière*. Paris: Albin Michel.

Lewin, K. 1948. *Resolving social conflicts: selected papers on group dynamics*, 1973. London: Souvenir Press.

Marx, K., and Engels, F. 1845. *Die Deutsche Ideologie*. In *Marx-Engels Werke,* vol. 5. Berlin: Dietz (English trans. *The German Ideology*. Prometheus Books).

Montebello, P. 1992. Axiomatique de l'individuation. *Cahiers Philosophiques* 52: 59–91.

Rabaud, É. 1951. Sociétés humaines et sociétés animales. *L'Année Psychologique* 50: 263–272.

Simondon, M. 1994. L'ambiguïté de l'opération technique. Le geste technique et la vie. In *Ordre biologique. Ordre technologique*, ed. F. Tinland. Seyssel: Champ Vallon.

Toscano, A. 2007. Technical culture and the limits of interaction: a note on Simondon. In *Interact or die!* ed. J. Brouwer and A. Mulder, 198–205. Rotterdam: V2_Publishing/Nai Publishers.

Chapter 7
Social Homeostasis and the Exceeding Normativity

What we commonly define as 'human' only appears in the transindividual regime of individuation; it can be neither foreseen and predetermined, nor even taken for granted. Although one can define and partially calculate its conditions of possibility by analysing the tension of metastability in a system, the process is undetermined and exceeds the structural and functional configuration of the system, both in the beginning where it has been triggered and in its 'transductive' course. In short, the threshold conditions of the transindividual cannot be set within definitive boundaries, nor can transindividual processes develop along a predetermined course. That is why Simondon does not present the concept of the transindividual as a solution to the epistemological problem of psychic and collective individuation, but rather as the 'field' in which a whole series of social, biological and technical dimensions can be made converge towards a science of anthropogenic processes. The transindividual allows him to illustrate the space of human relationships by avoiding a direct reference to any presupposed anthropology, whether grounded on human biology or on a supposed metaphysical human nature.

Simondon can thus challenge in his own way the question of human origins in the nature/culture threshold through the concept of the transindividual: he can avoid both the structuralist solution of simply abandoning it, and the phenomenological solution of situating origin within the pre-constituted horizon of (the consciousness of) a subject, i.e. sense. This is the perspective from which Simondon studies societies in the third part of *Individuation*, without dissolving the 'double bind' which ties him to his main critical references: Bergson and cybernetics. According to Simondon, Bergson's philosophy cannot grasp the discontinuity proper to transindividual individuation, because it tends to reduce psychic and social processes to the model of life processes:

> The psycho-social is part of the transindividual [*est du transindividuel*]: it is the reality which the individuated being carries with itself, as a charge for future individuations. It cannot be called *élan vital*, because it is not in continuity with vital individuation, although it prolongs life, which is a first individuation. (I 303, italics added)

A. Bardin, *Epistemology and Political Philosophy in Gilbert Simondon*,
Philosophy of Engineering and Technology 19, DOI 10.1007/978-94-017-9831-0_7

On the other hand, in Simondon's mind cybernetic schemas do not offer an appropriate grip on social systems, because the concept of information there elaborated can only explain processes in terms of homeostasis, thus tending to reduce both social and technical processes to the model of mechanical automatic processes. On the contrary, in *Individuation* Simondon defines all systems by the simultaneity of different kinds of processes; hence, in the third part of it, he proceeds to draw the composite field of the 'human'. Thus he dismantles the concepts that ground the sociological analysis on the 'individual' and/or its structural relations (e.g. organisms, species, work, sacredness, language), bringing them back to their constitutively twofold sources: tendencies, instincts, technicity, belief, signification.

Consequently, to explain the very relation between what Simondon names 'biological community' and 'ethical society' it is necessary to challenge the problem in terms different from the human/animal contraposition. The peculiar phase-shift of the social system must be understood in terms of tension between different kinds of normativities. For this aim it is worth analysing the relation between biological and social homeostasis, a theme Simondon inherits directly from cybernetics and from the analysis Canguilhem had devoted to it a few years before in *Le problème des régulations dans l'organisme et dans la société* [The Problem of Regulations in the Organism and in Society] (1955), where he explicitly referred to Bergson's *Les deux sources de la morale et de la religion* [The Two Sources of Morality and Religion] (1932).

As already shown, the third part of *Individuation* can be entirely interpreted in light of the Bergsonian contraposition of 'open society' and 'closed society'. The closed/open paradigm is precisely the conceptual tool which Simondon employs to connect the structural metastability of both biological and social systems, and the transductive operations taking place there. If the reform of the cybernetic notion of information is, as previously stated, the key for the understanding of Simondon's project of the unification of human sciences, in the same direction we shall look for his solution to the problem of providing a unitary and consistent paradigm for biological and social systems.

The possibility of a science of the 'intermediate zone' between the individual and society – i.e. transindividual individuation – is elaborated by Simondon in direct confrontation with Norbert Wiener's project of a cybernetic science of the social system, grounded on the concepts of homeostasis and self-regulation through *feedback*. I will therefore provisionally take for granted the ontogenetic dimension of the 'transindividual' displayed in Chap. 6, and I shall delve into the analysis of the dynamics of social systems, i.e. – in Simondon's jargon – of 'the collective' as a metastable system. Along this way, my argument will lead to a discussion of the notion of symbol and introduce the function of culture as a regulatory apparatus of social systems.

7.1 The Critique of Wiener's Social Automaton

Once again cybernetics is the reference in opposition to which Simondon defines his own stance, at the same time differentiating himself from Canguilhem. Therefore, it is worth sketching a brief account of Wiener's model for social systems, using the filter of Canguilhem's thought.

Wiener's aim is to provide a 'general theory of regulation', extended to natural, artificial and social systems: a project depending on the supposed universality of the processes of 'communication' (or 'information exchange'). Starting from this assumption, and against any possible humanism or vitalism which would erect an ontological barrier between humans and animals, or between organic and inorganic matter, cybernetics extends its basic information-based model to all systems, the social system included.[1] In *The Human Use of Human Beings. Cybernetics and Society* (1950) Wiener engages with a variety of thinkers who theorised society as an organism, and tries to renew this idea by integrating the concepts of homeostasis and communication. As already seen, it was Cannon who offered the classical definition of homeostasis, which cybernetics applied to society, as the set of organic processes which contribute to maintain the organism's morphology and internal stable state, despite external perturbations (Le Roux 2007: 114).[2] Now, the notion of communication enriches this concept of homeostasis with feedback exchange mechanisms, thus making of regulation the functional unity of the system itself in its open relation with the milieu. In this way the cybernetic concept of homeostasis includes all the functions characterising a system both in its internal relations with its own metabolism and in its external relations with the environment.

Such a system – whether an organism or a society – can be defined according to its rate of resistance to entropy, which ultimately depends on the amount of circulating information. As far as information exchange augments, the negentropic tendency of the system grows along with its capacity of interaction with the environment and, consequently, of adaptation and lifespan (Wiener 1950: 102). Society is a quite complex system where nevertheless, as Wiener suggests, a direct ratio is given between 'the degree of organization of society and the amount of information socially available':

> The existence of an efficient language and, in particular, the existence of a long-time store of written or oral tradition vastly increase the amount of communal information and the possible complexity of the commune. (Wiener 1946: 217)

---

[1] See also Wiener's paper on *L'homme et la machine* (1962), introduced by Simondon at Royaumont.

[2] For a schematic triangulation of Bergson-Wiener-Simondon, see Le Roux (2009). It is worth recalling that Wiener begins his early book on *Cybernetics: Or Control of Communication in the Animal and the Machine* (1948) with a chapter on '*Newtonian and Bergsonian time*'.

Systems and organisms thus last and expand thanks to information exchange. They last, insofar as information is the 'memory' grounding the persistence of their identity (Wiener 1950: 90). They expand, insofar the more information circulates the more the system increases its order of magnitude:

> It is even possible to maintain that modern communication, which forces us to adjudicate the international claims of different broadcasting systems and different airplane nets, has made the World State inevitable. (Wiener 1950: 82)

This point of view entails a politics of education aiming at the steady augmentation of social homeostasis, consistent with the hypothesis that communication is the regulator of the 'body politic' *par excellence*, such as information exchange is naturally integrated in the functioning of any complex machine, whether physical, biological or social (Wiener 1951: 67). In fact, although this model frees systems from the idea of finality, entirely absorbing teleonomy in the mechanisms of functioning, the complexity of their feedback mechanisms of regulation remains exclusively homeostatic. It escapes neither the postulate of the tendency of systems towards adaptation for self-preservation, nor the postulate of a deterministic nature of the information exchanges involved, may or may not they be 'captured' by mathematics. In short, adaptationism and determinism equally mark the functioning of all cybernetic systems, and also in the study of social systems the model remains therefore narrowly targeted for the understanding of deterministic and homeostatic functioning.

If we adopt Canguilhem's perspective, it is possible to appreciate how, through an apparently biological model, cybernetics actually draws from mechanistic conceptual tools and applies them to the social sciences. As Canguilhem explains, the concept of milieu originated from the physical concept of 'field' (Canguilhem 1952: 96; 133–35): the concept of regulation 'after being a concept of mechanics' was for a short period a biological concept, before 'becoming a concept of cybernetics, thanks to the mediation of the concept of homeostasis' (Canguilhem 1977: 99).[3] In this sense, the whole theoretical apparatus through which cybernetics tries to extend the use of the concept of regulation over all systems, is still inscribed in a mechanistic conceptual framework which eventually proves inadequate not only for the study of society, but also for the study of organisms.[4]

---

[3] See also the entry '*Régulation*' in the *Encyclopédie Universelle* (1972), where Canguilhem explains how the concept passed from the biological to the sociological field through the mediation of Malthus and Comte.

[4] Canguilhem formulates his hypothesis concerning the origin of the concept of adaptation within the same conceptual horizon of Simondon's hypothesis on the 'technical' origins of hylomorphism (see above, Sect. 2.1): 'After a quarter of a century, this concept has received such an application in psychology and sociology, often inopportune, that it can only be used in the most critical spirit, even in biology. The psychosocial definition of the normal in terms of adaptedness implies a concept of society which surreptitiously and wrongly assimilates it to an environment, i.e. to a system of determinisms. On the contrary, it is a system of constraints which, already and before all relations between it and the environment, contains collective norms for evaluating the quality of these relations [...] It is a popular concept describing technical activity. The human being adapts his tools and indirectly his organs and behaviour to this material, or to that situation' (Canguilhem 1943: 213–14).

Simondon himself addresses his criticism to the society-machine model presented by Wiener in the *Note complémentaire*, where he tries to supplement the specific limitations of the cybernetic analysis of social systems just as in *Individuation* he grounds his own 'reform' of the concept of information on a general critique of the conceptual apparatus of cybernetics. He attacks the model of the automaton, which entails a conception of regulation as a dynamic stability which would allow adaptation according to the changing milieu, subsequently preserving the system structure in the course of time. This model is perfect for the understanding of communitarian mechanisms ('The automaton is communitarian [...] a pure community functions as an automaton' NC 519) but totally inadequate to grasp the open processes characterising society:

> Norbert Wiener analysed how the powers of rigidity in a community ground its homeostasis. The community tends to automatise its individuals, assigning to them a purely functional significance. [...] Now, these capacities of direct adaptation through assimilation and of structural stability define the perfect automaton. Every civilisation requires a certain rate of automatism to grant stability and cohesion. It needs also the dynamism of societies, a constructive and creative adaptation, in order not to close itself within a stereotypical, hypertelic and inevolutive adaptation. (NC 519)

In fact, homeostatic regulation perfectly explains – and not only for machines – the mechanisms of conservation and reproduction in which all processes are reduced to a minimum effort thanks to a 'stabilising negative reaction' effected by the network of feedbacks which keeps the system functioning on from the thresholds of any possible self-destructive 'positive' reaction (MEOT 79–80). Furthermore, Simondon acknowledges the pedagogical efficacy of the cybernetic model. Through the understanding of these mechanisms of social regulation, cybernetics would 'free from the unconditioned prestige of the idea of finality' and therefore contribute to overcome the 'minority' of a pure and simple submission to authority (MEOT 103, 151). And nevertheless he warns that the fundamental inadequacy of the model limits the efficacy of this political 'pedagogy'. In fact, the homeostatic functioning of the automata can explain neither the genesis of the system nor the processes which do not concur to stabilise it. And therefore, from this perspective, both the emergence of the system and the processes destabilising it are ultimately conceived as anomalies.

Simondon's attack to the notion of 'perfect automaton' is first of all at the theoretical level. According to him the notion itself is contradictory. A perfect automaton would exclude any 'margin of indeterminacy', since all the processes involved in its functioning would proceed according to predetermined patterns. Now, an automaton needs to exchange information with the outside in order to regulate its inner processes according to the changing milieu. If this exchange is predetermined by a fixed code, the automaton cannot escape reactions and goals predetermined by the relation between its inner code and the variations of the milieu, that is, it entirely depends on a stringent, although complex and systemic, necessity. For the interaction to acquire a 'meaning [*signification*]', i.e. to produce – in Simondon's terms – true information capable of modifying the code and the programmed functioning itself, the automaton *must* function according to some margins of indeterminacy.

It must be able to interrupt the established procedures in order to renegotiate its regulatory patterns according to the changing relation with its milieu:

> If the margin of indeterminacy is zero, there is no more possible variation: the functioning repeats indefinitely, and this iteration has consequently no signification. (MEOT 140)

Simondon is applying Canguilhem's conception of the living as a normative being to technical objects. In the end, the notion of a 'perfect automaton' does not correspond any real being: such a 'perfection' would be an ideal isolation of the system and the orientation of all its processes to a perfect homeostasis. This would limit the system functioning to a passive adaptation, i.e. to a fatal tendency towards the entropic perfect equilibrium of death. From Simondon's perspective, the (impossible) cancellation of any margin of indeterminacy from the functioning of the system is therefore a kind of naïve contradictory dystopia, which clearly illustrates the process of the self-destruction of a perfectly closed system.

In short, the cybernetic model cannot explain the (partially) indeterminate functioning of real machines, because it implicitly reduces indeterminacy to a sequential causality, however complex and utterly 'subtle' it is for our instruments.[5] This model of an entirely deterministic machine is precisely the one Wiener adopts for the understanding of social systems. Against his move in MEOT Simondon synthetically presents a double counter-move: (1) he criticises the validity of homeostasis for the complete understanding of social processes; (2) he re-introduces a biological model, in order to conceive social regulation as an operation exceeding homeostatic processes:

> Nothing obliges us to consider society the domain of unconditioned homeostasis. Norbert Wiener seemed to admit a postulate of values which is unnecessary, i.e. that a good homeostatic regulation is the ultimate end of societies, the ideal which should guide any act of government. In fact – such as the living being relies on homeostases for developing and becoming, instead of perpetually remaining in the same state – in the act of government there is a power of absolute event, which, although resting on homeostases, uses and exceeds them. (MEOT 151)

7.2 Society as 'Machine and Life' in Canguilhem

Despite this direct reference to living beings, Simondon does not mean to go back to any – although renewed – biological model for understanding social regulation. He reformulates here the problem posed by Canguilhem in the well-known lecture on *Le problème des régulations dans l'organisme et dans la société* (1955).[6] In his

---

[5] For instance, in the process of the synchronisation of two oscillators, the emission and the reception of information are activities whose actual contemporaneity cybernetics cannot explain through a sequential theory – although complex – of *feedback* (MEOT 140–41).

[6] Lecture held at the *Alliance Israélite Universelle* (1955) (later collected in *Ecrits sur la médecine* without the final discussion). The lecture presents many themes that will appear in the section on the *Nouvelles réflexions concernant le normal et le pathologique* (*1963–66*), the contents of which

paper Canguilhem radically prevented any possible extension of the concept of organism *sic et simpliciter* to the study of social systems. The tacit assumption 'of the idea of social care, of social therapy' (Canguilhem 1955: 106) would be in fact a political act: 'the starting point for a political and sociological theory which would tend to subordinate the social to the biological [...] and become, in fact, an argument for political practice' (102). On the contrary, according to Canguilhem one cannot identify the functioning of the two systems, since for organisms 'the norm or the rule of its existence is inborn' (107), while 'for society, its ideal state or norm is precisely in question' (108). The thesis Canguilhem elaborated in *Le normal et le pathologique* plays a central role in his argument. In organisms existence and 'normative activity' ('life' itself in its proper sense) coincide, while in society they are structurally disjoint: this explains the apparently paradoxical conclusion that 'a society's life is not inherent to it' (109).[7]

An analysis of the different regulation mechanisms in organisms and societies is provided by Canguilhem in relation to the concept of homeostasis, as it was simultaneously treated by Cannon and Bergson at the beginning of the 1930es: 'It is quite interesting that in the period 1930–32, Cannon and Bergson face the same problem, respectively starting from their own biology and philosophy' (117).[8] According to Canguilhem, Cannon cannot avoid the analogical transposition of his concept of biological homeostasis on society, thus conceiving the body politic as a whole naturally endowed with 'wisdom' (Cannon 1932), whose parts are subject to different tendencies ultimately well balanced by a series of self-regulatory mechanisms. On the contrary, in Bergson's *Les deux sources de la morale et de la religion*, the hypothesis of a dichotomy of tendencies (society is for Bergson 'closed' and 'open' at the same time) prevented the analogical trap. For this reason Bergson could not therefore maintain a conception of social homeostasis which directly corresponded to the homologous biological function. The analogy of the homeostatic processes with the regular oscillation of a pendulum applied to society simply does not function: society would irremediably have a 'memory' affecting its oscillation (Canguilhem 1955: 119). According to Canguilhem, the repeated differing of its internal

Simondon therefore partially knew when writing *Individuation*. Canguilhem will only sketch there a possible extension of the concept of organisation to fields different from the biological one: 'The correlativity of social norms – technological, economic, juridical – tends to make their virtual unity an organisation. It is not easy to say what the concept of organisation is in relation to that of organism, whether we are dealing with a more general structure than the organism, both more formal and richer; or whether we are dealing with a model which, relative to the organism held as a basic type of structure, has been singularised by so many restrictive conditions that it could have no more consistency than a metaphor' (Canguilhem 1943: 185–86).

[7] Again in the *Nouvelles réflexions*, Canguilhem draws a scheme concerning the relationship between rules of adaptation: 'external to the adjusted multiple' in the social field, and 'immanent, presented without being represented, acting with neither deliberation nor calculation' in organisms (Canguilhem 1943: 186).

[8] Also in the *Nouvelles réflexions* Canguilhem does quote Bergson: 'One philosopher, at least, has noticed and brought to light the organic character of moral norms, much as they are first of all social norms. It is Bergson in *Les deux sources de la morale et de la religion* analysing what he calls "the totality of obligation"' (Canguilhem 1943: 185).

'oscillations' from the norm marks the irreducibility of the social to the biological. Different theoretical models are required for the understanding of society.

Canguilhem's first proposal is an evident provocation. Society – he says – is to be conceived as a 'machine', insofar it is deprived of inner finality: 'a society has no inner finality, it is like a machine or a tool, rather than a organism' (120). He eventually clarifies his view by complicating the metaphor. Society – he states – is 'both machine and life: it 'presupposes and also calls for regulations', and nevertheless it is 'deprived of any specific regulatory apparatus' (121–122). Therefore any regulation is 'superposed' to society, i.e. out of control of its homeostatic processes, and nevertheless crucial to its existence. This paradoxically 'external' regulatory apparatus characterising society is what Canguilhem names 'justice': 'although there are institutions of justice in society, justice, the supreme regulation, does not appear in the form of an apparatus produced by society itself', and it must therefore arrive 'from elsewhere' (122).

This stance smoothly recalls Bergson's metaphor of pendulum, according to which society is characterised by alternate periods of – so to speak – tendency of the rate of wisdom to fall, and occasional heroic 'invention':

> This is the reason why I see an essential link between the idea that justice is not a social apparatus and the idea that, until now, no society has been able to survive on its own without going through subsequent crises and without some exceptional beings we call heroes. (Canguilhem 1955: 124)

Now – relying on the problem posed by Canguilhem – Simondon aims at deriving, beyond the mechanistic conception of cybernetics, a general theory of the social system which would provide a new foundation for 'human sciences'. It is from this perspective that, at the level of psychic and collective individuation, Simondon both criticises the homeostatic adaptationism and determinism of cybernetics and the Bergsonian postulate of a structural exteriority of regulation, which would definitively subtract the question of justice to any possible theory of society. Again, Simondon seems to play the two stances one against the other. Against Wiener's technocratic universalism, he accepts the criticism to cybernetic mechanicism implicit in Canguilhem's peculiar 'vitalism'. Against Canguilhem's 'Bergsonian' postulate of the *exteriority* of the function of regulation, he connects Wiener's theory of metastability with a quantic model entailing thresholds of indeterminacy. This double move allows him to assume the possible emergence of justice *within* the social system, as a process exceeding its exclusively homeostatic regulation.

7.3 A Biological Model for Social Regulation?

On the basis of Canguilhem's definition of society as 'machine and life', it is now possible to question which kind of *model* is provided in Simondon's theory of social systems, and, in particular, which kind of *regulation* do we face, according to

Simondon, where society is concerned. Simondon sometimes actually adopts the distinction between the natural and the artificial, organism and machine, situating regulation *inside* the organism and *outside* the machine (MEOT 49). Nevertheless he tends to abandon the ontological opposition *tout court* between artificial and natural *structures*, and rather to question the *processes* of regulation. Thus in *Individuation* he goes straight to the point – the knowledge of the different *processes* that traverse different systems – in order to ground his theory of the social system on a model for the understanding of *all* systems as internally discontinuous, and whose functioning exceeds the conservative dynamics typical of homeostatic processes.

It is not surprising, then, that when discussing the regulation of machines in MEOT Simondon choses to draw on the Bergsonian heritage; he does so, moreover, in a twofold manner. On the one hand he seems to accept the vitalistic irreducibility of the organism to a machine. The machine differs from the organism because – as the 'Bergsonian' Simondon states – it does not have 'the sense of time'; it cannot 'modify itself in the function of the virtual' thus posing problems concerning an unforeseen future, and therefore attempting to provide an anticipated solution to those problems. The technical object would not really 'live', since it cannot *invent* information: starting from a given piece of information it can only substitute one form for another, i.e. receive and transmit information (MEOT 143–45). On the other hand, Simondon simultaneously revives the Bergsonian closed/open paradigm in order to redefine also mechanical regulation in terms of processes rather than structures:

> It might seem too easy to oppose open machines to closed machines according to the meaning Bergson gives to the two adjectives. And nevertheless this is an actual difference: the existence of a regulation in a machine leaves the machine open as far as it localises critical periods and points, starting from which the energetic channels of the machine can be modified thanks to the existence of a certain degree of indeterminacy. (MEOT 141–42)

As shown in Sect. 6.4, this basic Bergsonian closed/open schema corresponds to the way Simondon displays in *Individuation* a peculiar counter-topology of the regulation of the individual-milieu relationship, leaving aside the substantialist contraposition between internally self-regulated and externally hetero-regulated structures. According to this perspective, the regulation of the individual is open insofar as its code is open, modifiable according to the information exchange between the system and its internal and external milieu (Sect. 2.2). On the contrary, it is closed (and in fact hetero-regulated) insofar as it functions by merely reacting to the same information exchange according to pre-established patterns and procedures entirely determined at the moment of its emergence: 'construction' in the case of machines, 'birth' in the case of organisms, 'collective individuation' in the case of society.

In short, regulation is no more to be defined as 'internal' or 'external', but rather as partially undetermined or entirely determined, and in this sense 'self-regulation' and 'hetero-regulation' would identify processes rather than structures. As already explained, for Simondon the concept of a perfectly closed 'automaton' is itself a contradiction. On the contrary, an open machine, like an organism, is a 'transductor' which assimilates information, accumulates potential energy and releases it according to a temporal schema which is in interaction with its milieu. This is in fact a machine very close to an organism, capable of incorporating in its own (partially undetermined) functioning the regulatory mechanisms emerging from its relation to the milieu.

From this point of view what differentiates this machine and an organism is only the impossibility of non-organic matter to function beyond the threshold of 'a certain degree of indeterminacy'. In this sense, the difference natural/artificial is instituted by Simondon at another level: what characterises an organism, and what an automaton lacks, is the kind of 'divergent' adaptation defined by transductive invention:

> The automaton cannot but adapt converging towards a set of conditions by steadily reducing the shift between its action and the predetermined goal. It neither invents, nor discovers new goals during its action, because it does not enact any actual transduction, i.e. any expansion of a domain initially reduced which acquires more and more structure and amplitude. (I 161)

Of course, this does not entail just any *identification* of organism, machine and society. Simondon is shifting thresholds, defining individual beings not on the basis of their structures, but, on the contrary, on the basis of the different modes of functioning conceived as the effects of differently simultaneous processes. On these bases, rather than stating that the functioning of machines is *under* a biological degree of indeterminacy, one might more consistently assume from Simondon's philosophy that it is not worth trying to apply a distinctive functioning to different classes of objects, but rather to define different kinds of 'operations' as a criterion to classify and distinguish the different kinds of processes which cross – although with different configurations – the different systems.[9] And it is precisely on these different processes that Simondon grounds his explanation of the peculiar regulation characterising collective individuation, as Merleau-Ponty does not fail to notice in his unpublished notes:

> The notion of *regulation* should be broadened: there is the regulation of an organ and the regulation of an individual – There is the regulation of a society of bumblebees and the regulation of the true "collective" (Simondon) and history, which presupposes a new individuation and which (Lorenz) is not realized in animal societies –. The concept of

[9] The whole of Simondon's argument is built against the hypothesis of Ashby's *homeostat*, that is against the idea that the functioning of an organism can be entirely explained and reproduced through apparatuses of homeostatic regulation. In fact invention, as far as it is made possible thanks to the presence of thresholds of indeterminacy within the system, *is not* an entirely homeostatic regulation. The fact that only at the level of organisms the functioning of the system crosses the threshold of invention entails the irreducibility of life to purely deterministic laws. One can speak of 'life' only when the functioning of a system overcomes the threshold of invention (i.e. the trigger of processes which compel the system itself to call into question the configuration of its own internal and external relations). This cannot happen to machines precisely because of the high degree of determinism which makes them always depend on an external regulation. Or better, if invention took place, this definition of the machine would not fit anymore. On these grounds we might perhaps abandon today a substantialist logic of structures for a logic of operations when trying to conceptualise a machine endowed with biological features.

regulation should not be treated "objectively", as indicating a process in the third person, as in the habit of science – Nor believe that regulation is each time the {operation} of *the same* Nature. (Merleau-Ponty 1959b: 42)

Shall we therefore believe that Simondon's aim is to integrate the cybernetic model for social systems, which is mechanistic and deterministic, with a phenomenological-biological paradigm built on the assumption that 'homeostasis is not the whole of vital stability' (I 161)? If this is true, such an operation is possible not because society actually functions *as* an organism, but because *any* system functions thanks to an excess defined in a differential relation to the multiplicity of homeostatic processes the system hosts, an excess appearing with striking evidence in organisms. In this sense, society can be indifferently modelled as an organism *or* as a non-automatic 'machine': a system the regulatory apparatus of which is defined by both closing mechanisms of homeostatic regulation and by opening processes the partial indeterminacy of which allows the system to continuously invent a new compatibility between the configurations emerging from the internal-external relation. In this sense a process of structural re-configuration of a metastable system can be triggered by a singular structure emerging both from the outside and from within. And this should be true for both physico-chemical and social systems:

In his remarkable study P. Auger explains that in certain cases a seed crystal can be provided by a random encounter, by a fortuitous correlation between molecules. Similarly, in certain pre-revolutionary situations a resolution might occur either for the fact that an idea falls out of nowhere – and immediately a structure arises that spreads everywhere – or through some random encounter, although it is quite difficult to admit that chance might create a good form. In any case [...] we would need to ask ourselves why societies transform, why groups are modified according to different conditions of metastability. (FIP 550)

From Simondon's perspective regulation cannot be defined by any specific internal homeostatic processes, but rather consists of discontinuous processes *between* different homeostases. It corresponds to the capability of the system to integrate in its functioning exceeding factors which could *not* be integrated, but only refused or reduced, without a radical restructuring of the system, i.e. its further individuation. In the social system this is the function of what Simondon calls 'invention'. Now, within the social system invention cannot be classified according to the opposition internal/external, and it rather depends on a 'mixed' set of processes which crosses both 'milieus'. Since Simondon links invention to the term 'subject', one might expect him to locate it in the transindividual domain (MEOT 248). On the contrary, social invention must be explained by going back to Simondon's concept of individuation, focusing on the particular transductive function the individual fulfils in the social system.

7.4 The Transductive Function of the Individual

The image chosen by Simondon in *Individuation* to illustrate the individual 'transductor' is the classical example of the colony of coelenterates. In the regular temporality of colonial life, an elementary individuation marks the discontinuity in which life 'makes itself an individual' (I 169).[10] The appearance of a 'pure individual' entails a double sided discontinuity: an 'internal' separation of the individual from the vital system, and an 'external' relation to a new system. On the one hand the individual appears as a '*quantum* of living existence' against the background of the development of the colony (I 168); on the other hand, it is the very relation through which a new living system, a new colony, emerges from a partially aleatory process:

> The colonies of coelenterates occasionally lay eggs which become the Jellyfishes which provide reproduction. In other cases a complete individual detaches itself from the colony which, after carrying its life on independently, lays an egg far from the original colony and dies; thus a new colony is founded through sprouting from the individual-progenitor resulting from that egg. In this sense, between two colonies capable of indefinite growth, a free mortal individual exists, which plays for them the function of transductive propagation. At the moment of its birth the individual is issued from a colony, at the moment of its death, after moving in time and space, it is the starting point for a new one. The individual is not part of a colony, it is rather inserted between two colonies without being integrated in any of them, and its birth and death are balanced since the individual is issued from a community and it generates a new one: *it is relation.* (I 169)

It is worth noting that Simondon uses the terms 'colony' and 'community' here interchangeably, since they describe the background of the same circular regularity against which the individual appears as a singular and unforeseeable trace: in fact the condition – even though not the grant – of further (possible) transduction. This transductive function played by the individual within the system is the same in colonies and in the collective, in biological and in transindividual individuation:

> The individual as such, distinct from the colony and from the collective, is the result of a singularity and marks a discontinuity; but this is an amplifying discontinuity which tends towards continuity, through a changing of orders of magnitude. (I 331, n. 12)

This change of 'order of magnitude' is not without effects, both for the individual and for the system. In the collective it is the individual itself who incorporates the two distinct functions of 'social' discontinuity and 'communitarian' continuity originally co-implied in the colony-individual system:

> *The individual, in the individuated forms of life systems, is a mix.* It resumes in itself two aspects: a *pure individuality*, comparable to what one can see operating in the relationship between two colonies, and a *continual life*, which corresponds to the function of organised simultaneity that we can see operating within a colony. (I 169).[11]

---

[10] 'The vast domain of coelenterates shows a transitional zone between non-individuated and totally-individuated vital systems; the study of these mixed systems allows for the establishment of precious functional equivalences' (I 169). Duhem (2008) makes of the 'thanatological character' of the individual the centre of his analysis. He poses the problem in terms of finitude and creative force, referring to Jankélévitch and Nietzsche. Making the whole problem converge into the category of the 'pure individual' Duhem concludes that the power of the pure individual contrasts the set of social functions, thus differentiating the creative force characterising the individual from a derivative or 'secondary power' in which command would consist (Duhem 2008: 16–18). The idea is further developed in Duhem 2013a, where the author explores the limitations and opportunities offered by Simondon's 'thanatological' thought.

[11] 'The alternation of the individual and the colony leaves its place, in the superior species, to the simultaneity of individual life and society. This complicates the individual, by putting in it a double cluster of individual (instinct) and social (tendencies) functions' (I 171).

The individual thus conceived is the transductive excess of the system itself, which can be qualified neither as an internal function of the system nor as an external relation between systems: it is not exactly a structure but rather the 'unity of a system' with the functions of 'amplifying transfer and self-regulation' (I 192). Needless to say, this conceptual shift is made possible by the notion of information: 'the individual is not a being but an act [...] it condenses a piece of information, transports it, and eventually modulates a new milieu' (I 191). Again, the functioning of a system is shown to both exceed its internal regulation and institute an 'external' relation with other systems. At this level of abstraction the concept can be easily extended to individuals at all levels, since every individual is a transductor as far as it can 'accumulate energetic potentials and suddenly discharge them'. Also on this point Simondon does not avoid a criticism to the way Bergson's 'vitalism' confines within the domain of living beings a notion which could be profitably referred to an operation, and hence generalised:

> What Bergson was concerned with here was to show a function of temporal condensation which would be essential to life. Now, the relation between the slow accumulation of potentials and the sudden instantaneity of actualisation is not always present [...] the living being intervenes as a transductor [...] it is what *modulates*, and it is also where modulation itself takes place. (MEOT 143)

A few years later, in *Le relais amplificateur* [The Amplifying Relay], Simondon explicitly proposed to generalise the technical model of the transductor-amplifier he had already presented in MEOT: 'is it necessary to push further the research of models, in order to try to understand the phenomena of growth and metabolism as processes of amplification?' (MEC 139).[12] But in *Individuation* he was rather trying to describe transduction in terms of information, thus establishing a direct relationship between the transductive function of the individual and the 'internal resonance' of a system in the course of individuation. This 'recurrent causality' Simondon conceives as the condition of a system on the point of transductively exceeding its homeostatic regulation, an event that can take place *also but not exclusively* through an individual. In fact, within the given conditions for the emergence of a process of individuation, the individual plays a role precisely *as* information: it is in this sense that 'the individual becomes amplified in the collective' (I 328–30).

This explains why, although the notion of the individual remains in *Individuation* so central that it risks absorbing the function which it should just represent,[13] it would be wrong to assume that the individual carries out *all* the transductive dynamics of social systems. In fact, as early as in *Individuation*, the transductive function

[12] 'This notion of transduction can be generalised. Presented as pure in different kinds of transductors, it exists as a regulative function in all machines with a certain degree of indeterminacy [...] human being, and more in general the living, are essentially transductors' (MEOT 143–44). In a discussion following his paper at the conference on *Mécanologie*, Simondon thus responds to an intervention which invites him to expand his model: 'an event is not closed in itself; in the psychic domain it is relevant mainly for its repercussions. Now, the word "repercussion" is not correct, it would be better to employ the term amplification' (MEC 143).

[13] It is perhaps in this sense that Petitot (2004) claims it is possible to detect a 'superiority of the individual over the collective' in Simondon's thought.

at the level of the collective can be assumed not only by individual organisms, but also by different kinds of individuals or 'elements' which behave as the normative 'vehicles of the affective community': thus the *in-group* results from 'symbolic and effective elements of group life: the regime of sanctions and rewards, symbols, arts, objects culturally valorised or devalued' (I 249). These are different concretisations or 'crystallisations' of the activity of signification, which, at the level of the collective, carry out the same transductive function which the individual carries out at the level of biological individuation. This becomes evident in MEOT, where the technical object itself has the function of transindividual individuation: 'through the mediation of the technical object, an inter-human relationship emerges, which is the model of *transindividuality*' (MEOT 248). Also this function does not pertain to the technical object *as an individual*, it rather depends on the 'technicity' it carries on within itself. For this reason in the next chapter I shall treat technicity both in its emergence from the biological domain and in its transindividual function, as the vehicle of a normativity which cannot be simply confined to any of the two domains.

References[14]

Bergson, H. 1932. *Les deux sources de la morale et de la religion*. Paris: Alcan (English trans. *The Two Sources of Morality and Religion*. London: Macmillan & Co., 1935).

Canguilhem, G. 1943. *Le normal et le pathologique*. Paris: PUF, 1966 (English trans. *The Normal and the Pathological*. New York: Zone Books, 1998; First edition: Essai sur quelques problèmes concernant le normal et le pathologique. Paris: Les Belles Lettres).

Canguilhem, G. 1952. *La connaissance de la vie*. Paris: Vrin (English trans. *Knowledge of Life*. New York: Ordham University Press, 2008).

Canguilhem, G. 1955. Le problème des régulations dans l'organisme et dans la société. In *Ecrits sur la médecine*, 101–125. Paris: Seuil, 2002. (First edition: *Les Cahiers de l'Alliance Israélite Universelle* 92:64–81).

Canguilhem, G. 1972. Régulation. In *Encyclopaedia universalis*, vol. 14. Paris: Encyclopaedia Universalis.

Canguilhem, G. 1977. *Idéologie et rationalité dans l'histoire des sciences et de la vie*. Paris: Vrin.

Cannon, W. 1932. *The wisdom of the body*. New York: Norton.

Duhem, L. 2008. L'idée d' "individu pur" dans la pensée de Simondon. Revue Appareil 2. http://appareil.revues.org/583. Accessed 30 Dec 2014.

Duhem, L. 2013a. Engins de mort. Individuation, technique et thanatologie. In *Technologiques, la pharmacie de Bernard Stiegler*, ed. A. Jugon. Paris: Cécile Defaut.

Le Roux, R. 2007. L'homéostasie sociale selon Norbert Wiener. *Revue d' histoire des sciences humaines* 16(1): 113–135.

Le Roux, R. 2009. De Wiener à Simondon: penser l'invention avec et sans Bergson. *Cahiers Simondon* 1: 91–114.

Merleau-Ponty, M. 1959b. Notes de travail inédites. Cours 1959. *Chiasmi International 7: Merleau-Ponty. Vie et Individuation*: 39–42.

Petitot, J. 2004. Elogio della modernità. Il reincanto tecnico-scientifico del mondo nel pensiero di Gilbert Simondon. In *Realismo, illuminismo ed ermeneutica: percorsi della ricerca filosofica attuale*, ed. F. Minazzi, 89–107. Milano: FrancoAngeli.

[14] *Simondon's complete bibliography and a list of abbreviations are provided in the* Appendix.

Wiener, N. 1946. Time, communication and the nervous system. *Annals of the New York Academy of Sciences* 50: 197–220 (Also in N. Wiener, Collected Works IV: 220–243).

Wiener, N. 1948. *Cybernetics: or control of communication in the animal and the machine.* New York: Wiley.

Wiener, N. 1950. *The human use of human beings. Cybernetics and Society.* London: Sphere Books, 1968.

Wiener, N. 1951. Homeostasis in the individual and society. *Journal of the Franklin Institute* 251/4: 65–68 (Also in N. Wiener, Collected Works IV: 380–383).

Wiener, N. 1962. L'homme et la machine. In N. Wiener, *Collected works* IV 824–827 (First edition: In Le concept d'information, 5ème colloque philosophique de Royaumont. Cahiers de Royaumont. Philosophie 5. Paris: Minuit 1965, 99–100).

Chapter 8
Biological, Technical and Social Normativity

The reference to Leroi-Gourhan is central to Simondon's conception of the relation between biology, technology and the social system. It is on the basis of Canguilhem's idea of a 'general organology' and Leroi-Gourhan's palaeoanthropology that Simondon understands the different kinds of normativities implied by the biological and technical processes which structure and frame what he names the transindividual. On this background I shall try to read Simondon's conception of culture, particularly as he presents it in the *Note complémentaire*, as the regulatory mechanism through which the social system makes the different normativities it emerges from and is crossed by compatible. Hence it will be possible to grasp the ethical and political function Simondon attributes to the figure of the 'technician' as dependent on the kind of collective normativity it embodies rather than on some kind of individual heroic features.

In the essay *Machine et organisme*[1] Canguilhem declares his intention to 'inscribe the mechanical within the organic' through the assumption of technics as a 'universal biological phenomenon' (Canguilhem 1952: 126), thus conceiving the 'human being in continuity with life through technics' (127). Canguilhem hopes for a science of the living being which would extend the concept of 'organ' in the direction of a conjoint study of the production of organs (biological evolution) and instruments (technical evolution), including the analysis of machines. He refers again to Bergson's heritage, both in recalling the pre-scientific roots of the '*esprit* of mechanical invention' displayed in *Les deux sources* and in revamping a project of 'general organology' he estimates already implicit in *L'évolution créatrice*:

> Bergson is also one of the rare French philosophers, if not the only one, to have considered mechanical invention as a biological function, an aspect of the organization of matter by

---

[1] Originally – along with *Aspects du vitalisme* and *Le vivant et son milieu* – it was part of a cycle of lectures Canguilhem held at the *Collège philosophique* in 1946–1947. All these lectures were later published in *La connaissance de la vie* (1952).

A. Bardin, *Epistemology and Political Philosophy in Gilbert Simondon*,
Philosophy of Engineering and Technology 19, DOI 10.1007/978-94-017-9831-0_8

life. *L'évolution créatrice* is, in a sense, a treatise of general organology. (Canguilhem 1952: 125, n. 58)[2]

Canguilhem associates this project to a new field of research to which Simondon's MEOT could be directly connected. In France, he says, ethnography is attracting scientific research into the 'philosophy of technics', thanks to the compared analysis of existing primitive societies and pre-historical archaeology (122). In particular, Canguilhem refers to the last chapters of *Milieu et techniques* (1945), the second volume of André Leroi-Gourhan's *Evolution et techniques*, as 'what is today the most striking example of a systematic and rigorously detailed attempt to bring biology and technics together' (Canguilhem 1952: 124).[3]

8.1 Leroi-Gourhan: Tools and Technical Milieu

The development of Leroi-Gourhan's research, from *Evolution et techniques* [Evolution and Techniques] (1943, 1945) to *Le geste et la parole* [Gesture and Speech] (1964, 1965), is well traced in the essay *Technique et société chez l'animal et chez l'homme* [Technics and Society in the Animal and in the Human] (1957), published when Simondon was completing his two theses.[4] The biological originality of homo sapiens resides, according to Leroi-Gourhan, in the coincidence of technical and social abilities. The evolutionary history of these groups is strongly conditioned by the instrumental mediation characterising their relation to the environment. Leroi-Gourhan's thesis relies on the crucial evidence offered by 'human palaeontology': 'the only biologically undeniable criterion for defining humanity is the presence of the tool [*outil*]' (Leroi-Gourhan 1957: 69).

His whole work moves from this basic assumption to the study of human evolution both on the biological and the socio-cultural side. From that single starting point, Leroi-Gourhan reconstructs the hominisation process through the analysis of the biological grounds of 'gesture' and of 'speech' as human features resulting from the progressive 'exteriorisation' of – respectively – (technical) tools for acting on

---

[2]According to Canguilhem, such a project, first sketched by Alfred Espinas (1897), would continue through Bergson (1907) and along the path covered by Leroi-Gourhan (Canguilhem 1952: 122–25). In fact Bergson (1932) conceives the tool as a prolongation of the same vital function of the organ, but adds a remark concerning the peculiar acceleration of human history due to the shift from one to the other. It is worth underlining that the meaning attributed by Simondon to the expression 'general organology' in MEOT 65 – as already highlighted by Stiegler (1994) – is limited to the study of technical elements; but what is at stake here is the concept, not the expression: and to this concern it must be recalled that technicity is carried by the element precisely because it is detachable from the set and therefore capable of transduction (MEOT 73).

[3]As I will explain, Leroi-Gourhan's work is part of the same French tradition – 'sociological' in the wider sense – the influence of which, although not always evident, is crucial for the understanding of the political aspect of Simondon's work.

[4]As previously clarified, at the time two theses were scheduled for a PhD, which preluded to the entering of French academia (see p. 1, n. 1).

the environment and (linguistic) tools for instituting a relationship with other members of the species (Leroi-Gourhan 1957: 69 ff.). And not only technology and culture are studied from the point of view of their biological origin, also the functioning of human groups is understood, according to the same approach, through the functional analogy displayed by organs and instruments:

> The human group behaves in nature as a living organism. Such as for animals or plants, natural products cannot be immediately assimilated without organs which prepare the elements, thus the human group assimilates its milieu through an envelope (of tools or instruments) [...] With this interposed layer it feeds itself, protects itself, rests and moves. Differently from animal species, which own a fixed capital of means of acquisition and consumption, humans are all equal in their nudity, and they augment through conscious acts the efficacy of their nails and fur. (Leroi-Gourhan 1945: 353–54)[5]

The relation of human groups with their external milieu is further complicated by a double characterisation of the latter, since we have both the relatively stable relation with a geological, climatic, vegetal and animal milieu, and the eventual encounter with objects and ideas coming from different human groups. Correspondently, the internal milieu is itself 'doubled' by the tension between different kinds of tendencies: it is framed in a tendency towards stabilisation carried on by the convergence of language, religion, socialisation, and a tendency towards innovation typical of the 'technical milieu'.

This internal dynamic is explained by Leroi-Gourhan in terms of a topological relation between two components of the human group, the ethnical group and the technical group:

> The ethnical group is the material expression of the internal milieu; the technical group is the materialisation of the tendencies which cross the technical milieu. (Leroi-Gourhan 1945: 369)

In fact, the 'ethnical personality' of a group is defined by the set of elements constituting the internal milieu in general, while the 'technical group' is the part of the ethnical group which 'puts in contact the internal milieu and the external one' and therefore it is, in fact, a subset and 'a partial expression of the ethnical group'. Now, each encounter of the group with the external milieu triggers a reaction in its technical milieu, following which the 'technical group' imports from the external milieu a certain amount of discontinuity compelling the entire group to react either with a refusal or with an internal restructuration (367–70).

It is clear that, since all biological groups can be defined, at least in principle, by the relation between their internal and external milieus, the originality of human groups resides in the particular complexity of the internal milieu, which ultimately depends on the technical milieu functioning as a kind of 'interposed

---

[5] Simondon follows Leroi-Gourhan (via Canguilhem) in differentiating the tool [*outil*] and instrument [*instrument*]: the one conceived as a means of action on the environment (e.g. hammer), the other one as a means for gathering information from the environment (e.g. lenses). Where this distinction is not relevant, I will use the term tool. This distinction is, of course, precisely what an epistemology of subatomic physics calls into question according to Bachelard (1951); but this is not relevant at the scale of the hominisation process.

membrane' between the internal and the external milieus. The successful transposition of a technical innovation in the structure of the group is called 'loan' [*emprunt*]. The success of a loan depends of course on complex conditions which do not rely exclusively on the technical milieu, but on the whole of the internal milieu, although it is in the technical milieu that the 'materialisation' of the object takes place. In this sense Leroi-Gourhan can assume that 'it is the group itself who invents' (401), because the invention of an object is 'the point on the surface of the internal milieu in which materialisation emerges' (401). In time, the subsequent accumulation of objects forms a layer which constantly mediates between the internal and external milieus. The system of technical objects is thus the 'technical envelope [*enveloppe*] of humans', the product of a crystallisation of subsequent inventions (353).[6]

But this envelope is not at all inert. Of course 'the technical envelope of humans is not in itself provided with energy, it just fixes the creative tendency' (353). And nevertheless, it is precisely in this way that it transmits information in the form of functional schemas, eliciting systemic effects. In Simondon's words, the technical envelope develops a transductive function. It is in this sense that the event of the materialisation of an object can be indifferently described as the action of the internal milieu on the external milieu, or vice-versa:

> Depending on the point of view assumed by the observer, the object appears as either the answer to a stimulation of the external milieu *or* the attack of the technical milieu on matter. (Leroi-Gourhan 1945: 393)

In short, the technical milieu is the zone of maximal permeability of the internal milieu: it is the mediator between the internal milieu of a group and the part of the external milieu in which it meets the material productions of other groups.

This explains the transductive function of the technical milieu in relation to other groups, but does not explain its function in relation to the natural environment. In fact, the technical gesture and the object it produces clearly depend on a factor the stability of which tends to invariance: matter. In *L'homme et la matière* (1943), the first volume of *Evolution et techniques*, Leroi-Gourhan asserted the universal normativity of the elementary forms of the technical gesture. Being tied, on the one hand, to the structural and functional anatomy of human body and, on the other, to the kind of matter concerned, the technical gesture undergoes an evolutionary tendency which allows a very limited number of variations. This means that, given a determinate organic configuration and a determinate physical milieu, the actions that a determinate kind of body can accomplish on a determinate kind of object

[6] It is worth noticing that such a characterisation of invention, in which the active and the passive functions are indiscernible – quite close to the Simondonian indeed – was reformulated by Leroi-Gourhan 20 years later in terms of a direct connection between 'favourable milieu' and 'impersonality': 'in my *Milieu et techniques* I stressed the importance of a "favourable milieu" in the phenomenon of invention, and the fact that this phenomenon is usually impersonal in character' (Leroi-Gourhan 1964: 223). In fact, from this perspective loan and invention become almost indiscernible, at least in relation to the mechanisms of their emergence (Leroi-Gourhan 1945: 461).

respond to a strictly combinatory logic. It follows that a physico-biological necessity dominates the technical milieu, at least at the level of the *elementary* actions on matter.

It is on these grounds that, in *Milieu et techniques*, Leroi-Gourhan connects the universalising tendency of the development of techniques to the fundamental physiological needs of human being, demonstrating that the number of variants they present across all cultures is limited. Thus the 'technical milieu', actively operating according to a tendency based on a limited number of operative schemas, is conceived as a zone of direct contact between the biologically determined 'continuous tendency' and an external milieu essentially discontinuous because of the physico-chemical structure of matter, climatic variations, scarcity of materials, etc.:

> It is evident that, if the technical milieu is continuous, the technical group assumes from the external milieu a certain amount of discontinuity. Each technical attempt has to model itself on more or less rebel bodies. (Leroi-Gourhan 1945: 369)

The 'envelope [*envelop*] of objects' becomes thus the stable 'support on which the conflict between the human being and matter is traced' (353).

Why then should the technical milieu – in constant tension between a rigid 'internal' normativity and the utmost rigidity of physical laws – result in being particularly 'permeable' to innovation, or even be the peculiar zone of the internal milieu in which processes of innovation in human group structures begin? Being the 'subset' of the internal milieu in contact with natural and biological determinisms, the technical milieu is also the zone of maximum and common universality within any human group, partially detached from the factors that ground cultural identity and determine cultural differentiation. In short, the technical milieu is for all human groups a 'free zone' of exchange with any other group, since it condenses the same universal relationship between the biology of the species and natural world. The technical milieu is therefore a factor of innovation for human groups *precisely because* it links the internal milieu to the variations of the external milieu rather than to the homeostatic mechanisms of stabilisation typical of ethnical identity, the normativity of which is, on the contrary, blind to the variations of the external milieu.

Thus the technical milieu carries on the force of normative invention, a single 'tendency' deployed in a variety of forms (a 'layer of objects') which displays, according to Leroi-Gourhan's vision, the basic trend of human history:

> The tendency – which is essentially universal and charged of all the possibilities that can be expressed with general laws – crosses the internal milieu immersed in the mental traditions of each human group. There it acquires particular properties – as a ray of light acquires different properties according to the different bodies it goes through. Then it encounters the external milieu, which offers to these acquired properties an irregular possibility of penetration. And finally, at the point of contact between the internal and external milieus, a layer of objects is materialised which constitutes the most general set of human's material goods. (Leroi-Gourhan 1945: 361)

From this perspective the history of human groups is marked at the outset by the presence of technical objects. The objects generated within the technical milieu of a group penetrate into the internal milieu of other groups, putting the two groups in relation. But the technical loan also carries on some ethnically 'particular properties' which challenge the peculiar forms of ethnical stability of the contaminated group, thus constituting the vehicle of possible ethnical innovation. To be more direct, the object of a technical loan, because of its capacity of penetration, is the very privileged vehicle of social innovation: 'the progressive accumulation of successive loans results in a changing of the internal milieu' (386).

It is the peculiarly mixed status of technical objects – at the same time technical and cultural[7] – to determine on the one hand the efficacy and on the other hand the pregnancy of the transformation. According to the given set of conditions, the technical milieu can be 'conquered', and in that case the transformation 'expands to the whole of the internal milieu' (389–90) of the conquering group as a deeper cultural 'contamination':

> Each object is permeated with traces of the *whole* of the internal milieu. The first principle one can derive from this dependency of all elements on the internal milieu, is what follows: when an object is loaned from a foreign group, it can escape the internal tendency, avoid the effect of refraction and remain, in short, the proof of its origin. (Leroi-Gourhan 1945: 364–65)

8.2 Biological and Technical Normativity

The greater part of Simondon's reference to Leroi-Gourhan –, at least since MEOT[8] – can be possibly read as a study of 'general organology' centred on the way humans have developed relationships with the natural milieu through the technical elements, objects and systems:

> The technical object [...] is the first detached object, since the world is a unity, a milieu rather than a set of objects. In effect, there are three kinds of realities: the world, the subject and the object, mediator between the world and the subject. The technical object is the first form assumed by the object. (MEOT 170)

It is difficult not to grasp in this passage the reference to Leroi-Gourhan's conception of technics as the study of the 'artificial envelope' which is part of the internal milieu of human groups. This reference is essential in order to understand Simondon's hypothesis concerning the relationship between psychic and collective individuation. As we have seen, for Simondon the vital activity of an organism

[7] 'When assimilated, the object is marked by two conditions: it receives the personal footprint of the new group [...] and was bended to the exigencies of the raw materials present in its new *habitat*' (Leroi-Gourhan 1945: 382). After establishing the difference between solutions concretised in 'universal objects [...] shared by all humanity', the emergence of which depends on the 'powerful' influence of the external milieu, and 'complex objects [...] linked to a determinate ethnical group', Leroi-Gourhan adds: 'for each object one should balance the two causes: that is why there are no pure examples' (Leroi-Gourhan 1943: 293).

[8] The two volumes of *Evolution et technique* first appear in the bibliography of MEOT, but only in the *Entretien sur la mécanologie* (1968) does Simondon explicitly declare his debt towards Leroi-Gourhan. The recently published fragment "Anthropo-technologie" (1961a) provides further evidence of Simondon's debt.

is intrinsically normative. This normative activity of the organism crosses the normativity of the natural milieu, and therefore the relationship between the organism and the world is not a relation of adaptation, it is a metastable relation the terms of which are constantly reconfigured through the mediation of organs, according to rules which sketch – so to say – the gamma of possible developments without determining them. Now, the same dynamics can be found at the level of human social groups, with the further complication of the emergence, there, of a technical milieu.

In MEOT, analysing the *Evolution de la réalité technique; éléments, individu, ensemble* [The Evolution of Technical Reality: Element, Individual, Set], Simondon explains that the technical object actually evolves not when it adapts to the context and to the goals it is produced for (this 'hypertelic' functionality tends, on the contrary, to be fatal), but rather when it institutes a dynamic relationship between two milieus, the technical and the geographical (MEOT 53). Organs and tools are in this sense the result of the same process of 'systematic and multifunctional convergence', whether it be of an organism or of a group (MEOT 56). But the effects of this same inventive activity does vary according to the scale.

From the moment they are invented, the different lineages [*lignées*] of technical objects tend to converge towards the production of a 'techno-geographical milieu' which becomes in turn the condition for further processes of collective invention and technical evolution. The invention of technical objects elicits a phenomenon of 'recurrent causality' in which the hominisation process itself is involved:

> It is, in effect, neither a progress conceived as a predetermined movement, nor a process of the humanisation of nature: this process might conversely appear as a naturalisation of humans. Between humans and nature a techno-geographical milieu emerges thanks to the intelligence of the human being. (MEOT 56)

A peculiar configuration of the collective relationship characterises human groups, in which invention is not limited to the relationship between single individuals and the natural milieu. By surviving the act of invention through its concretisation in the object, invention is integrated into the patrimony of a group.

The way in which Simondon reformulates the concept in his course on *Imagination et invention* [Imagination and Invention] (1965–1966), allows us to appreciate Simondon's analogical application of the same function to the entire 'system of created objects' (and not only technical objects). In social species the latter functions as a principle of organisation both of the individual-milieu relation and of the relation between individual and the social system:

> The system of created objects, in the double perspective of the relation with nature [...] and with the social [...] is the envelope [*enveloppe*] of the individual. (IMIN 186)

Translated in the jargon of Simondonian cybernetics, this means that the progressive production of a milieu of objects significantly modifies the collective regime of information exchange. Posing the problem in terms of information not only allows Simondon to cross the dynamics of technical objects and symbols as different 'materialisations' of gesture and speech, i.e. to explain within the same paradigm material and intellectual invention. This also allows him to conceive social systems as made of organisms and machines.[9] In human social systems, in effect, the biologically determined information exchange crosses the normativity of the technical milieu which is its partial sedimentation. This modifies trajectories and rhythms of human biological individuation. Thanks to its peculiar tendency towards universalisation, technical normativity is therefore the true condition of possibility for historicity.

As I will show in the third part of this book, during the 1960s Simondon will insist on the self-constitutive nature of technical normativity partially sketched in MEOT, by conceiving technicity as the key factor for the building of the social bond. But in *Individuation* the question of technicity never refers, as it happens in Leroi-Gourhan, to the hominisation process. Furthermore, its function is quite marginal in the explanation of the ontogenesis of the collective: it rather appears as the mark of an opening of the social system through a normative invention exceeding homeostatic regulations. In general, in *Individuation* technicity is still inscribed in a general theory of the functions of discontinuity between systems, according to dynamics of phase-shift and metastabilisation which can be drawn in the closed/open paradigm presented in Chap. 7, corresponding to the oppositions between community and society, and structure and process of invention.

And nevertheless, in the *Note complémentaire* technical normativity is conceived by Simondon as essentially linked to social invention in a way quite close to Leroi-Gourhan's:

> Technical normativity is intrinsic and absolute; it is worth underlining that it is through technics that the penetration of a new normativity in a closed community is possible. Technical normativity modifies the code and system of values of a closed society. Each closed society which admits a new technique introduces the set of values entailed by that technique, and operates in this way a restructuration of its code of values. Since all communities use a technique or introduce a new one sooner or later, there are no completely closed and non-evolutionary communities. (NC 513)

As already explained, every group understood as a metastable system is crossed by a twofold tendency. On the one hand, as cybernetics theorised, it is characterised by a main tendency made of homeostatic regulation through internal information exchange: this is for Simondon the tendency of 'community' in which the whole interplay of *inter*individual relations take place. On the other hand, and simultaneously, the collective is such (and not only 'community') precisely because it is crossed by processes constituting *trans*individual relations, which cannot be forecasted within the 'normal' functioning of the system, but can introduce new normativity for further configurations. The transindividual processes of invention depend on the indeterminacy of the pre-individual milieu and on the persistence of biological individuation, *plus* the feedback effect exerted over the system by the whole set

[9] Although grounded on a different conception of information processes (Sect. 2.2), the concept is already in Wiener: 'It is the thesis of this book that society can only be understood through a study of the messages and the communication facilities which belong to it; and that in the future development of these messages and communication facilities, messages between man and machines, between machines and man, and between machine and machine, are destined to play an ever-increasing part' (Wiener 1950: 18; see also 68 ff.).

of the on-going processes of psychic and collective individuation. 'Culture' is how Simondon names this whole set of processes, among which technics plays a crucial role of connection with the pre-individual milieu and with biological individuation. Thus, if it is not possible to conceive technical activity as *tout court* biological, at the same time the integration of technical normativity into what we call 'culture' entails a radical reconfiguration of the concept of culture itself.

8.3 Values and Norms: The Reflexivity of Culture

The *Note complémentaire* is a key text to delve into Simondon's concept of culture, there defined by the conceptual couple value-norm. But a propaedeutic enquiry is needed on the intertwining of the different kinds of normativities characterising the transindividual regime of individuation in *Individuation*. In the conclusion of his book Simondon aims at clarifying the concept of 'transductive series of metastable equilibriums' through the notions of norm and value:

> Norms are lines of internal consistence in each of these equilibriums, and values are the lines along which the structures of a system are translated into structures of the system substituting them. (I 331)

Evidently the normative apparatus is attributed here to the homeostatic function of maintaining the internal consistence of the system, while values refer to the transductive operation from one system to another through a structural change: 'values' – Simondon writes – 'allow the transduction of norms', 'they are the power of amplificatory transfer within the normative system', since they are norms themselves transformed into information (I 331).

As usual, Simondon attempts to grasp the process. In this case psychic and collective individuation, takes place as the resolution of a normative disparation: 'a system of norms is problematic as two images in a state of disparation; it tends to be resolved into the collective through constructive amplification' (I 331, n. 13). 'Value' is in short the name Simondon attributes to the formula of normative conversion. This 'formula' cannot be entirely formalised, of course, because in that case a determinate normative system would be absolutised, and nevertheless it is not reducible to a kind of undifferentiated becoming. The concept of value refers to a trend which can be defined *only* in relation to the system of established norms. It is not out of the normative system, but rather its transductive tendency itself:

> Norms might be conceived as expressing a defined individuation, and having consequently a structural and functional meaning at the level of individuated beings. On the contrary, values can be conceived as linked to the emergence of norms, expressing the fact that norms emerge with an individuation and only last for the duration in which this individuation actually exists. (I 332)[10]

[10] Norms and values are therefore the becoming of the social system, its 'double' historicity: 'there is a historicity of the emergence of values as there is a historicity of the constitution of norms'. A historicity characterising the opening and closing of social systems and, with it, of their ethics (I 333).

To grasp and to follow such a transductive movement is the task of an ethics of individuation. For being ethical, an act must fulfil a double condition of open potentiality and clear determinacy: 'in each act resides both the movement to go further and the schema that will be integrated into other schemas' (I 333). Only the tension of this double condition will make an act ethical, i.e. at the level of transindividual individuation, without reducing it to a fanciful aspiration or, on the contrary, to a supposed absolute norm in which the in-temporality of transduction is only mimed (I 332).[11]

The point is that – as already explained in Sect. 7.4 – the agent of this transductive function can only be an individual, both at the biological and at the social level:

> Values are the pre-individual of norms. They express the link to different orders of magnitude: issued from the pre-individual, they push towards the post-individual, both in the colony phase form, and in the transindividual form for superior species. They derive from continuity and go back to continuity through the individual as a discontinuous transfer. (I 332, n. 14)

This attribution of a key role to the individual is maintained in the *Note complémentaire*, on this point fully consistent with *Individuation.* It is here that the term 'community' is widely used in contrast with the term 'society' to highlight the normative function of the latter, peculiarly related to invention and to the technical individual. In the first part of the *Note*, titled *Values and the Research of Objectivity*, the term value is assumed – says Simondon – as a 'symbol' of the possible complementarity of individuals, more precisely as 'actions' through which 'complementarity can emerge'. Simondon differentiates here three kinds of values. Two of them are related to what institutes a kind of relationship functional to the individuation of the living: 'value as organic or technical condition' as food or medicine. The third kind of value, called 'absolute' by Simondon, is the 'beginning or trigger' of the collective relation (NC 503–4). This 'absolute' value is culture.[12]

The peculiar status of culture in contrast with biological and technical normativity is to be grasped by going back to the concept of the transindividual. Among the conditions of possibility of the transindividual there is a kind of inventive technicity which is a process of individuation in progress at different levels: biological, technical and collective. At this level, 'human nature' is not a given fact – neither biological nor cultural – but a biological-technical becoming which repeats itself, instituting norms and posing problems to the social system: it is a factor of structuration as

[11] In this circumstance Simondon sketches the figure of the sage: 'this directive force which continues cannot be a norm. The research of an absolute norm [mimes] the eternity and in-temporality within the becoming of a life: in the meantime vital and social becoming continues and the sage is reduced to the image of the sage' (I 332). An analysis would be required of the notion of 'wisdom' he presents when evoking Zarathustra (I 280–82). For a first account of the theme of ethics in Simondon's philosophy, see Hottois (1993).

[12] In fact, Simondon asserts that 'among these values one can include culture'. And nevertheless he does not provide any hint on what these 'absolute values' would be based. My hypothesis is that the concept of 'Culture' expresses the absolute pervasiveness of 'value' as a transductive force in social systems.

much as of destabilisation of the system in which it takes place. Biological and technical normativities are thus both the conditions of the social system and what menaces any stable configuration it might assume, since, as former phases of individuation, they carry instability: 'a society is characterised by factors of discontinuity the circumstances of which are organic or technical' (NC 508). Culture, instead, is the normative dispositive 'capable of manipulating in some way the symbols representing such a technical gesture or such a biological drive' (NC 504). The efficacy of such a 'manipulation' is always partial.

'Human nature', thus conceived, is a biological-technical feature which doubly exceeds the normative stability of the social system. On the one hand, 'biological' normativity appears as 'instinct' at the moment of the birth of the organism, and is prolonged as a phase within the individuated subject: this makes of each living being a new problem of symbolic integration for the social system. This biological surplus can be (always partially) integrated in the communitarian normativity through education, a symbolic capture into an 'elementary normativity suffered by the individual', without which society would be impossible (NC 506). On the other hand, 'technical' normativity is always dominated by its relation to the natural milieu, and therefore it cannot be entirely absorbed within the social normativity established by symbolic practices, which depend on the singular history of the social system. Rather, it forces the social system to repeat the effort of symbolising its achievements.

Now, as far as what is organic and technical threatens and, *simultaneously*, produces and maintains the social system, these processes and the respective normativities cannot be suppressed, but they have to be continuously 'manipulated' in order to be functional to the maintenance of group cohesion. The collective integrates the normativities exceeding the functioning of the social system by 'enveloping them' – Simondon says – with significations. Culture is therefore to be conceived as the transindividual milieu in which social systems emerge thanks to a tendential homeostatic stabilisation of their *constituting* biological and technical processes. In fact, biological and technical processes are the condition of the possibility of culture, i.e. of the collective process which makes them compatible through the production of a system of symbols:

> Culture is like a set of beginnings of actions rich of schemas, which attend to be enacted. Culture allows us to solve problems, but it does not allow us to build or to live. It presupposes that the possibilities of organic and technical life are already given, although unconnected and therefore sterile: culture creates then the system of symbols which allows them to enter a relation of reciprocity. (NC 504)

In this sense Simondon can assert that culture is not simply the superstructural or mythological expression of the technical and biological basis (it is not the 'means of expression' to which both Marxism and Freudianism would reduce it, NC 504), it is rather 'reflexive' insofar it resolves the problems posed by the biological and technical normativity to the social system in which it emerges. For this reason its relation with biological and technical normativities is essentially phase-shift and instable.[13] Hence the ambivalence of any culture: a closed system of rigidly normed practices, or the constant restarting of symbolic production which comes to terms with different forms of normative excess.

Only this perspective allows the puzzle posed by Simondon's assertion that the demagnetisation of the collective depends on organic or technical circumstances to be solved (NC 508). In fact, biological and technical normativity both exceed the homeostatic stability of the social system, but in quite different ways. It must be clear by now that technical activity is provided with a peculiar nature, suspended – so to speak – between nature and culture, and therefore it shifts the social system in a different way compared to the 'natural' tendency of biological individuation. While biological individuation introduces in the social system a 'bi-polarity of values' typical of a community, technical activity introduces a 'mono-polarity of values' typical of society (NC 509). As aforementioned, technical activity pertains to society, while biological life pertains to community. Thus the function of technical activity is differentiated from the corresponding function of biological normativity also in relation to the dynamics *between* diverse social groups. The primitive categories of inclusion and exclusion directly correspond, in fact, to the biological acts of assimilation and nourishment and the rejection of what is harmful, and therefore 'external communities are thought of as bad for the mere fact of being external' (NC 509). On the contrary, technical activity is the vehicle of the opening of a community as far as it can constitute a domain of transductivity, a regime of information exchange *between* different social groups which can metastabilise one or both of them.

Technical activity provides the social system with the tools and instruments the adoption of which can be potentially extended to any human group precisely because it is rooted both in biology and in physics, which are universal. At the biological level, the exercise of technicity is conditioned by 'schemas of action' functional to the satisfaction of needs shared by the whole species[14]; at the physical level it depends on its efficacy on the kind of matter it is concerned with. Thus the adoption of a technique can be retarded or even refuted on the basis of the existing culture, but once it has occurred, it becomes irreversible on the long term.[15] This allows Simondon to conclude that

---

[13] Furthermore, the risk run by culture is ambivalent. In fact, on the one hand it can simply adapt to the biological and technical normativities, and on the other hand it can totalise their symbolic capture, thus reducing them to mere homeostatic functions. In the first case culture would be reduced to 'the promotion of the organic or the expression of the technical' (NC 504), while in the second case it would determine the exclusion or recruitment of individuals in the social system according to symbols 'of organic or technical nature' (NC 509).

[14] 'Only technics is absolutely universalisable, since what resonates in it of the human being is so primitive, so close to the basic conditions of life, that each human being owns them in itself' (LPH 272).

[15] Furthermore, according to Simondon technical thought would benefit from a 'direct universality' and the utmost communicability, thanks to the use of images that would avoid a 'detour' through the institution of language (MEOT 97–98). But this idea of images as codes mainly subtracted to cultural conditioning because directly linked to the perceptive apparatus seems frankly valid only within the boundaries of the mechanical techniques, i.e. within the project of the *Encyclopédie*.

> Technical norms are entirely accessible to the individual without relying on social normativity: the technical object is valid or non valid according to internal criteria that translate the schema inherent to the effort through which it has been constituted. (NC 513)

It is precisely out of these premises that Simondon speaks of a 'pure individual' to indicate the technician who 'brings together in itself *the two conditions of reflexive thought*: organic life and technical life'. (NC 512, italics added).

8.4 The Political Function of the Technician

According to Simondon the 'pure individual' is in the last instance the critical true point of social innovation, the 'event' exceeding the simple homeostatic regulation of the system. In the *Note complémentaire*, Simondon individuates a few exemplary figures which can fulfil this function in the social system: the magician, the priest, the engineer and the physician. All of them 'have succeeded in detaching from the community and instituting a direct dialog with the world' (NC 512). Capable of observing reality and of moving in it according to a relationship which exceeds the normativity established by the 'groups of action', i.e. of work, these 'pure individuals' are 'technicians': 'mediators between the community and the hidden, inaccessible object' (NC 512). Thus their invention, although grounded on communitarian relationships, exceeds them: the technician 'is not only a member of the community, it is as if he belonged to another species' (NC 511).

The debt towards Bergson is here more evident than ever, as if Simondon were trying to reformulate in non dualistic terms what Bergson could only explain in the terms which grounds the *élan vital* of an exceptional individual:

> An *élan*, which had ended in closed societies because it could carry matter no further along, but which later on – in place of the species – looks for some privileged individual. This *élan* is thus carried forward through the medium of certain men, each of whom thereby constitutes a species composed of a single individual. (Bergson 1932: 285)

But in the *Note complémentaire* the 'pure individual' cannot be simply identified with the Bergsonian hero, although it seems to carry on its ethical grandness. In fact, in order to maintain its transductive function, the individual must act *between* communitarian belongingness and openness towards nature, because on the one hand the complete absorption of individual activity in communitarian automatism produces 'stereotypical, ipertelic, non-evolutionary' adaptation, on the other hand a 'purely individual enterprise' would risk to destroy the basic interindividual relationship which grounds any possible further social invention.

It is in this very direction that in the final part of *Individuation* Simondon stigmatises the reduction of the individual to an 'absolute individual', isolated in itself, fixed by the constraints of the exclusively internal normativity of a self-referring '*acte* fou'. The becoming of the subject would thus be reduced to a single, sterile, closed individuation. In fact the normative invention which opens the social system is possible *thanks to* the individual, on the condition that it can continue its 'role of transfer', by maintaining its ethical act within 'a measure both activating and inhibiting' (I 335).

It might be noted that this 'measure' still recalls the Bergsonian 'measure' between mystical invention and fabulatory function. Should we then conclude that a Bergsonian – or even Nietzschean – ethics of heroism is taken over by Simondon in the technical-inventive power of the pure individual? By unpacking the concept of the individual, I aimed to show why this is not the case, at least if we assume the theoretical stance implicit in Simondon's epistemology. Thus the question concerning the status of the pure individual has to become a question concerning its function: What does the individual amplify? Simondon is quite clear: *through* the individual and *within* the social system (*in-between* them, indeed), what is amplified is the system of values and norms which constitute the dynamism of the system. It is clear that, as far as the social system is concerned, there is no predetermined typology of the individual to carry on transduction, because in its metastable dynamics collective invention is essentially transindividual. In fact, the ethical *act*, which opens the social system, can be defined by its conditions of emergence and its function, rather than by the individuality in which it is – so to speak – embodied.[16] The centre of transindividual individuation is to be situated – as explained in Sect. 6.3 – in the exchanges between emotion and signification rather than in the dynamics of language, in the belief implicit in the in-group processes rather than in myth and opinion, and in technicity rather than in work. All these transindividual acts of invention exceed the individuals that carry them out and the structures within which they take place.

Simondon oscillates, in *Individuation*, between a perspective centred on the individual as a transductor and one concerning the systemic relations from which it emerges. This oscillation is structural to his thought, and traverses also the transindividual domain. It can be grasped by observing the symptomatic additions and cuts Simondon himself made to his oeuvre between *Individuation* (1958) and IGPB (1964). When publishing IGPB in 1964, Simondon had to exclude the whole 'third part' of his original doctoral thesis,[17] but he added some sentences and notes, among

[16] It is in this way that Simondon can for example detach the concept of 'moral consciousness' from any direct reference to the individual as such, and use it for indicating the shift between the exclusive and closed community and the transductive and open society (NC 509). It is precisely because it is grounded on 'other than the vital necessities of a community', that moral consciousness (the 'sense of values') entails the transductive opening characterising not only, but *also* the individual, whose transductive effort is suspended between the double risk of a solipsistic closure and a regressive (re)absorption into community, an 'interioristic or communitarian deviation of transindividual spirituality' (NC 508–509). It is precisely against the constraints of the normativity typical of the *acte fou* that Simondon evocates Zarathustra's act of 'going beyond' (I 330 ff.). Is then the rupture of social normativity, the normative invention of the technical individual in itself political? Is the pure individual political as such, insofar it is a possible germ of collective individuation? This hypothesis of a basic political 'power' of the individual as such is not compatible with the results of Simondon's thought and it can be assumed only through a partial reading of *Individuation*, in which the extension of the identification *tout court* of individual and political function would end up making of being itself a political issue, i.e. to make politics coincide with ontology. On the '*acte fou*' interpreted from a political perspective, see Aspe and Combes (2004).

[17] In fact, the second and third subsections of the second section (see above Chap. 1, n. 28).

which the concluding lines in which he celebrated the ethical role of the individual as an 'amplifying transfer' for social systems:

> Ethics expresses the sense of perpetual individuation, the becoming of the pre-individuated being, in course of individuation [and tending towards the continuum which rebuilds in the form of organised communication a reality as vast as the pre-individual system. Through the individual, amplifying transfer issued from Nature, societies become a World]. (I 335)[18]

Twenty-five years later, IPC presented the same conclusion of the original thesis of 1958, in which the amplifying role of the individual was, on the contrary, clearly attenuated.[19] On the other hand, the *Note complémentaire*[20] was published for the first time in IPC (1989) as a kind of appendix, whose final was devoted to the way *the machine* opens the social system:

> Between the community and the individual isolated in itself there is the machine, and the machine is open to the world. It outstrips communitarian reality and institutes the relationship with Nature. (NC 527)

As I will show, outside of *Individuation* the social system is rarely analysed by Simondon through categories which directly link it to the transductive function of the individual. In fact, this transductive function is prevalently reserved to the technical object (or better, to the technical 'element') or to the symbol. Furthermore the problem is mainly posed in terms of a phase-shift between different tendencies and processes which cross, metastabilise and finally reconfigure systems. Shifting the focus from the human individual to the machine and, furthermore, to the system in which transduction takes place makes more evident what was already implicit in the consideration of the individual *as* a process of individuation, which determines both the terminological ambiguity and the epistemological complexity of the concept of individual presented in *Individuation* (see Part 1, in particular Sect. 4.4).

In conclusion, one must assume that in a philosophy of individuation there can be no other invention (social, artistic, technical etc.) except the one related to the openness of the social system and therefore entailing the transindividual constitution of a subject, since – as one must keep in mind – 'it is not the individual who invents, it is the subject' (MEOT 248). Consistently with this view, there can be no individual invention *ex nihilo*, but only transindividual invention of actions and significations, both technical and symbolical. This is why in the third part of this book we shall follow Simondon transforming the problem of social ontogenesis into the political problem of the diagnosis and the possible solution of the conflict between the phase of technicity and other phases of culture, if not between technicity and

[18] In conformity with the typographical choices of the 2005 complete edition of *Individuation*, I insert between squared brackets the parts added in IGPB and later omitted in the IPC edition. It is worth noting that some variations, even important ones, have not been indicated as they should have (and this is precisely the case here). For a brief and clear account of these variations, see Carrozzini 2011: 156.

[19] 'Ethics expresses the sense of perpetual individuation, the stability of becoming of being as pre-indivuated and individuating' (IPC 246).

[20] It is worth recalling that the *Note* was written in same period as the two theses. See the Appendix to this volume.

culture as such. For this purpose the analysis of the inventive power of technicity has to be extended to other phases of culture that – if not explicitly presented as a function of social closure – rarely appear as processes of invention.

References[21]

Aspe, B., and M. Combes. 2004. "L'acte fou." *Multitudes. "Politiques de l' individuation. Penser avec Simondon"* 2004/18. http://www.multitudes.net/category/l-edition-papier-en-ligne/Multitudes-18-Automne-2004/. Accessed 30 Dec 2014.

Bachelard, G. 1951. *L'activité rationaliste de la physique contemporaine*. Paris: PUF.

Bergson, H. 1907. *L'évolution créatrice*. Paris: Alcan (English trans. *Creative Evolution*. London: Macmillan & Co., 1954).

Bergson, H. 1932. *Les deux sources de la morale et de la religion*. Paris: Alcan (English trans. *The Two Sources of Morality and Religion*. London: Macmillan & Co., 1935).

Canguilhem, G. 1952. *La connaissance de la vie*. Paris: Vrin (English trans. *Knowledge of Life*. New York: Ordham University Press, 2008).

Carrozzini, G. 2011. *Simondoniana. Commento storico-critico analitico de L'individuazione alla luce delle nozioni di forma e d'informazione*. Milano: Mimesis.

Espinas, A.V. 1897. *Les Origines de la technologie: étude sociologique*. Paris: Alcan.

Hottois, G. 1993. *Simondon et la philosophie de la "culture technique"*. Bruxelles: De Boeck.

Leroi-Gourhan, A. 1943. *Evolution et techniques I. L'homme et la matière*. Paris: Albin Michel.

Leroi-Gourhan, A. 1945. *Evolution et techniques II. Milieu et techniques*. Paris: Albin Michel.

Leroi-Gourhan, A. 1957. Technique et société chez l'animal et chez l'homme. In *Le fil du temps. Ethnologie et préhistoire*, 68–84. Paris: Fayard 1983.

Leroi-Gourhan, A. 1964. *Le geste et la parole. Technique et langage*. Paris: Albin Michel (English trans: Gesture and Speech, I. Cambridge-London: MIT Press, 1993).

Stiegler, B. 1994. *La technique et le temps. 1. La faute d'Épiméthée*. Paris: Galilée.

Wiener, N. 1950. *The human use of human beings. Cybernetics and society*. London: Sphere Books, 1968.

[21] *Simondon's complete bibliography and a list of abbreviations are provided in the* Appendix.

Part III
Technicity, Sacredness and Politics

De la société close à la société ouverte, de la cité à l'humanité, on ne passera jamais par voie d'élargissement.

(Bergson, *Les deux sources de la morale e de la religion*)

Il convient de rechercher si ce qui, dans l'individu, dépasse l'individu ne lui viendrait pas de cette réalité supra-individuelle, mais donnée dans l'expérience, qu'est la société.

(Durkheim, *Les formes élémentaires de la vie religieuse*)

Simondon was at the height of his philosophical creativity when, at the end of the 1950s he wrote his two doctoral theses: *Individuation* and MEOT (see p. 1, n. 1). The conceptual innovations of the former and the more strictly sociological and political issues challenged while developing his philosophy of technics in the latter hereafter converged in the plan for a reform of the social sciences he presented in *Forme, information et potentiels* [Form, Information and Potentials] (1960). Some of the following writings – the course on *Psycho-sociologie de la technicité* [The Psycho-Sociology of Technicity] (1960–1961), the short essay on *Culture et technique* [Culture and Technics] (1965) and the course on *Imagination et invention* [Imagination and Invention] (1965–1966) – are the most relevant traces of Simondon's reflection on the social sciences after the two theses, and the adequate ground for a complete understanding (and a critique) of the connection between his philosophy of technics and his political thought.

In order to have at our disposal all the elements for an adequate analysis of these texts, it is worth going back for a moment to the general theory of the social system Simondon derived from a twofold approach, which I named – assuming Merleau-Ponty and Canguilhem as his main references – epistemological and phenomenological. Simondon's theory combines analyses of the processes and structures constituting social systems: on the one hand the ontogenetic analysis of biological normativity allows him to trace the threshold conditions of transindividual individuation; on the other hand the exercise of an individuation of knowledge provides the understanding of the peculiar 'oscillating' temporality which characterises social relations.

The theoretical tool Simondon adopts in *Individuation* for connecting the two perspectives is the concept of information, reformed in light of the quantic paradigm: this concept would bring an understanding of the discontinuity of processes and the partial indeterminacy of relations within and between systems. According to Simondon, what exceeds the automatic functioning of a social system depends on the emergence *within* the system of destabilising factors which are *at the same time* constitutive of it: (1) pre-individual potentials, (2) phases relative to former individuations still taking place at the same or at different scales; and (3) the results of other processes of individuation. In effect, the social system – beside being in constant relation with its own associated pre-individual milieu and with the still present potentialities of the physical and biological individuations – institutes a peculiar relationship with the distinct milieu that emerged from psychic and collective individuation, the mixed milieu of technics and culture, composed of objects which are symbols and vice-versa.

In order to understand the genesis, nature and effects of this peculiar milieu, and in order to understand how the collective produces, undergoes and regulates its processes of structuration and of destabilisation, it is necessary to traverse another cultural referent which, although less evident, is crucial to Simondon's oeuvre: the French sociological tradition. This reference is the key for challenging the third part of MEOT, the epicentre of which is what Simondon calls the 'primitive magic relationship' (Chap. 10), and for the understanding of the relationship between technicity and sacredness (Chap. 11). After having explained the nature of technicity by referring to Leroi-Gourhan's 'prehistorical ethnology', I will then discuss themes typical of French sociology which will then bring me to trace the genealogy of Simondon's political thought to Marcel Mauss and Émile Durkheim.[1] This will bring to a reconstruction of Simondon's own pedagogical-political project based on the direct confrontation with the positivistic theme of the connection of human progress with the efficacy of technological education (Chap. 11). On this basis I shall provide an interpretation of his political thought that aims at disclosing the critical and constructive potentialities implicit in his epistemology of individuation for political philosophy (Chap. 12).

[1] This tradition was born within Positivism. It started from Durkheim's inscription of ethnological studies within a Kantian paradigm, moved through Mauss to arrive, in one of its ramifications, at Lévi-Strauss's structuralism. Mauss, nephew and recognised successor of Durkheim, was the professor of Leroi-Gourhan at the *École des Hautes Études* and at the *Institut d'ethnologie* at Sorbonne, and in the end his *directeur de thèse*. My analysis will follow this perspective through the very few quotations authorising it (in Simondon's oeuvre Durkheim does not appear, and Mauss only once in IMIN 26), but convinced that the weight of this cultural heritage has been underestimated. Besides the references to Leroi-Gourhan, in the bibliography of MEOT one can find *Le rite et l'outil* (1939) of Charles Le Cœur (a Durkheimian sociologist) and the book of Georges Friedmann, *Le travail en miettes* (1956), whose critique of Durkheim Simondon abundantly draws from.

Chapter 9
Techno-Symbolic Function

Simondon's study of the ontogenesis of the social system in *Individuation* places his theory at distance from both phenomenology and structuralism. This is evidenced by his choice of elaborating the concept of the 'transindividual' by way of refuting the phenomenological concept of 'intersubjectivity' and the structuralist concept of 'symbolic system'. In his main oeuvre he describes the transindividual regime of individuation differently from both the way in which the relation 'to the Other' (*à Autrui*) is – for phenomenology – the (transcendental) institution of intersubjectivity, and from the way in which the symbolic structure is – for structuralism – the functioning of society itself out of any 'natural link'. His seminar *Imagination et invention* [Imagination and Invention] (1965–1966) provides a completely different approach. Its analysis will allow me to situate Simondon's theory of symbolic function in relation to the 'fight between two kinds of originary [*deux originaires*]' which took place during the 1950s between phenomenology and structuralism (Bimbenet 2001: 163). In distinction from structuralism, Simondon does not avoid posing the question of the origin of society, but at the same time he detaches himself from phenomenology, since he avoids challenging such an origin as the supposed horizon of sense. In fact Simondon avoids both the *disappearance* and the *hypertrophy* of the question concerning the origin of society. What he aims at is rather a description of the relational structures of the social system, seen as the result of dynamic and differential processes in which the origin of society plays the role of an active 'phase'.

The problematic hesitation between the concepts of 'signification' and 'symbol' characterising Simondon's lexical choices from *Individuation* (1958) to *Imagination et invention* (1965–1966), shows a path which can be compared to Merleau-Ponty's attempt to overcome the opposition between phenomenology and structuralism in

A. Bardin, *Epistemology and Political Philosophy in Gilbert Simondon*,
Philosophy of Engineering and Technology 19, DOI 10.1007/978-94-017-9831-0_9

De Mauss à Lévi-Strauss (1959a).[1] But the outcomes of Simondon's research undisputedly show the influence of Canguilhem, whose conceptualisation of the relation between organism and milieu allowed Simondon to provide an original connection between the processes of symbolic production and the technical-cultural milieu.

A close analysis of the course *Imagination et invention* shows that for Simondon the symbolic field cannot be defined in opposition to a supposed senseless real which would mark its boundaries (whether it be conceived as a blind natural determinism or as a pure noumenal limit of the phenomenal horizon). The symbolic field is, according to him, a determinate regime of production of relations and of processes of information exchange in which the individuation of object-symbols accelerates and amplifies these processes to the point of making them inconceivable through categories *exclusively* biological *or* cultural.

Thus Simondon explains the emergence of the symbolic function starting from the relations *between* organisms and their mixed milieus made of nature, other organisms, and symbols derived from the subsequent sedimentations of such relations. This is an approach according to which 'sense' is neither produced by organisms nor by *homo sapiens* but emerges from the relations of communication through which groups of organisms and the organism itself, at different levels and through different milieus, are structured. From this perspective a theory of the symbolic function should firstly explain the 'natural link' between symbols and the reality they emerge from, and secondly the way in which, starting from the feedback effects of this milieu of symbols, the relations between organisms change their order of magnitude and the collective emerges.

9.1 The 'Cycle of the Image' in *Imagination et invention*

Simondon's course on *Imagination et invention* (IMIN) is an analytic study of the activity of symbolic production. It shares the same structure, context and academic finalities of the course on *L'imagination* held in 1962–1963 at the Sorbonne by

[1] A rapid glance through the discussion on the symbolic function at the time is offered by the paradigmatic form assumed by the debate that took place during the 1950s between structuralism and phenomenology on the heritage of Mauss's *Essai sur le don* (1923–1924). According to Lévi-Strauss, *Introduction à l'œuvre de Marcel Mauss* (1950), in Mauss the concept of symbol would be partially compatible with structuralism insofar as it becomes more and more detached from the Durkheimian theory of the collective subject and representation. The following year Claude Lefort, *L'échange et la lutte des hommes* (1951) (published in Sartre's review *Les temps modernes*), contested from a phenomenological perspective Lévi-Strauss's supposed cancellation of the individual in the whole of the social system and in the impersonal abstraction of symbolic function. At the end of the decade, on the occasion of Lévi-Strauss's candidature to the *Collège de France*, Merleau-Ponty wrote *De Mauss à Lévi-Strauss* (1959a), where he looked for a convergence between the two perspectives.

Simondon's colleague Juliette Favez-Boutonier.[2] In IMIN Simondon aims at showing a continuity between the concepts of image and symbol in order to explain the process of symbolic production and invention according to the 'cycle of the image', the status of which is mixed: both individual and collective.

In the foreword he immediately clarifies that 'image' is in fact the name of an operation. It is the 'local activity' of a subject that leads to invention in a system subject-milieu made of three subsequent *and* simultaneous phases: motor, perceptive and affective-emotive. Firstly, at the biological level, the motor schemas of the organism are the 'primordial source' of the *a priori* (IMIN 42): these potentialities of 'the present of the subject' (i.e. of the *present relation* in which the subject is involved) are amplified through the variety of cultural contents (IMIN 57). Secondly, there is perception and the cognitive content of the image that emerge through 'the relation to the milieu according to primary categories of valence and signification' (IMIN 63). Once again Simondon denies any identification of such a threshold with the organic/psychic or even animal/human differences (IMIN 42–43, 64). He speaks of a '*coupling* of two systems, subject and world', where the category of object emerges (IMIN 75). The resulting system is clearly metastable, although Simondon does not use the term: 'this image is not given; it does not result from a state of stable equilibrium. It is the act of a subject who finds sense at all orders of magnitude in the perceived reality' (IMIN 92). Thirdly, the affective-emotive content of the image concerns the 'a posteriori images of symbols'. At this level is situated the non-dialectical system of subsequent (although not necessary) phases of the process from which the 'world of the imaginary' emerges as the ground for any possible invention: (1) *imprinting* as a biological event; (2) its fixation into memory-images; (3) the formalisation of images in symbols.

The whole 'cycle of the image' spins around the local activity of the 'subject living-being [*être vivant sujet*]', and from this perspective it might appear as a 'problem of general psychology' concerning the complex biological activity of superior organisms (IMIN 3). But this is not the case. In the introduction Simondon in fact clarifies that the image is to be intended as an 'intermediate reality between subject and object' neither perceptively nor conceptually graspable in its entirety (IMIN 7–8) which necessitates, as such, a phenomenological analysis.[3] Furthermore,

---

[2] These courses of general psychology were attended by a mixed group of psychologists and philosophers. In the organisation of studies in force until 1966–1967 the courses for the *Certificat de psychologie générale* were the base for both the degree of psychology and for one of the four qualifying courses for the teaching of philosophy. Favez-Boutonier is part of the generation of Daniel Lagache, in 1953 cofounder of the *Société française de psychanalyse* with Lacan, with whom he also founded the *Association Psychanalytique de France* in 1964, after the scission from which Lacan's *École freudienne de Paris* had emerged. A quick glance at the authors quoted in Favez-Boutonier's course is sufficient to give evidence for a connection to IMIN, in particular in the third part where Simondon relies on the problem posed about the relation between image and symbol.

[3] 'The existence of different categories of image-objects [*objets-images*], the third reality between the subjective and the objective, requires a particular kind of analysis which one might call, in the proper sense of the term, phenomenological, since it is peculiar to this kind of reality to manifest and impose its nature of image' (IMIN 15).

the triple connotation, motor, perceptive and affective-emotive (in the terms of *Individuation* we might translate: biological, psychic and collective) makes of the image the common universal background, the 'basis of cultures' (IMIN 28). That is why its analysis must also traverse the study of mythological systems and of what we might understand in general terms as 'material culture':

> Part of the reality of human groups [is] made of images materialised as drawings, statues, monuments, dresses, instruments and machines, and also of rhetorical figures, formulas, as the proverbs which are actual verbal images. (IMIN 18)

In short, the 'cycle of the image' is to be studied as the universal model of the process at the end of which 'cultures disorganise, change their structure and are revived according to new principles' (IMIN 28).

By defining such an approach as 'phenomenological', Simondon neither intends to refer to any consciousness nor to an original 'intersubjective cogito' as does Lefort (1951).[4] Simondon is simply pursuing the mixed study of structures and operations seemingly authorised by the form Merleau-Ponty's thought had assumed since *La structure du comportement* (1942), an approach forbidden by strictly synchronic structuralism. Simondon's course is thus divided into three parts which correspond to three different phases of the 'cycle of the image' as three different relational modalities between the organism-subject and the world. Yet nevertheless an exclusively phenomenological reading of the course centred on the activity of the symbolic production of the organism-subject simply does not work.

In fact the influences of the external milieu and of the biological functions of the organism are crucial of symbolic production. First of all it is the external milieu that, both as a natural and as a symbolic world, introduces elements of discontinuity in relation to the continuous tendency of the cycle (IMIN 30, 88, 92): this prevents at all levels the closure of the cycle and obliges the subject to a series of quantum leaps which the homeostatic cyclical activity would not entail in itself (from motor tendencies to perceptions, form memory-images to symbols), through which the subject succeeds in 'organising the world of the imaginary' (I 138).

Thus conceived, the 'production of sense' resulting from the cycle of the image might be simply counted among the activities of the living being. This systematic study of biological activity and of the organism's relation to its milieu is the 'general praxeology' Simondon defines, when closing his course, through Alfred Espinas's words: 'a science of the most universal forms and the most elevated principles of action in living beings' (IMIN 191). Simondon seems to have abandoned here the hypothesis of a possible 'allagmatic' theory unifying the human sciences, in order to adopt a wider non-substantialist biological theory unifying the phase-shift system of all activities of 'the living being, considered in its most primitive forms as a self-kinematic system in interaction with the milieu' (IMIN 191). Not a science of the organism, indeed, but of the living system, a concept which – understood in the light

[4] See above, n. 1.

of the subject-milieu relationship – might explain also the activity of symbolic production as a mixed, biological, technical and historical, function.[5]

It is from this perspective that the evidence of different linguistic choices reflects the conceptual shift from *Individuation* to IMIN. The key concepts of *Individuation* almost entirely disappear in IMIN: the terms 'transindividual' and 'transduction' are never used (for transduction Simondon always uses 'amplification'); other terms, like 'individuation' and 'analogy', although they sporadically appear (we are talking of one-figure numbers in about 200 pages) loose their theoretical centrality.[6] These terminological choices partially derive from Simondon's exigency for a rhetorical reconfiguration of his teaching to the peculiar audience of IMIN,[7] and possibly stem from the difficulties he experienced in trying to introduce his jargon to the philosophical community of the time (Sect. 3.3). But this also manifests a partial conceptual re-elaboration – rather a shift than a radical abandonment – of the theoretical horizon of *Individuation*, charged with tensions between the epistemological and the phenomenological approaches that inhabit it. If this were not enough, in IMIN the elaboration of the concept of symbol marks an important point, which it is not possible to grasp without reference to a set of problems of a structuralist sort.

9.2 Imaginary/Symbolic

Simondon follows Favez-Boutonier in questioning the relationship between the 'imaginary' and the 'symbolic' as respectively individual and collective. His aim is to theorise the cycle of the image as a symbolic function producing the metastable 'organised world' of symbols, which is the condition of possibility for individual *and* collective invention.

The two main references in Favez-Boutonier's course are Sartre and Lacan. Favez-Boutonier intends to show that Lacan's theory includes and widens the theory of Sartre, who erroneously 'mixes up the imaginary and symbolic functions' (Favez-Boutonier 1962–1963: 102):

> Sartre (*L'imaginaire*) did not differentiate the imaginary from the symbolic [...] he situates both of them at the level of what in Lacan's theory is the imaginary, a relationship that the subject produces and in which it is lost, because it neither moves there towards the others nor towards the real. (Favez-Boutonier 1962–1963: 96)

---

[5] The reference to Espinas (1987) relies on Canguilhem 1952: 122–23. The program was also, although in a different way, Bergsonian, insofar as it allowed 'the term biology the large sense it should have, and which it will probably acquire one day' (Bergson 1932: 103) (see Chap. 8, n. 2). Once established that in IMIN Simondon 'refuses to return to the anthropological presupposition of thought as the privilege of man' (Lefebvre 2012: 65–66), I am trying here to push this hypothesis towards the 'living system'.

[6] As I will show, the term metastability represents a notable exception: its partial usage is still carried on by Simondon in further writings, along with the relative concept of a state far from equilibrium full of potentials.

[7] IMIN was a course for both psychologists and philosophers. See above, n. 2.

Favez-Boutonier inherits the problem posed by Lacan's early article (already dated at the time) on *Les complexes familiaux dans la formation de l'individu* (1938)[8]: a rare product of a phase in which Lacan's thought addressed the problem of the genesis of the symbolic order within the subject through the *institution* of the law rather than through a more genuinely structuralist analysis of its *functioning*. Lacan introduces there the concept of '*imago*'. This makes the structuralist distinction between the imaginary and symbolic more problematic, because it poses the problem of the function of mediation of the family complexes, conceived as 'collective archetypes' situated between the world of images of the infant and the oedipal institution of the symbolic world. In short, the family complexes would be situated on the Freudian threshold between the principle of pleasure and the principle of reality: 'between the two there is an ambiguous zone [...] where narcissism entails the calling upon a myth' (Favez-Boutonier 1962–1963: 92)

Although attenuating it, Simondon fundamentally adheres to Favez-Boutonier's critique to Sartre, admitting that the latter 'denies the distinction between imaginary and symbolic functions' (IMIN 130). Yet he borrows and uses the Lacanian notion of *imago* precisely to understand the transition from the mental image to the object-symbol, a transition in which 'the imaginary' would emerge as an 'organised world' of symbol-objects, the status of which is *both* individual *and* collective (in Simondon's words 'transindividual'). But in the paragraph *La notion d'*Imago*; en quel sens l'*Imago *est un symbole* [The Notion of *Imago*: in Which Sense the *Imago* is a Symbol], Simondon also distances himself from Lacan. For Lacan there is a difference between the image and the symbol, since the latter only appears over the threshold of the complexes 'where three terms are involved (Oedipal complex), while images express a duality'. On the contrary, for Simondon 'the Imago as a kind of organiser is already an elementary symbol', and therefore there is no essential difference between the two (IMIN 128).

In this role of mediation the *imago* acquires the paradoxical status of an 'unconscious representation', which can be grasped through the concept of metastability: 'the Imago emerges as a figure of a tense equilibrium' related to two different orders of magnitude in which two different regimes of activity take place (IMIN 127). On the one hand we have the discontinuous and partially aleatory spectre of the memory-images organised according to an elementary binary logic, on the other hand we have the continuously structured net of the socially instituted symbolic relations that provide the subject with access to reality. The differentiation between binary and ternary structures allows Simondon to describe the process of the development of the individual in classical psychoanalytic terms (from the imaginary-mother to the symbolic-father, IMIN 127–28) precisely because this difference of

[8] Published in the volume VIII of the *Encyclopédie Française*, entitled *La vie mentale*. In it Lacan reformulates the Jungian concept of 'family complex' in the terms of his 'mirror stage', thus anticipating the structuralist development of his thought during the 1950s, but still maintaining a 'genealogical' approach to the problems of law and the institution of the symbolic order. On the tormented writing of this text and its different versions, see Roudinesco 1993: 193 ff., where the author also delves into Lacan's (quite shared indeed) debt towards Uexküll's concept of *Umwelt* (see below, n. 19).

structures (not 'of nature') defines the phase-shift of the social system between two regimes of relational activity or communication which simultaneously take place at different scales, individual and collective:

> The ternary structures actually allow individuals in the same group to communicate, insofar as they formalise the experience of interaction and provide a universal ground corresponding to the intellectualised, adult, vigilant and conscious expression. But also binary structures allow communication, according to modalities less collectively universal, less inserted in the group action and not implying the same degree of vigilance: tales, legends, myths, present sometimes binary structures [...] Finally, a certain connection links within culture the individual binary structures and the ternary structures that entail the presence of Society, Law, Divinity. (IMIN 129)

That is how the shift from the imaginary to the symbolic (from binary to ternary), i.e. the integration of the subject in the collective, is made possible by the *imago*: a 'structure of conversion' that Simondon defines (with more than simply a Lacanian nuance) 'with a double entry [*à double entrée*]', since on the one hand it is related with the I [*moi*] and on the other to the symbolic order (IMIN 130). Thus Simondon can claim that the world of the imaginary 'prepares access to what is usually called the symbolic' (IMIN 130), precisely because the latter has since the beginning contributed to the emergence of the former.

In this sense the concept of *imago* explains the collective as a system of relations between the individual imaginary and the symbolic milieu. But what does this entail concerning the relation between the process of integration of the subject in the collective and the very institution of the collective? Simondon does not treat here the second side of the process and, as it is typical of IMIN, he seems to adopt an approach of general psychology. Therefore we have to go back to the ontogenesis of the individual imaginary to discover that it has a double source. On the one hand – as binary – it derives from a kind of archetypical elaboration of the universal conditions of the existence of the species, the basic normativity concerning 'life and death, sanity and illness, pleasure and pain'. On the other it emerges from an 'asymmetrical structuration' of experiences. Although linked to an individual history, this 'symbolic power' tends towards universality, since experiences emerge out of the relation with a partially shared natural and cultural milieu (IMIN 130).

Thus the 'formalisation' of individual experiences takes place *between* the biological-archetypical duality and the triadic relationship [*relation de trialité*] of the collective, so that there is no problem of compatibility between the individual imaginary and collective symbolism. The former is always already involved in a process of symbolisation entailing a relation of constitutive reciprocity between memory-images in the subject and what Simondon names here the 'situation':

> In the study of the genesis of images, we will call symbols the memory-images resulting from an intensive exchange between a subject and a situation. Taking part in an action, in a situation, the subject has given something of itself to it. In exchange it preserves a sufficiently intense image which is a kind of fragment of the reality of the situation, and up to a certain point it allows for its reactivation. (IMIN 5)

A further step is the reactivation of the two 'pieces' of the symbol, which entails the information exchange between the individual imaginary and the symbolic world, the condition of possibility of which is the fact that the symbol is *actually* a part of the complex relation between the subject and the milieu: 'the symbol can never be a *flatus vocis*; it presupposes an implicit realism' (IMIN 6). In short, because of its ontogenesis within a metastable situation, the symbol is rooted in a 'natural linkage' with the reality it refers to.

In the foreword to IMIN Simondon accepts Favez-Boutonier's terminological distinction between 'sign' and 'symbol', making of the concept of symbol the centre that connects the two milieus between which it emerged: the 'internal' psychic activity of the organism and the 'external' natural and social environment.[9] This situates the symbol beyond the distinction between the mental image and the physical object (i.e. beyond the subject/object schema). Consequently the symbol will be able to activate new (transductive) processes in both milieus, processes of knowledge included. Thus 'theory' simply has to follow 'the natural tendency of the symbol to become action' (IMIN 126), accordingly to what Favez-Boutonier stated in the conclusion of her course:

> The image, whether expressed or created, even the simpler one, does not only tend to possess, but also to transmit, to communicate, each time that the subject marks with a sign the *milieu* that surrounds it. (Favez-Boutonier 1962–1963: 114)[10]

In order to deny the structural closure and the contraposition between the individual and the collective, Simondon thus refutes the structuralist break between the image and the symbol. And, consistently, he refuses to explain the collective regime of information exchange in terms of 'sign' and does not consider language as a system of signs capable of explaining the symbolic function. It is precisely the concept of *imago* that allows him to avoid the structuralist identification of language and symbolic function, maintaining the anteriority and independence of the *imago*. Thus Simondon can understand the symbolic function as a process anterior to language, of which the latter would be the result 'stabilised through conventions' (IMIN 131). And this demonstrates the radical incompatibility of Simondon's philosophy with the structuralist project.

Furthermore, since he acknowledges biological and physical roots to symbols, Simondon cannot adopt the phenomenological *escamotage* of dissolving them into the indefinite production of a pre-linguistic intersubjective relationship. In this sense his preference for the concept of 'symbol' and his abandonment of the term 'signification' is possibly the symptom of a deeper refusal, although a simple

[9] 'The sign is, in relation to the thing, a supplementary term; the black table exists and is in itself complete without the word that refers to it [...] the symbol, on the contrary, entertains an analytic relationship with the symbolised. Symbols are couples: this means that a symbol is the fragment of a primordial whole which has been divided following an accidental line' (IMIN 4–5). IMIN 4 refers to Favez-Boutonier 1962–1963: 92–93. Her argument takes its start from Ortigues 1962: 203, where the author differentiates the image which tries to make the object present and the sign which accepts its absence: the symbol, instead, is an image employed as a sign, that is – one might say – an image which does not pretend to provide 'presence'.

[10] In the conclusion, the author reaffirms the distinction *and* the natural continuity between image and symbol (Favez-Boutonier 1962–1963: 107–19).

terminological choice does not entail in itself an epistemological shift. In fact, in Merleau-Ponty 'the *symbol* is still conceived according to the model of *signification*', i.e. it 'depends from the position of a consciousness for which it would have a meaning' (Karsenti 1997: 298). But it has been already demonstrated that in *Individuation* Simondon detaches 'signification' from 'sense' precisely against Merleau-Ponty and precisely because signification does not depend there on the position of any subject (Sect. 3.2).

In the period which runs from *Individuation* to IMIN, Simondon increasingly abandoned the term 'signification' and frequently adopted the term 'symbol', which had the advantage of 'keeping more domains together',[11] as the concept of information did in his former work. In fact, moving from a theory of the social system centred on the concept of signification to one revolving around the concept of symbol, Simondon continued to work on the concept of information as the way out of the subjectivism still implicit in the phenomenological horizon, although without adhering to the destitution of the subject of which structuralism was supposedly the name. That is why, despite the different linguistic choices, in IMIN Simondon's philosophy seems to prolong the former conceptual effort to grasp a transductive (i.e. quantic) relation through the different fields of being and knowledge. Differently from significations, which result 'from the assimilation of the real to the *moi*' (IMIN 131), the symbol – affective-emotive crystallisation of an image – has thus in IMIN the same paradoxical status which characterised in *Individuation* the transindividual production of significations, i.e. collective invention.[12]

9.3 The Problem of Symbolic Invention

It should be clear now that – from the doctoral theses to IMIN – a basic continuity characterises Simondon's effort to define the internal processes of phase-shift within social systems, in which the mediation of an individual organism, an object or an element, allows for the amplification of the potentials present in the milieu. In *Individuation* the individual is a support for schemas, who triggers the development of processes internal to social systems, by 'traducing [itself] into signification, implicit or explicit, vital or cultural' (I 217). And 'culture' is this milieu of potential schemas of action (NC 504) in which the individual intervenes as an amplifying

[11] See here the 'suggestion' of Favez-Boutonier 1962–1963: 94, relying on Ortigues 1962: 61–62.

[12] See Sect. 4.1. Such as the transindividual, also the symbol is what really crosses the 'inside' and the 'outside' as the 'mental symbol-image' in the subject and as the materialisation of the 'symbol-object'. Also in IMIN a double perspective necessarily follows on the same processes, the description of which depends on the side one choses to describe them: it is possible both to claim that the symbol-image elaborated within the subject can 'borrow [*emprunter le secours de*] the materiality of the objects' (IMIN 5), and to maintain that the 'organising power' of the '*a posteriori image*' 'continues when the situation [from which it emerged] ceases to exist' (IMIN 20). Thus significations can circulate as linked to their origin and actually take part to the productivity of the symbolic function, *only* when they continue to 'adhere to symbols' (IMIN 132).

device through its (discontinuous) symbolic existence, the continuity and efficacy of which depends on the milieu:

> If there is an external reality, it is the individual as a transductive being: not as a subject-substance or a body-substance, consciousness or active matter. During its objective existence the individual, as far as it is experiencing, is already a connected being. Perhaps something of the individual is eternal, and reintegrates itself in some way to the world in relation to which it was an individual. When it disappears, the individual is annihilated only for what concerns its interiority. For it to be annihilated objectively, the milieu itself should be annihilated. It is as an absence in the milieu that the individual continues to exist and also to be active. When dying the individual becomes an anti-individual: it changes sign, but it still perpetuates its own being as an individual absence. The world is made of actually living individuals, which are real, but also of 'holes of individuality': true negative individuals composed of a knot of affectivity and emotion who exist as *symbols*. When an individual dies, its activity is unaccomplished, and it will be unaccomplished as long as there will be individual beings capable of re-actualising this active absence, this germ of consciousness and action. (I 250, italics added)

As already explained in Sect. 8.4, the function of the individual is the central and most problematic issue in Simondon's philosophy of individuation, but further texts show a major concern with the flip side of the same problem: the theme of the milieu. The third part of MEOT concerns the way humans derive their biological and cultural specificity from the technical and symbolic milieu they build. There the transductive function is in fact not proper to the individuals or to the technical object, but rather distributed among technical sets, objects and elements, although the ultimate transduction of technicity occurs at the level of elements, 'the true bearers of technicity', in which 'technicity exists in its purity' (MEOT 73).[13] As the bearer of a simple operational schema, the element is more easily detachable from the context and therefore transferable according to unforeseen trajectories, the vehicle of the technicity which, 'model of collective relationship' (MEOT 245), expresses itself in invention. The process of invention – to which Simondon will devote many of his courses in the following years – is only partially analysed in MEOT, where the theme of symbolic invention does not appear if not occasionally. And nevertheless, we can already find there clearly formulated the way in which the transductive function can be carried on by an image which assumes the status of a symbol, according to the same scheme we will find in IMIN: the image must become 'an object containing a structure to be analysed by the activity of the individual being' in order to transductively function as a 'still and radiating symbol' (MEOT 99).

The fact that in the 1965–1966 course it is the image as a symbol-object – and not only the signification or the technical element – the privileged vehicle of information around which the collective emerges, entails a perspective more and more centred on the system rather than on the individual. According to my hypothesis,

[13] 'Technicity at the level of the element is concretisation: it is what makes the element the actual product of a determined set, but not itself a set or individual. This characterisation makes the element detachable from the set ['*de l'élément*' is an evident editorial mistake] and liberates it for the constitution of new individuals' (MEOT 73).

this oscillation between the transductive function of the individual and the transductive dimension of the collective as a system reflects in Simondon's thought the presence of a double and conflicting paradigmatism, biological and phenomenological. The result is that from his perspective the social bond can be reduced neither to a dynamic interaction among human organisms, nor to the social institution that would precede and form it. These are major complementary risks for a theory equally far from any physico-biological or structural determinism. And nevertheless the risk had to be run. Although this methodological oscillation between the two approaches remains unresolved in Simondon, as the inventive and 'transductive' function of internal excess of the system shifts more and more from the variously intended individual to the relation individual-milieu, a biological paradigm becomes more and more prevailing, and the concept of organism gains a central place in the explanation of the mechanism related to the 'human' function *par excellence*: the symbolic function.

It is for this reason that concepts otherwise central for the explanation of psychic and collective individuation and of invention – 'transindividual' in *Individuation* and 'technicity' in MEOT – leave their place to the systematic treatment of the 'cycle of the image' in IMIN. In effect, the 'cycle of the image' is described there in terms of the relationship organism-milieu, where the milieu is a mix of nature and culture (technicity included), which explains the symbolic function from which the social bond emerges. As it happened with 'values' in *Individuation*, in IMIN the symbol has the function of 'guaranteeing the cultural continuity of groups' (IMIN 18), of building (*with* action and perception, the status of which is mainly biological) the 'basis of cultures' that changes according to a cycle the trigger and result of which is invention: 'after each cycle cultures de-organise, change their structure, and are born again according to new principles' (IMIN 28). In this sense, in IMIN 'image' and 'symbol' must be read as moments of the same discontinuous process – of which invention is part – that characterises both the ontogenesis and becoming of the organism and, on a larger scale, the *isomorphic* ontogenesis and becoming of civilisations: 'The act of invention is not essentially different from the modalities of organised growth characterising organisms' (IMIN 162).[14]

In *Individuation* the theme of invention was only treated in a general sense, as related to the transductive heuristic of analogical thought, while the concept of the transindividual traduced its structuring force in the psychic and collective domain. In the *Note complémentaire* only the theme of technical invention emerged as crucial to transindividual individuation. Similarly, in MEOT and in great part of his subsequent work, Simondon is mainly concerned with invention in techniques. In particular in MEOT it is clear that technical invention is possible thanks to an analogic transposition of the paradigm of the individual-milieu relationship. Technical invention is only possible for the living being as far as it is 'an individual being who carries within itself its associated milieu': it is precisely this capability of conditioning oneself that allows for the production of objects that condition themselves

[14] 'The process of growth, maturation and decline, directly correspond to the common ground of images that constitute cultures, as norms for individual knowledge and action' (IMIN 27).

(MEOT 58). What is not at all clear in MEOT is the role played by invention in the fields of symbolic production such as culture and, in particular, the phase he defines here as 'religion' and later as 'sacredness'.

I suppose Simondon's resistance to treat symbolic production as a fruit of invention derives from the defensive stance he adopts in order to 'liberate' technicity from the minoritarian position it had in relation to both the dominant humanist culture and the 'structuralist turn'. The way he addresses structural linguistic in the course *L'invention et le développement des techniques* [The Invention and Development of Techniques] (1968–1969) is emblematic in this sense: 'structuralist formalism generalises a classificatory and categorical [*catégoriale*] thought which is only one of the aspects of human relationship', the synchronic, static and formalised aspect (IT 85). On the contrary, the study of the technical object would provide 'conceptual models different from those provided by linguistics' (IT 84–85), which would allow an adequate understanding of the diachronic characterisation of such relations. The strength of this model, capable of becoming the paradigm for a new 'systematic vision of the human world' (IT 85), resides precisely in the fact that the subject-object relation would not be 'neutralised' there, but rather seen through the prism of the organism-milieu relationship:

> Technics provides the basis for a representation more powerful than formalism, because it absorbs the subject-object relations through the reversible mediation between the *tool* and the *instrument*, which is the third reality of technical objects, suturing the human being and the world, and the paradigm of the relation between the living and its milieu. (IT 85)

As a result, the theme of symbolic invention, related to a cultural field dominated first by the literary tradition and then by structuralism, is very marginal in Simondon's oeuvre. Therefore it is quite difficult to find a direct analysis of it, with the exception of IMIN.[15] The course of 1964–1965, although pointing to technical invention, does not limit itself to this subject. It aims at demonstrating that for technical invention to be understood, it has to be situated at the background of a process – the cycle of the image – of which it is only one of the moments and of the possible outcomes.

As aforementioned, according to Canguilhem invention consists in the solution of problems posed within the relationship between organism and milieu. Now, at the elementary biological level the solution can occur thanks to an action which allows for the bypassing [*détour*] of the obstacle or – but only in the second instance – thanks to the production of a tool: in both cases the 'detached' images of desire provide the necessary and sufficient milieu for 'individual' invention (IMIN 152–53). But in order to function at the higher scale of the collective – notably human – invention must rely on a formalised milieu of signs and symbols, because only 'a symbolic formulation [of the problem] can solve […] general and theoretical problems in relation to which actual difficulties appear as particular cases' (IMIN 153 ff.). The social factors of the milieu are therefore crucial for symbolic invention,

[15] According to Van Caneghem (1989) the systematic absence of a theory of language in Simondon would depend on the 'extreme and certainly excessive respect Simondon paid to the "territory" of his colleagues' (Van Caneghem 1989: 816). Although academic opportunism might confirm Simondon's resistance to treat the theme, it does not provide a definitive explanation.

which counts among its conditions both a determinate development of social relationship and the partial sedimentation of processes of problem solving through practices, technical objects, images and symbols.

This mixed milieu, first of all organised according to a biological goal – the survival of the group – is the very ground of any symbolic invention whether it is the objective 'metrological formalisation' which moves from techniques to the constitution of the sciences, or the subjective 'axiological formalisation' on which normative invention (ethical and political) and artistic invention depend. Technical and symbolic inventions are in this sense moments of the same process, along which the 'basic' technical invention of the tool follows and oversteps the organic *detour*, thus introducing symbolic invention. Signs and symbols are themselves tools (and instruments) that, within a regime of collective individuation, groups produce in order to solve problems deriving from their relation to the natural milieu, thus creating a further milieu, the symbolic one, which would support the relaunch of invention – *also*, but not exclusively, technical – at yet another level.

'Practical and symbolic invention' *can* finally take place once the 'actual field of accumulated finality and experience' in which the milieu of collective invention consists emerges (IMIN 162). When the symbolic invention becomes part of the milieu, a cumulative causality is established. Thus the circulating symbol-object can trigger further cycles of images, thanks to a systemic causality taking place in a mixed milieu made of organisms, symbols and objects (technical, sacred or artistic), where collective processes eventually appear which can define – although not exclusively – the human field:

> The process of invention is better formalised when it produces a detachable object or an artwork independent of the subject, which can be transmitted and therefore put in common and constitute the support for a relation of cumulative participation. Although I am not denying the theoretical possibility or the actual existence of cultures in certain animal species, it is worth noting that the main limit of such cultures resides in the poverty of the means of transmission, because of lack of an object detachable from the living beings that produced it [...] what animal societies lack is not the capacity of creative spontaneity, but rather the capacity of creating objects. (IMIN 163–64)

The technical-symbolic milieu peculiar to humans is therefore defined by a 'cumulative causality' in which the production of 'objects' crosses an established culture both as its condition and as its effect:

> These effects of 'cumulative causality' only appear in a definite and decisive way – in the form of created objects having a sense for a culture – with human species [...] the created object is thus an element of reality organised as detachable, produced according to a cultural code which allows it to be utilised far from the space and time of its creation. (IMIN 164)[16]

[16] For Leroi-Gourhan the symbolic function, language in the wider sense, generally differentiates groupings based on instinct (species) from other based on language (ethnos). In analogy with the 'exteriorisation of the organs involved in the carrying out of technics', the 'exteriorisation' of memory makes of language a 'particular form of memory' which differentiates humans from other animals (Leroi-Gourhan 1965: 63–65). On Leroi-Gourhan's influence on Simondon, see in par-

9.4 Between Phenomenology and Life Sciences

Indeed Merleau-Ponty had already focused on the organism-milieu relationship in order to reconfigure the problem of the symbolic function and the emergence of culture within phenomenology. A few years before Simondon started working on his doctorate, Merleau-Ponty had assumed the concept of 'institution' as a 'remedy to the difficulties of the philosophy of consciousness' (Merleau-Ponty 1952–1960: 59). His attempt was to conceive the intertwining of what is biological and historical on the common ground of an intersubjective process in the course of 'institution', rather than idealistically inscribed in the horizon of a subject or scientifically fixed in a structure without subject. The 'institution' in fact exists already in animal groups as what introduces stability in the individual sequence of experiences, as far as it is the intersubjective ground of experiences and thus, thanks to its historical dimension, can grant the continuity of the collective (Merleau-Ponty 1952–1960: 61).

In short, the concept of 'institution' is for Merleau-Ponty the mark of a program of research: phenomenology would find in biology the point of departure for the understanding of the process through which organisms reconfigure their milieu as the 'web [...] of history', i.e. as the 'sphere of symbolism' (Merleau-Ponty 1955: 94). The origin and effects of the symbolic field remain thus always defined in relation to the historical dimension of an 'intersubjective' relation.

Conceiving culture as a 'field' seems to bring Merleau-Ponty and Lévi-Strauss closer, as the former seems to explicitly concede when closing the notes to his course quoting the latter's *Les structures élémentaires de la parenté* [The Elementary Structures of Kinship] (1949; Merleau-Ponty 1954–1955: 154, n. 121). But this is not enough to cover the distance separating the structuralist concept of symbolic function, with its 'triadic' structure, from the 'dyadic' structure implicit in the phenomenological concept of intersubjectivity (Merleau-Ponty 1954–1955: 103). And this is evidenced by the way the concept of 'institution' was created by Merleau-Ponty precisely in order to *overcome* structuralism through a theory of perception:

> The perceptive orientation of the social field is to take literally what Lévi-Strauss offers as a metaphor. As the thing perceived is a principle of lived cohesion without being an essence, thus the symbolic system, the *pattern*, would be a social thing. (Merleau-Ponty 1954–1955: 121)

Is it therefore possible to attribute to Simondon the aim of developing such a program of research? Can we read his philosophy of individuation as a partial continuation of Merleau-Ponty's effort to oppose to structuralism a dialectic of subjectivity? In the already quoted working notes it is Merleau-Ponty himself who denies that Simondon might be situated along this line of research:

> Simondon's point of view is trans-perceptive: perception is for him on the order of the inter-individual, unable to account for the true collective [...] For my part, the philosophy of

ticular Sects. 8.1 and 10.4. On this trajectory opened by Leroi-Gourhan are also based the 'grammatologies' of Derrida (1967) and Stiegler (1996) (see also Chap. 10, n. 28).

> brute (or perceptive) being [*être brut*] takes us out of the Cartesian *cogito*, of Sartrean intersubjectivity [...] reveals to us institutions beneath the flux of *Erlebnisse* and the fulgurations of the decision, – but for it, the nexus [*foyer*] remains the perceptive field, insofar as it contains everything: nature and history. (Merleau-Ponty 1959b: 42)

It is not my aim here to discuss whether this statement is for Merleau-Ponty the path for a possible convergence of structuralism and phenomenology, or the mark of his resistance to the possible absorption of phenomenology within biological categories. This perspective is what allows me to situate Simondon's philosophy in relation to the phenomenological heritage. In effect, since the beginning of his research (1953–1954) Simondon aims at the destitution of the primacy of consciousness (and *therefore* of intersubjectivity), he refuses to conceive the symbolic as a *cut* in relation to the biological, and he substitutes a 'natural' history of the emergence of the subject to the phenomenological questioning of the origin of human societies. But Simondon's 'natural' history is in fact made of the crossing of the cultural heritage with individual trajectories in singular 'situations':

> If it is true that culture can be conceived as the *non somatic* heritage the species provides to the individual during its formation [...] this mental nature is not constituted by the pure presence of symbols [...] since no concept can by itself, without an emotive emphasis, form a subject. (PI 116)

Human societies, although grounded on a biologically coded 'original sociality', cannot be understood on the simple bases of the biological features of the species, because they emerge from the circular causality triggered by the institution of their techno-symbolic milieu. This conception of the human milieu as a mix of life and history, perfectly matches the characterisation of the transindividual in *Individuation*: a limit concept which tries to hold together within the field of the 'human' the duality of the technical and symbolic functions and their respective normativities. It is in the form of this constitutive twofold nature that – even when the concept of the transindividual has disappeared – the double function of opening and closing characteristic of technicity and symbolism repeatedly returns in Simondon's oeuvre. In this sense Simondon's trajectory may be understood as a continuation of Merleau-Ponty's work on the concept of institution along the line of development suggested by Canguilhem's vitalism. But this may be accepted only with the premise that the quantic discontinuities Simondon injects into vitalism cancel any pre-determined 'substantialist' distinction between the living and the non-living, and therefore nature and culture.[17]

---

[17] This is not Merleau-Ponty's stance, of course. According to him the systemic effects of the symbolic field on the relationship between organisms ground the institution of human societies (Merleau-Ponty 1954–1955: 49–50). Yet this is a necessary but not sufficient condition to grant a shift which always remains, for Merleau-Ponty, unbridgeable: 'animal institution as "imprinting" [...] does not have the value of a symbolic matrix', since it does not own a force of 'indefinite productivity', due to the lack of any storage [*mise en réserve*] of historicity (39). Culture thus conceived as a field of 'cultural knots' (103) still marks a threshold in relation to 'nature' conceived as a kind of noumenal ground: the 'non-instituted' (Merleau-Ponty 1956–1960: 20).

In fact, observed from the perspective opened by IMIN, Simondon's oeuvre essentially seems to follow Canguilhem's footprints rather than Merleau-Ponty's. In *Le vivant et son milieu* [The Living Being and its Milieu][18] Canguilhem derived from Uexküll the assumption that all milieus are 'subjectively centred, humans being's included' (Canguilhem 1952: 153).[19] This reference allowed him to directly connect the notions of subject and organism, extending the paradigm of the organism-milieu relationship to scientific knowledge itself: if no milieu can be included in a hypothetical universal milieu, this is true also for the milieu of human beings and for the 'objective reality' of science as well. Hence Canguilhem's philosophical challenge:

> From an authentically biological point of view, a general theory of the milieu of the human being as technician and scientist [*l'homme technicien et savant*] – one like Uexküll's theory for the animal and Goldstein's for the sick – remains to be elaborated. (Canguilhem 1952: 96)

There is a permanent shift between the milieu as a 'subjectively centred world' and the 'objective reality' produced by scientific knowledge, which 'dissolves living beings' – the centres of organisation, adaptation, and invention – 'into the anonymity of the mechanical, physical, and chemical environment' (Canguilhem 1952: 153). The crucial point is the 'reflexivity' of life sciences entailed, in Canguilhem's view, by the fact that the distinction between the objective environment [*environnement*] and the subjective milieu is an operation of life itself, insofar as science is 'a sort of adventurous enterprise of life' (Canguilhem 1952: 153). Considering scientific objective reality as a milieu entails a *reflexive* conception of biology, and the transposition of themes and problems typical of the human sciences – such as the emergence of the (collective) subject of knowledge – to the epistemology of life sciences. In this sense 'the specificity of [Canguilhem's] vitalism' was – as Lecourt poses it – the intellectual demand to resist 'subordination to a philosophy of Being' [i.e. 'a substantialist ontology'] and move towards 'the elaboration of the new – non-Aristotelian – notion of form' (Lecourt 2012: 183–84).[20]

---

[18] One out of three conferences held in 1946–1947 at the *Collège philosophique*, subsequently published in Canguilhem (1952).

[19] The German biologist Jakob von Uexküll made the concept of *Umwelt* a tool for ethologists and philosophers; Heidegger, for instance, used the concept to highlight the difference between human beings and animals which 'do not have a World'. Uexküll's little book *Streifzüge durch die Umwelten von Tieren und Menschen,* was translated in French, according to a clearly French Heideggerian inflection, as *Mondes animaux et monde humain* (1934). Not only does the original title not suggest any difference between human beings and other animals, but Uexküll also uses in it the term 'subject' when referring to organisms, whether *homo sapiens* or not. Von Uexküll was a key author for more than a generation of French philosophers from Canguilhem to Deleuze (e.g. they both derive from him the well-known example of the milieu of the tick), and also including Merleau-Ponty (Merleau-Ponty 1956–1960: 220–234) and Simondon.

[20] It is worth recalling Canguilhem's quick reference to Simondon's IGPB when he hinted at 'a new kind of Aristotelianism, on the condition, of course, that Aristotelian psychobiology and the modern technology of transmission would not be confused' (Canguilhem 1943: 277–278).

It is from this perspective that more than an echo of Canguilhem's vitalism can be found in Simondon's philosophy of individuation. In fact, Canguilhem's reference to the original relational meaning of the term 'milieu' as what is '*between-two centres* [*entre-deux centres*]' in order to demonstrate the 'fecundity' of the concept 'for a philosophy of nature centred on the problem of individuality' (Canguilhem 1952: 130–31), clearly segues into Simondon's concern for the 'central zone' of individuation. Starting from the crucial reference to quantum physics, in *Individuation* Simondon aimed at extending the same philosophical operation to all the domains of individuation through the concepts of form and information. In effect he deepened Canguilhem's criticism of mechanicism, by shifting his questioning of the relation between the subject-organism and its milieu to all regimes of individuation, physical and technical individuation included. Beyond the epistemological reflexivity Canguilhem attributed to biology, Simondon thus opened the possibility of extending to the sciences of matter the same 'encounter between history and its object' (Macherey 1998: 179). As a result, in Simondon's philosophy *all sciences* share the same reflexive, and therefore problematic, nature, since their objects cannot be seen as pre-inscribed in any (inter)subjective horizon.[21]

It is in this sense that Simondon makes the concept of milieu more and more crucial for defining a relational regime of information exchange which can be reduced neither to the objectivity of the natural environment nor to the subjective view of an organism-subject. All his writings from *Individuation* to IMIN can therefore be read in the light of the reflexive circularity he derives from the non-homeostatic and inventive relationship between the diversely conceived individuals (from the subatomic particle to the social group) and milieus (from the pre-individual to the techno-symbolic milieu). This is what can be considered the true invariant in Simondon's oeuvre, the permutations of which originated the whole series of concepts aimed at explaining the very dynamics of the biological, technical and social processes of invention. This is not only the backdrop of his conceptual effort around the symbolic function in IMIN, where the shift to a biological paradigm apparently allows him to abandon phenomenology. It also retroactively explains why Simondon had to conceive the symbolic invention as connected – if not subordinated – to technicity, as both concurring to the emergence of the collective of which they are the 'two sources', inseparable although irreducible to one another. And this is what also explains the decisive and otherwise unintelligible role of magic in Simondon's oeuvre, particularly in the third part of MEOT.

[21] This connection between Canguilhem's and Simondon's epistemologies has been developed in Bardin (2015). Simondon's necessity of overcoming Canguilhem's 'vitalism' is possibly reinforced by the latter's constant reference to perception, which evidences something more than a phenomenological blend in his approach. One should notice, for instance, the centrality of the 'world of perception' in determining the concept of milieu: 'moreover, as a living, the human being does not escape from the general law of living beings. The milieu proper to it is the world of perception [...] the environment to which the human being is supposed to react is originally centred on it' (Canguilhem 1952: 152). Canguilhem concludes his essay stating that 'if science is a fact in the world at the same time as it is a vision of the world, then it maintains a permanent and obligatory relation with perception' (154).

References[22]

Bardin, A. 2015. On substances and causes again: Simondon's philosophy of individuation and the critique of the metaphysical roots of determinism. In *Morphogenesis and individuation*, ed. A. Sarti et al. Cham: Springer International Publishing.

Bergson, H. 1907. *L'évolution créatrice*. Paris: Alcan (English trans. *Creative Evolution*. London: Macmillan & Co., 1954).

Bergson, H. 1932. *Les deux sources de la morale et de la religion*. Paris: Alcan (English trans. *The Two Sources of Morality and Religion*. London: Macmillan & Co., 1935).

Bimbenet, É. 2001. L'échange consenti: la phénoménologie face à l'anthropologie structurale de Lévi-Strauss. In *Phénoménologie et sociologie*, ed. J. Benoist and B. Karsenti, 157–184. Paris: PUF.

Canguilhem, G. 1952. *La connaissance de la vie*. Paris: Vrin (English trans. *Knowledge of Life*. New York: Ordham University Press, 2008).

Derrida, J. 1967. *De la grammatologie*. Paris: Minuit.

Espinas, A.V. 1897. *Les Origines de la technologie: étude sociologique*. Paris: Alcan.

Favez-Boutonier, J. 1962–1963. *L'imagination*. Les cours de la Sorbonne, Centre de Documentation Universitaire. Paris: BNF.

Friedmann, G. 1956. *Le travail en miettes*. Paris: Gallimard.

Karsenti, B. 1997. *L'homme total. Sociologie, anthropologie et philosophie chez Marcel Mauss*. Paris: PUF.

Lacan, J. 1938. Les complexes familiaux dans la formation de l'individu. In *Autres écrits*. Paris: Seuil, 2001.

Le Cœur, C. 1939. *Le rite et l'outil*. Paris: PUF, 1969.

Lecourt, D. 2012. The question of the individual in Georges Canguilhem and Gilbert Simondon. In *Gilbert Simondon. Being and technology*, ed. A. De Boever et al., 176–184. Edinburgh: Edinburgh University Press.

Lefebvre, A. 2012. Realism of the image and general praxeology in Simondon's course *Imagination et invention. The Warwick Journal of Philosophy* Special Volume. Deleuze and Simondon:57–67.

Lefort, C. 1951. *L'échange et la lutte des hommes*. In *Les formes de l'historie,* ed. Lefort, C. 1978, 15–29. Paris: Gallimard.

Leroi-Gourhan, A. 1965. *Le geste et la parole. La mémoire et les rythmes*. Paris: Albin Michel (English trans. *Gesture and Speech*, II-III. Cambridge-London: MIT Press, 1993).

Lévi-Strauss, C. 1949. *Les structures élémentaires de la parenté*. Paris: PUF (English trans. *The Elementary Structures of Kinship*. London: Eyre & Spottiswoode, 1968).

Lévi-Strauss, C. 1950. Introduction à l'œuvre de Marcel Mauss. In *Sociologie et anthropologie*, ed. M. Mauss. Paris: PUF.

Macherey, P. 1998. Georges Canguilhem's philosophy of science: epistemology and history of science. In *In a materialist way. Selected essays by Pierre Macherey*, 161–187. London: Verso.

Mauss, M. 1923–1924. *Essai sur le don*. In *Sociologie et anthropologie*, 141–279. Paris: PUF, 1950.

Merleau-Ponty, M. 1942. *La structure du comportement*. Paris: PUF (English trans. *The Structure of Behaviour*. London: Beacon Press, 1967).

Merleau-Ponty, M. 1952–1960. *Résumés de cours (Collège de France, 1952–1960)*. Paris: Gallimard, 1968.

Merleau-Ponty, M. 1954–1955. *L'institution. La passivité. Notes de cours au Collège de France (1954–55)*. Tours: Belin, 2003.

Merleau-Ponty, M. 1956–1960. *La Nature. Notes. Cours du Collège de France*. Paris: Seuil, 1995 (English trans. *Nature: Course Notes from the Collège de France*. Illinois: Northwestern UP, 2003).

[22] *Simondon's complete bibliography and a list of abbreviations are provided in the* Appendix.

Merleau-Ponty, M. 1959a. De Mauss a Lévi-Strauss. In ed. Merleau-Ponty, M. 1960. Signes, 154–168. Paris: Gallimard.

Ortigues, E. 1962. *Le discours et le symbole*. Paris: Aubier.

Roudinesco, E. 1993. *Jacques Lacan – Esquisse d'une vie, histoire d'un système de pensée*. Paris: Fayard.

Stiegler, B. 1994. *La technique et le temps. 1. La faute d'Épiméthée*. Paris: Galilée.

Van Caneghem, D. 1989. Hommage à Gilbert Simondon. *Bulletin de Psychologie* 42(392): 815–836.

Von Uexküll, J. 1934. *Streifzüge durch die Umwellen von Tieren und Menschen – Bedeutungslehre*. Berlin: Springer (French trans. *Mondes animaux et monde humain. Suivi de La théorie de la signification*. Paris: Denoël, 1965).

Chapter 10
Magic, Technics and Culture

As explained in the previous chapter, Simondon refuses to symbolic function the task of defining the boundaries of the human field, whether it is conceived as the intersubjective horizon of sense (phenomenology) or a universal set of unconscious structures (structuralism). Rather, thanks to a strategy widely experimented in *Individuation*, in IMIN he formulates the problem of the emergence of the symbolic function in terms of a process situated *between* nature and culture, reviving the topic of magic previously developed in MEOT. A close analysis of Simondon's theory of magic in the third part of MEOT clearly shows its reliance on Hubert and Mauss's *Ésquisse d'une théorie générale de la magie* [A General Theory of Magic] (1902–03). This will demonstrate that a considerable source of Simondon's philosophy of technics is sociological, and explain how he conceives the symbolic function as essentially 'techno-symbolic'.

The analysis of the double phase-shift of the original human-world relationship displayed in the third part of MEOT also pushes the enquiry further into Simondon's sociological heritage. This allows us to consider two important texts of the 1960s, *Psycho-sociologie de la technicité* [The Psycho-Sociology of Technicity] (1960–61) and *Culture et technique* [Culture and Technics] (1965), as the key to explore the relationship between religious rituals and technical instruments in the production of culture. Among these writings one can appreciate Simondon's consistent analysis of the different normativities that cross social systems, and fully understand the peculiar function he attributed to technicity – following, again, Leroi-Gourhan – in shaping the functioning of social systems today, thus making of technological progress a crucially political problem.

A. Bardin, *Epistemology and Political Philosophy in Gilbert Simondon*,
Philosophy of Engineering and Technology 19, DOI 10.1007/978-94-017-9831-0_10

10.1 For a *Théorie général de la magie* in *MEOT*

What marks the passage from nature to culture within the 'cycle of the image' displayed in IMIN is first the symbolisation of the body (tattoos, cuts etc.) and then the production of 'prosthetic objects' as external 'organs' (IMIN 135). In this sense 'the symbolic function is in continuity with the function of skin appendages [*phanères*]', the physical manifestation of relational capability (IMIN 134). It is precisely through the creation of a milieu of 'prosthetic objects' that an intermediate reality emerges, which begins the properly human modality of regulation of the organism-milieu relationship. Simondon describes this as 'a kind of *pandemonium* floating between the object-situation and the subject-situation, interposed between the living being and the milieu' (IMIN 137). As is easily noted, in IMIN symbolic formalisation is described, according to the model of magic, as a shift from the symbolised body (e.g. with tattoos) to the symbol-object, which culminates in the detached 'fragments' of the voodoo practices (IMIN 131–38). This magic-symbolic operation is what Simondon himself ritually invokes when presenting his edited version of the course:

> Symbols, the fragments of objects in which the part stands for the whole and communicates with it, are the basis of the voodoo practices [*voults*] serving magic operations: a simple lock, the strip of a vest taken from a person are fragments of their reality and allow for the possibility of acting on the person at a distance, through the symbolic relationship [...] the symbol can never be a *flatus vocis*; it presupposes an implicit realism. The present draft is presented here as a symbol of the course. (IMIN 6)[1]

When Simondon was probably still working on MEOT, in his course on *Le concept de nature. L'animalité, le corps humain, passage à la culture* [The Concept of Nature: Animality, The Human Body and the Passage to Culture] (1956–60) Merleau-Ponty made magic a key moment in the shift from instinct to symbolism. Following Konrad Lorenz (1937), Merleau-Ponty observes there that the animal's instinctive imagination prefigures symbolisation. Animal 'mimicry' is therefore to be considered a kind of 'natural magic' that begins the shift from the biological domain of instinct to the symbolic domain of institution. But Merleau-Ponty, although stating that 'the relationship between humans and animals is not hierarchical', concludes his reflections by redirecting his research to its main subject, 'the series *physis-logos*-History', and finally claiming for the human body a privileged function 'as the root of symbolism, as the junction between *physis* and *logos*' (Merleau-Ponty 1956–1960: 259).[2] Indeed, if we go back to Merleau-Ponty's

[1] It is probably superfluous to recall the etymology of the term 'symbol': it comes from the Greek verb *symballo*, an amalgam of *sym* (together) and *bole* (throw), with the approximate meaning of 'putting together' two different parts. This is something Simondon himself does not fail to remind in his writings.

[2] This phenomenological conception of the human body as 'symbolised' will encounter Lacan's criticism: 'the structure of the organism in Goldstein, the structure of behaviour in Maurice Merleau-Ponty [...] the soul has been presented for centuries as a spiritualised body, contemporary phenomenology makes of the body a corporalised soul' (Lacan 1962–1963: 237–38).

source, Uexküll, we find a completely different approach to the question. In fact in Uexküll's theory the phenomenon of magic is the model of an action that introduces in the subject (i.e. the organism) the first hints of a finalistic behaviour. This activity 'escapes all objectivity but acts on the milieu' thus making a 'magic milieu' emerge *both* for the animal and for the human, according to an approach quite similar to Simondon's in IMIN, but far from the approach he adopted in MEOT (Von Uexküll 1934: 73, 153).[3]

In the third part of MEOT magic is treated as the 'original phase' full of potentials where the passage from nature to culture originated through technical and symbolic invention. A reconstruction of Simondon's draft of this part of the book is really problematic. Relying on a gestaltic conceptual framework Simondon tries to make explicit the theoretical foundations of the first two parts. My hypothesis is that the draft of the first two parts of MEOT preceded the draft of *Individuation*, and the entire text of MEOT was later reviewed in its light, without modifying the original structure and adding the third part.[4] Through magic Simondon seems to sketch here a general theory of original human sociality, which takes into account human biological and technical features on the basis of the hidden reference to Leroi-Gourhan. But he carries on his enquiry in a partially phenomenological fashion, by assuming magic to be the model of an original relationship between humans and the world. In short, what seems to be at stake here is a 'phenomenological capture' of the prehistorical process of hominisation. However, for a full understanding of the theme of magic in MEOT it is necessary to go back to a primary source for the

[3] See in particular chapter XIII on *Les milieux magiques*, which clearly demonstrates Uexküll's influence on Simondon. From his interesting analysis of the concept of milieu, Petit also derives that 'Piaget's genetic interactionism and Simondon's genetic relationalism share the same ambition: to think the co-genesis of organism and milieu' (Petit 2010: 64). Although he is well aware that 'this answer is not a solution but a problem', I believe the path he opens here is a very profitable one, given that Piaget's profound influence on Simondon's thought is still to be understood.

[4] This would explain why, when Simondon thanks Canguilhem for his remarks which allowed him to find out the definitive form of his work, he particularly refers to the fact that the third part 'owns a lot to his suggestions' (MEOT 7). Furthermore, it would explain why the direct references to the concepts and terms developed in *Individuation* – in particular to the term transindividual – only appear in the introduction of the third part and in the conclusion, as seemingly later additions aimed at justifying the relation between the two theses. All this seems to evidence that MEOT was the result of subsequent drafts, while it is acknowledged by scholars that *Individuation* was the product of a continuous and rapid writing experience. As a result, the first two parts and the third one are often different in style, semantic choices and contents. Parts one and two – the only ones included in the first unofficial English translation – have interested the philosophers of technology since their publication. On the contrary the third part – genuinely speculative – has been rarely taken into consideration by the critique, as Hottois complains when referring to Michel Simondon's memory – Gilbert's son – that this section of the book was the one his father was most attached to (Hottois 1994: 118). The third part of MEOT is indeed, when seen from an exegetic and philosophical political perspective, the most interesting. It is full of insights that will be later developed by Simondon in subsequent works, and the ontological status of politics in his oeuvre can only be understood starting from the subsequent phase-shifts of the human-world system the original configuration of which is the net of relations called in MEOT 'magic primitive unity' (Sect. 10.2). For an attempt to interpret the third part of MEOT in relation to *Individuation*, see Barthélémy (2011).

whole of Simondon's generation: *Ésquisse d'une théorie générale de la magie* (1902–03), of Henry Hubert and Marcel Mauss, the latter being master and supervisor of Leroi-Gourhan when he wrote his doctoral thesis in 1945.

In the *Ésquisse* the authors pose first of all the problem of providing a clear definition of magic. As a social phenomenon, magic 'has no genuine kinship with anything apart from religion on the one hand and science and techniques on the other' (Hubert and Mauss 1902–1903: 134), and therefore magic can be easily confused with them. On the one hand magic resembles techniques in 'its practical goals, [and in] the mechanical nature of many of its actions' (79), while on the other, because of its rituality and its connection to belief, it is hardly distinguished from religion. Therefore the research of a clear definition of magic has to take its starting point precisely by a differentiation from techniques, sciences and religions:

> There are two types of special functions in society to which we already assimilated magic. They are, on the one hand, techniques and sciences, on the other religion. [The question is:] Is magic a kind of universal art or possibly a class of phenomena analogous to religion? (Hubert and Mauss 1902–1903: 82)

At a first glance, the link between magic and religion seems evident: as magic is not connected to individual invention, also in religion 'invention only emerges in the [collective] form of revelation' (83). But it is precisely because of the simplicity and universality of magic that a genealogical link appears quite evident in the case of techniques:

> Since magic is the most childish technique, it is possibly also the oldest. In effect, the history of techniques proves that there is a genealogical link between techniques and magic [...] techniques are like seeds which bore fruit in the sole of magic. Later, magic was dispossessed. (Hubert and Mauss 1902–1903: 135)

Hence the authors conclude that techniques and the sciences originated along the same process which progressively made of the 'collective representations' connected to magic the ground for individual thinking through the notions of force, cause, goal, substance (137). What about religion then? Is it possible to maintain also that religion similarly derives from the 'undifferentiated whole' of magic, which yet lacked the stability typical of (institutionalised) religions? Or do they come from a common source? In the *Ésquisse* this is quite problematic, and the doubt on the relation between magic and religion emerges from the anthropological evidence that the sacred, which grounds the phenomena of religion, is a social category while, on the contrary, magic practices are individual.[5]

A possible solution of the problem can be found according to the authors in the notion of *mana*. This shift of focus from magic to *mana* can only be understood if it is clear that Hubert and Mauss's work aims – in a purely Durkheimian orthodoxy – to grasp the a priori categories that define the collective. Demonstrating 'the *social*

[5]The conditions of possibility for magic to be not only the source of techniques, but also of religions and are set in the *Appendice* as follows: 'either magic is collective or the notion of the sacred is individual' (Hubert and Mauss 1902–1903: 140). The *Appendice* is not present in the English version.

characterisation of magic and of the notion of mana' means therefore to place the enquiry in the domain of religion – an analysis later developed in Durkheim (1912). According to Hubert and Mauss the notion of *mana* – the unconscious condition of magic – would allow us to 'dig deeper still, in order to reach those collective forces, which we claim to have produced magic, and of which the idea of *mana* is the expression' (Hubert and Mauss 1902–1903: 115). In distinction from Durkheim, they do not assume the sacred as the common origin of all societies. They rather derive from ethnographical studies that the social category of *mana* is not only provided of the standard degree of universality typical of all the categories of collective thought, it is so general as to be likely to substitute Durkheim's notion of the sacred:

> Our analysis shows that *mana* is an idea of the same order as the idea of the sacred [...] As a result, we find that not only is the idea of *mana* more general than that of the sacred, but the sacred is inherent in the notion of *mana* and derives from it. It would be probably fair to say that the sacred is a species of the genus *mana*. Beneath the sacred rituals we might have found something better than we were looking for: we might have found their origin. (Hubert and Mauss 1902–1903: 112)

From the originality of *mana* in relation to the sacred (which defines religion), Hubert and Mauss do not conclude that magic is prior to religion. Rather, they untie the concept of *mana* from its privileged link to magic, making it the common source of both magic and religion: 'the notion of *mana* cannot be said to be more magical than religious' and 'the original facts of magic [are] also the original facts of religion'. What they commit to future research is precisely the effort to demonstrate that magic and religion 'derive from a common source' (Hubert and Mauss 1902–1903: 130).

Although within the limits of the *Ésquisse* the relationship between magic and religion is strongly problematic, the basic claim is clear: there are forces of a collective nature that constitute the social bond, the universality of which is demonstrated by ethnological evidence. The authors of the *Théorie générale de la magie* decide to name these forces, the unconscious 'common source' of religion and magic, *mana*. But in a *mémoire* almost 30 years later, precisely opposing Durkheim,[6] Mauss does not restrain himself from restating the common foundation of religion and magic in the notion of *mana*, he also recalls that his former work with Hubert on magic derived from the exigency of proving the following hypothesis:

> We had to fix our ideas on magic which we considered the primitive "form": the pseudoscience that preceded religion [...] we believed we had primitively to do with nothing else than magic formulas. (Mauss 1930: 217–18)

[6] This text was redacted by Mauss on the occasion of his candidature to the *Collège de France*, and published with the title *L'œuvre de Mauss par lui même* (1930). 'We [Mauss and Hubert] found at the basis [of magic] and of religion, a vast common notion we named by a term borrowed from the Melanesian-Polynesian context, the term *mana*. This idea is perhaps more general than the one of the sacred. Hereafter Durkheim tried to sociologically deduce the idea of the sacred. We were never sure he was right and I still speak of a magic-religious background' (Mauss 1930: 218). In fact, when displaying his conception of the origin of the institution in his *Les formes élémentaires de la vie religieuse* (1912), Durkheim seemingly tried to 'expel' magic from the domain of religion (Karsenti 1997: 223).

Now, the possible conclusion that magic is the 'primitive form' in which techniques and religions are implicit, or at least present although in a partially undifferentiated embryonic state, is perfectly compatible with the hypothesis of a common origin of religions and techniques in the 'magic phase' maintained by Simondon in the third part of MEOT. In the concluding section of this work magic is treated both 'anthropologically' as the 'primitive mode of the relation of the human being to the world' (PR 267), and 'phenomenologically' as the 'primitive magic unity, the vital relation between human being and the world, which defines a universe at the same time subjective and objective, anterior to any distinction between subject and object' (MEOT 163). My interpretation of the third part of MEOT takes its starting part precisely from this double inspiration.

In a brief introduction Simondon explicitly adopts the conceptual framework of *Individuation* to recapitulate the ontogenesis of the 'human-world system' as a metastable one:

> The term genesis is taken here – in the sense previously defined in the study concerning *L'individuation à la lumière des notions de forme et d'information* – as a process of individuation in general [...] such a genesis opposes itself to the degradation of the potential energies contained in a system during the passage to a stable state in which no transformation is possible anymore. (MEOT 155–56)

The 'magic phase' has in the system thus conceived a key function: it is the energetic charge which maintains it in its metastability: 'magic, the primitive mode of the human being in the world, can supply without exhaustion an indefinite number of subsequent divisions'. These are characterised as quantic leaps entailing interaction between simultaneous phases and chronologically subsequent ones (MEOT 161). In short, it is immediately evident that in MEOT the magic 'phase' plays a phenomenological role: the 'magical mode' is 'a central, original [*originaire*] and unique mode of being in the world' (MEOT 160) beginning from which a 'generalised genetic interpretation' of the relations between human beings and the world would be possible (MEOT 154).

But this is not the only function of magic in Simondon's discourse: it also indicates the evolutionary transition between nature and culture – for which he often uses the term 'primitive'[7] – which can be situated 'immediately over the simple relationship between the living and its milieu' (MEOT 156). In the terms of *Individuation*, it is possible to describe magic as the regime of individuation which marks the threshold from a biological individuation to a psychic and collective one. This threshold has its own particular structure: an asymmetrical distribution of potentials and of key-points describes the composite field of 'the magic universe, structured according to a modality anterior to the separation of object and subject' (MEOT 164).

The study of magic is, in short, the study of the 'phase' that allows a better explication of the shift from nature to culture both as a primitive stage and as an original phase recurring in each new ontogenesis of the collective:

[7] See MEOT 156 e 161, and also MEOT 196, where he speaks of 'primitive magic thought'.

> The first stage [*étape*] of the relation to the world, the magic stage – in which the mediation is neither subjectified nor objectified, neither fragmented nor universalised – is the most simple and fundamental structuring of the milieu of a living being; [it is] the birth of a net of privileged points of exchange between being and milieu. (MEOT 164)

Against the background of the *Ésquisse* on magic it is thus possible to grasp the meaning of the human-world relationship Simondon calls the 'magic phase in the wider sense'. Just over the living being-milieu relationship, this phase is pre-technical and pre-religious (MEOT 156) (it precedes the 'primitive opposition of technics and religion' MEOT 212) and marks the unaccomplished shift from nature to culture. From this perspective a striking connection can be seen between Simondon's 'magic phase' and Hubert and Mauss's conception of *mana* as a field: a 'magic force-milieu' (Hubert and Mauss 1902–1903: 109) in which things act on one another because of their 'difference in potential' (114). The mixed status, biological and cultural, of both concepts directly refers to the emergence of the collective. The forces expressed by the concept of *mana* can be reduced neither to biology nor to society: although they have their condition in the biological and their development in the social, only their intersection – the 'instinct of sociability' – 'is the initial condition of all the rest' (120). Given the equivalence between what is original-primitive and what is collective, which is more or less implicitly assumed by the authors (124), the phenomenon of magic is in fact one of the first pieces of evidence for the emergence of collectivity (120 ff.), a force that brings together the individuals through collective belief, at the point that only when the whole group 'believes' in magic is the social body 'realised' (126–27).[8] It is central then to understand magic as the 'presence' of an origin that never ceased to produce effects: according to Hubert and Mauss the creation of society by magic is continual (132), and the result of the entire process is always a *partial* exit from the horizon of magic, the presence of which can never be definitively exorcised: 'neither techniques, sciences, nor the directing principles of our reason are quite free from their original taint' (148).

Similarly for Simondon magic continues to exist as an associated pre-individual milieu, with a double meaning: it is the permanent risk of regression to an archaic phase and the energetic charge grounding any further transindividual individuation.

10.2 The Second Phase-Shift and the Sacred Field of Politics

In the third part of MEOT magic appears as paradigmatic as quantum physics in *Individuation*: it is the basic theoretical model to be analogically transposed to all fields as a possible source of scientific explanation. In Simondon's oeuvre there are not many other concepts capable of grounding the explanation of the emergence of

[8] It is worth noting that Mauss's description of the crowd phenomena does not fit Le Bon's (1895) picture: they do not only threaten the social bond, they are also the moment of the possible *formation* of society.

the human field. As I have shown in Chap. 6, *Individuation* offers a quite fragmentary picture of this process through the concepts of biological community, personality and emotive-affectivity, and through the analysis of belief, work and language by means of the concept of the transindividual.

In MEOT, instead, Simondon develops a phenomenology of the social relationship, which is also a kind of reconstruction of a 'natural history' of civilisation through a sequence made of a series of phase-shifts starting from the 'primitive and original phase of the human's relationship to the world' (MEOT 156): 'the primitive magic unity' (MEOT 163). In a presentation characterised by a strong phenomenological emphasis, everything is orderly displayed according to the gestaltic model of the figure-ground relation (MEOT 164)[9] and everything occurs according to the same dynamics that in *Individuation* mark the metastability of a system: internal tensions due to differential potentialities go along with the series of subsequent phase-shifts, starting from the primary divergence of 'technicity' and 'religion', which resulted from the former universe of magic.

But in MEOT, Simondon's main treatise on technics, there is no general theory of the transindividual function of technicity, just sporadic hints. In fact, Simondon's book on technical objects is basically a book on the contemporary issues resulting from the problematic form assumed by the 'original' tension between symbolism and technicity, nowadays transformed in a conflict between culture and technology. Thus, the question of 'origins' is not posed there as a scientific question on the (pre) historical relation between technical inventions and the process of hominisation.[10] It is rather a question concerning the specificity of the technical 'phase' at the scale of the contemporary shape of the human-world system, and of the social systemic effect that the lack of the adequate knowledge of technicity entails.

Furthermore, the problem is formulated in a quite peculiarly fashion. Simondon speaks of these 'phases' of the becoming of the human-world system (MEOT 159 ff.) qualifying them as different 'modes of thought'. Now, what does 'mode of thought' mean here? What does Simondon mean with magic, technical, religious, aesthetical, philosophical – 'modes of thought'? And, in particular, if the magic phase is in-between nature and culture, *before* the subject/object differentiation, how is it possible to conceive a magic 'thought' *before* this earlier distinction of thought and action?

---

[9] As Simondon himself will claim three years later, the original hypothesis of MEOT concerning the magic 'net' 'presupposes an analysis of the structures of perception and action which discover in the word a certain number of key-points: an analysis which would follow the latest acquisitions of the theory of Form' (PST 327).

[10] The choice appears consistent with Canguilhem's suggestion: 'the problem about the origin of the tool, the problem about the origin of society, the problem about origins in general, are unsolvable problems; the problems concerning origins are not historical problems. Receding to some anterior state is in no way particularly clarifying' (Canguilhem 1955: 78; this part of the discussion only appeared in the original essay).

Although the expression 'mode of thought' is a singular exception in Simondon's oeuvre,[11] we might be partially aided by recourse to IMIN where he also characterises the 'magic universe' as made of 'noticeable points', 'extremes terms of reality' that express the intertwining of 'situations and beings [...] with the natural and social world, according to a "savage" mode of perception and action' (IMIN 134). This 'savage' *mode* expresses there – in a text centred, as shown, on the organism-milieu relationship[12] – the same common ground of thought and action Simondon refers to in MEOT as 'theoretical and practical thought' (MEOT 162) derived from the phase-shift of the original magic unity in what he names the 'active' and 'representative' modes of thought (MEOT 158). This conception of 'thought' relies on the hypothesis of a 'set formed by the human and the world' (MEOT 155): a set which, both in the theoretical and in the practical domain, relates to 'the figure/ground relationship taken as a complete reality' (MEOT 211). Now, since this system can be grasped only in the light of its phase-shift from different 'modes of being in the world' (MEOT 157), the concept of a 'mode of thought' theoretical *and* practical appears to be consistent with the approach characterising this part of Simondon's book. It is in particular the part devoted to technical activity that clarifies how Simondon's use of the term 'thought' points there to the phenomenologically concrete systematic unity from which theory and praxis emerge at each step:

> The emergence of these two modalities, one theoretical and the other practical, expresses the break of a primary unity which was of both knowledge and action: technical thought complete and concrete. (MEOT 203)

Consistently, primitive magic 'thought' – despite the term 'thought' being employed – *is not* conceptually connected to the category of subject. Magic is in this sense the prototypical field of the original identity of thought and action, seen through the actual process of phase-shift from which both subjects and objects emerge. Indeed, the third part of MEOT displays an ordered sequence of phase-shifts of the 'primitive magic unity'. Let us briefly go through the two subsequent phase-shifts.[13]

The first phase-shift is a process of distancing between the human and the world. Technicity and religion emerge there from the magic phase: this 'first wave' of phase-shift originates on the one hand from multiple techniques that apply to diverse parts (figures) of the natural world, and on the other hand from religiosity that situates the individual in the whole (ground). The 'primary modes of thought' correspond to a first configuration of the human-world relationship at the level of biological individuation – and continues to 'secondary' modes of thought which, on the contrary, 'entail communication and expression', the emergence of subjects and

[11] Probably due to the anomalous writing process of the third part of MEOT. See above, n. 4.

[12] The 'savage' mode of perception and action characterises there the passage from the imaginary to the symbolic and therefore to the collective, which carried the extraction from the milieu of 'pseudo-objects charged with the potential energy of a metastable system' which determine the emergence of thought: 'abstraction means to extract from' (IMIN 136) (Sect. 9.2).

[13] Clearly pictured in Hottois (1994: 72).

therefore collective individuation (MEOT 201–202).[14] In the first phase-shift – this is the crucial point:

> Mediation itself [...] acquires a certain density; it is objectified in technics and subjectified in religion, thus making appear the technical object as the first object and the divinity as the first subject, while before there was nothing but the unity of the living and the milieu. (MEOT 168)

The relation between the living and its milieu acquires a 'density' that, as the reading of IMIN has shown, is a new milieu made of objects that are also symbols, in which either the subjective or the objective components can prevail, without ever making one of the two exclusive. It is the 'mixed' milieu made of organisms, and also technical objects and symbols, the appearance of which marks the threshold of an 'inter-human' relationship. If through the technical object 'an inter-human relationship which is the model of *transindividuality*' takes place (MEOT 248),[15] this is possible because in this milieu the technical object also assumes the symbolic function it derives from the invention it originated from:

> The technical object, considered in its nature, i.e. as invented, thought and willed, assumed by a human subject, becomes the support and symbol of this relation we would like to name *transindividual*. (MEOT 247)

When the two tendencies to technical reticulation and religious totalisation cross each other, they generate the human and the geographical world. The 'geographical world' emerges from the application of the (religious) totalising function to the technical segmentation and reticulation of the natural world, while the 'human world' emerges from the application of the (technical) exigency of segmentation to religious affectivity, from which images, institutions, symbols and singular ethical actions emerge. At this stage reticular structures appear and techniques and religions display their full maturity:

> The maturity of techniques and religions tend towards reincorporation in the world, geographical for techniques, human for religions. (MEOT 182)

The second phase-shift arises, once the techno-symbolic milieu is constituted, from the exigency to 'reincorporate' the human in this new 'world' after aesthetics failed in its attempt to provide a reconciliation of the two tendencies and stabilise the system. Furthermore the charge of the 'magic phase' persists in the dynamical

---

[14] It is worth noting, however, that Simondon uses here the term 'intersubjective' to characterise collective individuation. This would confirm the hypothesis of an earlier phenomenological inspiration of Simondon reflected in the different layers of MEOT, with the later addition of the term – and elaboration of the concept of – transindividual. See above, n. 4.

[15] Simondon's explicit statements concerning the technical constitution of the transindividual – well highlighted by Stiegler – have to be read carefully, since in MEOT it is not technicity, but magic, the original phase from which humanity emerges. In fact, although it is 'the model of transindividuality', technical activity never becomes for Simondon the *only* explanation of transindividual individuation: 'technics and religion are the organisation of two symmetrical and opposite mediations, but they form a couple because they are each *a* phase of the primitive mediation and in this sense they are not endowed with a definitive autonomy' (MEOT 169).

differential relation between the technical and the religious phases, as a 'metastabilising' energetic source constantly supplying 'the drive of the primitive magic universe' (MEOT 161). Out of the immediacy of the religious and the technical primary phase-shift, in the 'second stage' the human milieu is not anymore the natural world populated with other organisms. Two milieus have emerged now: the 'geographical world' as the object of representation and usage, and the 'human world' as the collective subject of thought and action (MEOT 214).

It is with this second phase-shift that Simondon's systematic deduction abandons its previous symmetry: he characterises this 'second wave' by the unilateral operation of technics and religion on the 'human world'. The newly instituted collective milieu in fact amplifies the insufficiency of both the function of the simple craft techniques and the function previously carried on by religious thought that, at this level, 'charged with social inferences [...] cannot realise the mediation between the human and the world anymore' (MEOT 208). What is at stake now is not the relation to the natural milieu, but the relation to the 'human world', i.e. the technical-symbolic milieu. Thus from the second phase-shift, along with new techniques directed to the human milieu, 'political thought' also emerges.

In the whole of Simondon's oeuvre, the issue of politics is often implicit but always marginal. In *Individuation* religion only sporadically appears, but – at least in one case – with a strikingly decisive connotation: 'religion is the domain of the transindividual' (I 250). The assertion is not isolated if one considers the other two statements that, at the two extremes of his oeuvre, provide it with due relevance. In his first published article Simondon claims that 'every closed community secretes a form of the sacred' (PI 117), thus anticipating what later developed in the *Note complémentaire* concerning the relation between closed community and open society, and the juxtaposition of sacredness on the first of the two sides. On the other hand, at the end of his intellectual life, in a draft letter to Derrida entitled *Sur la techno-esthétique* [On Techno-Aesthetics] (1982), Simondon invites his colleague to consider, along with technics and aesthetics, the importance 'of religious thought and practice' as 'interfaces' for the regeneration of philosophical thought.[16] But the theme of religion becomes explicitly political only in the third part of the course *Psycho-sociologie de la technicité* (1960–61), in which Simondon presents a theory of the structures of 'technicity' and 'sacredness'. But before focusing on the course of 1960–61, it is necessary to follow briefly the 'deduction' of politics from religion sketched by Simondon in MEOT.

As already demonstrated, after the first phase-shift technicity and religion both emerge as 'heirs of magic' (MEOT 173) in structuring the human-world relationship, yet it is only religion that carries on an 'exigency of totality and unconditioned unity' (MEOT 208). With the second phase-shift – this time concerning the relationship between humans and the techno-symbolic milieu – new 'techniques of the human world' emerge along with other 'modes of thought also referring to the human world, but in this case grasped in its totality' (MEOT 214). It is precisely

[16] Pages are not numbered.

here, at the level of the second phase-shift, from within the field opened by the religious phase, that 'political thought' emerges:

> Techniques on the human being and political and social thoughts derive from a new wave that divides magic thought [...] when techniques on the human being have broken the former reticulation by considering the human being as technical matter – from this new break of a figure-ground relationship simultaneously emerges a thought that grasps human beings under the level of unity (the techniques of human manipulation) and another thought that grasps them beyond the level of unity (social and political thoughts). (MEOT 214–15)

Politics is, according to Simondon, a mode of religiosity: at least at the level of 'the great worldwide political movements' these 'modes of thought', are precisely the 'functional analogue' of religions (MEOT 214). In the second phase-shift techniques tend to fragment the human world: through the imposition of measure and control they 'pluralise and study [the human being] as citizen, worker, [or] member of a familial community'. In connection to this modality of technicity, politics emerges in MEOT as a compensative function in the social system: in short, it implements the religious 'function of totality' within the new milieu, thus conveying technical experimentation and the fragmentation of the human to a new unity (MEOT 215). Thus politics carries an evident analogy with religion: social and political thought classifies and evaluates, includes and excludes according to categorical dichotomies functioning as the sacred/profane, pure/impure couples featuring religious thought.

Centred as it is on a programmatic assumption of the structural social 'openness' of technicity, MEOT is not at all focused on the religious and the political phases and, wherever they are concerned, it is in view of a differential definition of the opening opportunities offered by technicity. Thus MEOT seems to conclude in favour of the 'cultural' regulatory function of politics in perfect analogy with the results of the theory of social systems developed in the third part of *Individuation* and in the *Note complémentaire* (Sect. 8.3). But not only is the text of MEOT much more nuanced (as I will show in Sect. 12.1), Simondon himself felt the need to better scrutinise these topics, giving them a new shape in a course he held at the University of Lyon a couple of years later, in the academic year 1960–61.

10.3 *Psycho-sociologie de la technicité*: Isomorphism and Asymmetry

The course on *Psycho-sociologie de la technicité* (PST) ideally continues and integrates the problems Simondon presented in MEOT, by projecting the sequence of phases displayed there onto the synchronic surface of the actual opposition of 'technicity' and 'sacredness'. The analysis of this course will provide a better scrutiny of the phase of 'sacredness', thus allowing us to test the diagnostic efficacy of Simondon's theory of a structural asymmetry of the social system from which he derived his project of the formation of a 'technical culture'.

In order to introduce the course we shall go through the only sociological text (a strongly ethnographic sociology indeed) presented in the whole bibliography of MEOT: *Le rite et l'outil. Essai sur le rationalisme social et la pluralité des civilisations* [The Ritual and the Tool: Essay on Social Rationalism and the Plurality of Civilisations] (1939) by Charles Le Cœur.[17] Approach and terminology are so impressive for a reader of Simondon that it is worth quoting an entire passage:

> *Two kinds of action and thought stand opposed.* The first have universal value. Maxwell's principles of electricity are true for everyone, just as Edison's lamps illuminate the whole world. Whether one is French or Arab, a driver pushes the same accelerator and pulls the same break, because there is only one way to make a car go on. The same causes produce the same effects, no matter in which society we live in. But the Arab bourgeois of Rabat who decorates his boutique on the day of the Feast of the Throne with the Sharifian red flag does not produce the same effect as the Parisian worker who on the first of May waves the revolutionary red flag. In order to greet, Muslims bring their arm to the heart, while Christians tip their hats. This second category of actions only makes sense in relation to a given society. And this is what sociology is concerned with. Two consequences result from this definition. First of all the *social is not opposed to the individual*. Many Moroccans have a dark skin: this is neither an individual fact, nor a social one. Brought away from their families since their birth and educated in a different society, they would not be Moroccans anymore, but they would still have the same skin. Conversely, a few years ago, a Moroccan Muslim converted to the Christian religion. This is a strictly individual fact, probably unique – an exception – in our epoch, but the emotion and indignation raised by this conversion demonstrates that this was a social fact. On the other hand the *social cannot be mistaken for collective*. The foundation of Rabat on the banks of the Bou-Regreg is a collective fact: it is not in itself a social fact. It is not necessary to be an Arab or Muslim to understand the advantage of an estuary for a city, and the Romans of Sala Colonia [Rabat] had been aware of that far before the contemporary inhabitants of Ya'qoûb el Mançour. (Le Cœur 1939: 9–10)

Le Cœur presents in this old-style the methodological grounds of his work establishing – according to Maussian teaching – what a 'social fact' is. The main features of his methodology are: (a) the opposition of a universal *technical* normativity to particular *cultural* normativities; (b) the consequent lexical choice that distinguishes the 'collective' and the 'social', by giving different semantic value to the two terms – the first refers to actions connected to the universality of technics, while the second to behaviours specific of cultures –; and (c) the refusal of any sociological relevance to the conceptual opposition of the 'individual' and the 'social'. As a result, the concept of the 'individual' is freed from *exclusive* dependence on the semantic area either of the collective-technical or of the social-cultural, and possibly connected to both. At the heart of the book this duality is condensed as the fundamental distinction 'between two types of action and thought' the symbols of which are the 'ritual' and the 'tool'.

---

[17] Charles Le Cœur, a former student of Malinowski, wrote his doctoral thesis in order to 'resume the lesson of 10 years of African life [in Morocco]' (Le Cœur 1939: 1). But the book is also a reflection on the ethical-political function of what today we would name cultural anthropology, through the criticism of two theories: Lévy-Bruhl's theory of the primitive mentality and what Le Cœur calls 'theory of economical rationalism', in which he stigmatises the technocratic approach shared, in his view, by classical liberal theories and Marxism.

Each action can be read, according to Le Cœur, as technical or symbolic according to the adopted point of view, internal or external: the 'human being appears to itself as technical, to others as a creator of rituals' (Le Cœur 1939: 4). Each action has in fact a utilitarian side linked to 'natural determinism', and a symbolic one linked to 'social obligation'. It is simultaneously technical and ritual, since it is rooted in the common source from which the 'technical effort' and the '*élan* of sensibility' emerge (15–19). Le Cœur's argument thus suggests that the hypothesis of the complementarity of ritual-symbolic invention and technical invention (what he calls in a clearly Bergsonian way the 'obligation' to invention) can be used for the explanation of social ontogenesis. In fact, Le Cœur's argument follows here, '*bon gré, mal gré*',[18] the Durkheimian path when, in partial contradiction with his premises, he reduces the tool/ritual dynamics to the individual/social opposition, thus risking the identification of the social and the collective:

> We will resume our study in the following formula: *human action, creative of sensibility is, in its essence, ritual, but the attention it addresses to things is basically utilitarian.* Goals suppose society; individuals arrange the means. (Le Cœur 1939: 32–33)

Since *De la division du travail* social [The Division of Labour in Society] (1893), for Durkheim the distinction between social and individual representations parallels the one between moral and technical norms, which defines the nature of society itself, as far as the two normativities follow different regimes of development and degrees of necessity. Now, in Durkheim the individual is a *Homo duplex* precisely because it embodies this asymmetrical duality: each individual of the human species is irremediably divided between *social* duties and *individual* needs and aspirations, among which is the force of invention. While the social reaction (*sanction*) to the immoral act (*infraction*) 'follows with authentic necessity', the only possible progress in morality is that 'collectively accomplished by society'; on the contrary, through technical acts (of the physician, the industrialist, the artist) 'individual variations can be produced in complete liberty, and also with success'. In short, technical 'infractions' are not immediately rejected by society, and therefore 'changes are easier and more rapid' through them (Durkheim 1893: 23–24).[19]

And nevertheless Le Cœur cannot be said to simply adhere to Durkheim's view as another source is driving his essay. In the *Evolution créatrice* (1907) Bergson tried to translate the Durkheimian opposition in an evolutionary sequence which would root technical normativity directly into the biology of *Homo faber*:

> If we could get rid of all pride, if, to define our species, we kept strictly to what the historic and the prehistoric periods show us to be the constant characteristic of man and of intelligence, we should say not *Homo sapiens*, but *Homo faber*. In short, intelligence, considered in what seems to be its original feature, is the faculty of manufacturing artificial objects, especially tools to make tools, and of indefinitely varying their manufacture. (Bergson 1907: 117)

---

[18] 'For better or worse, we are brought back to opposing the ritual character of society to the utilitarianism of those who are part of it' (Le Cœur 1939: 32).

[19] This part is not present in the 1930 edition of Durkheim's book (Paris: PUF) and, consequently, in the English translation. On the same subject see also Durkheim (1924: 60 ff).

Now, although working in the Durkheimian framework, when Le Cœur recalls the Bergsonian opposition he reforms it according to his programmatic theoretical distinction between the 'social' and the 'collective'. As a result, in his book the expressions *Homo faber* and *Homo vates* neither point to the (Bergsonian) distinction of two different moments in the evolutionary process, nor do they plainly correspond to the (Durkheimian) opposition between the individual and the social. They rather represent the twofold nature of human normativity, the concretisations of which are the ritual and the tool, respectively 'social' and 'collective':

> To the *Homo faber*, who considers and treats the world as a machine, the *Homo vates* is opposed, who makes of the world an artwork, a touching set of symbols. The human being draws from natural determinism the tools that extend its power on things, and from social obligation the rituals that make it profoundly vibrate. (Le Cœur 1939: 15)[20]

When in *L'homo faber: la main* (1950)[21] Leroi-Gourhan questioned Bergson's 'philosophical' expression *Homo faber* from a scientific point of view (Leroi-Gourhan 1950: 75–77), he sharply concluded that 'the distinction between *faber* and *sapiens* is deceiving and scarcely useful for the scientific understanding of human origins' (89). His point was that the distinction *faber/sapiens* does not serve the evolutionary enquiry into the origins of technics, it can only be a useful way of differentiating human operative modalities, and not a very serious one, as can be easily deduced from the joke concluding the debate: 'The notion of *Homo faber* has a certain utility. We are all *Homo faber* to a certain extent, and me – I am quite sure – more than most of you' (Leroi-Gourhan 1950: 98).

Considering all these sources, it is possible to conclude that Simondon's theory of the structural 'phase-shift' of both human beings and human society between two different kinds of normativities descended from Durkheim and that Simondon re-elaborated it through the lens of Le Cœur and Leroi-Gourhan. What this permitted him to accomplish was: (1) to implement Le Cœur's programmatic differentiation of the social and the collective, thus conceiving invention independently from the Durkheimian contraposition of the individual and the collective; (2) to follow Leroi-Gourhan's criticism, thus refusing Bergson's hypothesis on the evolutionary sequence *Homo faber* – *Homo sapiens*.[22] Simondon could thus assume the sociological reference without denying the biological one, while refusing to deduce the phase-shift between technicity and sacredness from a simple and naïve contraposition

[20] Also Mauss declared he would accept Bergson's concept of *Homo faber* only on the condition of revising the notion of invention out of any 'mystery'. Invention is not a 'creation' but a 'transformation' of matter, the subject of which is not an individual *élan* but a 'common effort' (Mauss 1948: 75).

[21] Paper delivered at the *Centre international de synthèse*. In a note Leroi-Gourhan adds: 'the object of this exposition had been initially treated in 1948–49 at the *École des Hautes Études*, and, during the same year, in two courses held at the University of Lyon' (Leroi-Gourhan 1950: 89, n. 1).

[22] Simondon's evaluation of Bergson's philosophy of technics is of course quite critical. According to him, although by connecting technical activity to *Homo faber* he had the merit to show the relation between technicity and intelligence. Bergson contributed by reducing technicity to utilitarianism, by situating it on the passive side of his 'axiological dualism of closed and open, static and dynamic, work and dream' (MEOT 254).

between individuals and society.[23] And, in effect, at the beginning of PST Simondon explicitly declares his intention to continue Leroi-Gourhan's oeuvre by extending his analysis to contemporary societies:

> It is necessary to promote this study, mainly concerning preindustrial societies, by the examination of the genesis of technical objects in industrial societies. (PST 130)[24]

The declared intent of Simondon's course on the *Psycho-sociologie de la technicité* is to reveal the normative function, in the wider sense 'cultural', of technical objects. The first two sections mainly repeat the analysis displayed in MEOT on the historicity of the technical object and the alienation of technicity in society: their professed aim is to demystify a social relationship exclusively based on the reduction of technical objects to their utility or symbolic value, alienating the technicity therein contained.

Finally, in the third part, entitled *Technicité et sacralité. Etude comparé des structures et conditions de genèse, de dégradation et de compatibilité* [Technicity and Sacredness. A Comparative Study of Structures and Conditions of Genesis, Degradation and Compatibility], Simondon explicates in few highly condensed pages his theoretical effort to analyse the structures of 'technicity' and 'sacredness'.

He takes from Mircea Eliade's *Images et symboles* [*Images and Symbols*] (1952) the working hypothesis of a 'structure of sacredness': a net in which sacred objects are the knots, '*centres* which make the fundamental regions of space communicate' (PST 129). According to Simondon the same approach has to be extended to technical objects, and therefore the existence of a 'structure of technicity' must be assumed (PST 324). In short, there are two isomorphic structures – technicity and sacredness – in which the objects-knots have the same function. In PST the technical object is analogous to the sacred object: in the structure of technicity 'each tool exists less and less as an *object* and more and more as a *symbol*' (PST 325).

Now, the isomorphism of technicity and sacredness depends precisely on their reticular structure. This authorises a conjoint analysis of their common origin, which Simondon calls 'primitive ritualisation', characterised by the reticulation of the natural milieu and the repetitive organisation of action. This feature clearly recalls magic as the original source that, in MEOT, was subsequently phase-shift in technicity and religion; and in effect Simondon quotes his complementary thesis to this regard:

> This is the hypothesis of a parallel genesis by splitting starting from an original reticular structure we presented in the third part of the oeuvre entitled *Du mode d'existence des objets techniques*. (PST 327)

[23] This is the same theoretical standpoint from which Simondon (particularly in *Individuation* and in the *Note complémentaire*) preferred to adopt the term transindividual – neither referred to the individual nor to the social system – to name the more fundamental dynamics of which these latter terms are only *parts*.

[24] It is worth noting that also Leroi-Gourhan pays a surprising tribute to Simondon by including PST in the bibliography of his *Le geste et la parole*.

In the 1960–61 course these original structures of 'ritualisation' produce the simultaneous genesis of the space and time of technicity and sacredness (PST 328–29), which is confirmed by the palaeoanthropological evidence of the intertwining between primitive techniques and rituals.[25] Sacredness and technicity in fact both entail the reference to the dimension that in *Individuation* was called transindividual: they are 'dimensions in which action exceeds itself' and cannot be grasped as the act of a subject (whether individual or collective), but rather as the effect of a flux of forces. These forces structure the net of objects and symbols in which individuals and groups build their identities operatively and symbolically: 'technicity and sacredness suppose that the individual in the technical operation and the group in sacralisation exceed their unity and identity: they form a consistent world of structures' (PST 332).[26]

And nevertheless, at the backdrop of their possible convergence, the common nature of technicity and sacredness does not cancel the asymmetry of their social impact:

> Through technicity action detaches, condenses and mobilises the aspects of the world it organises and utilises. On the contrary, through sacredness action merges itself with the space and time it penetrates, without detaching objects, without mobilising elements. Sacredness immobilises forces, displays them over the world, while technicity gathers and mobilises them. (PST 332)

Simondon confirms here that the difference between the two functions, as respectively 'representative' and 'operational' is due to their working at different orders of magnitude. Sacredness operates at the level of groups, while technicity 'oversteps the level of the vastest of human groups' (PST 343). This difference in the orders of magnitude corresponds to different regimes of 'cumulative causality' and, consequently, to a different degree of connection to the dynamics of single social groups. Sacredness 'recruits forces and energetic resources in the human world of motivation and faith', and thus it remains always local and determined according to the characteristics of a single group (PST 340). In short, the 'positive cumulative causality' of sacredness cannot escape the sacred/profane dualism, the binary structure that determines both its stability and structural rigidity:

> Sacredness is rigid and limited. Thus the tendency to ecumenism internal to the category of the sacred is an unattainable dream: each system of sacredness is virtually universal, but in fact concurrent with other systems of sacredness. (PST 341)

[25] Simondon refers to Eliade (1956a) when picturing the original phase. But it is clear that he identifies it with magic, while Eliade to religion: 'For religious man, space is not homogeneous; he experiences interruptions, breaks in it; some parts of space are qualitatively different from others' (Eliade 1956b: 25).

[26] These structures can also be conceived as 'codes' the psycho-social function of which is to 'decode everyday reality in order to know, interpret and implement it with a determined action' (PST 340).

On the contrary, since the historical moment in which it crosses the boundaries of different human groups, technicity

> Provides a system of reference the wide network of which relativises the particularities of human groups and the regionalisms of sacredness to their order of magnitude. (PST 341)

Thus technicity alone is endowed with 'an actual power of ecumenism' (PST 341) and can be the 'basis' of both 'relativity and universality' (PST 343), in opposition to the merely ideal, closed and exclusive universality entailed by the binary logic of sacredness (PST 340–41).

In conclusion, the hypotheses of the structural isomorphism and co-originality of sacredness and technicity do not cancel the difference between respective functions; on the contrary they highlight the epochal importance of the historical phase-shift manifested by the risky development of technology today. Therefore the hypothesis of a conciliation of sacredness and technicity cannot rely on any spontaneous tendency of the two phases to converge. And nevertheless the assumption of a common origin of sacredness and technicity preserves each phase from its reduction to a mere epiphany – or worse, degradation – of the other. Furthermore, once shifted the focus on the structural problem of a diagnosis of the actual opposition between the two phases, the original isomorphism grounds the project for their possible (re) convergence. In particular – what Simondon is mainly concerned with – it prevents considering technicity a mere modification or 'deviation' from an alleged original sacredness.

That is why Simondon attacks Eliade's conception of an opposition between 'the historicity of civilisation and the a-temporality of culture (PST 227). From Eliade's contraposition of 'the rational and conceptual contents' of civilisation and the (rationally non-representable) 'images, symbols and myths' of culture, a very simple diagnosis follows concerning the discontents of civilisation: the 'modern human being is characterised by the fact that, for him civilisation has prevailed over culture' (PST 319). Eliade's quite classical differentiation patently contrasts with Simondon's theory in the *Note complémentaire*, where the social homeostasis of 'community' opposed the excess of technical invention destabilising the social system, but this excess is *also* the condition for innovation and the restructuration of the system and in the last instance the condition for its vital and moral continuation. Thus when Eliade classifies 'the contents of representation and use of technicity among the contents of civilisation' (PST 319) against the eternity of the symbols of culture, according to Simondon he commits the error of transforming what is actually a concrete opposition of operations into an abstract and superficial one between structures, thus instituting an ontological opposition between the ahistorical world of sacredness-culture and the contingent and artificial world of technicity-civilisation.

In his attack against Eliade Simondon also involves Heidegger and Toynbee, and the whole French translation of the German dichotomy Kultur/Zivilisation (notably in existentialism and phenomenology) which would be in this sense a 'defence mechanism' through which culture itself produces 'defensive myths' resulting in an impotent technophobia, blind to the actual 'cultural' content of technical objects (PST 320). Simondon completely reverses the terms of the question, claiming on the one hand the historicity of culture and on the other a kind of 'a-temporality' of technicity, to which he attributes a sense compatible with the theory of technical evolution displayed in MEOT: a kind of 'eternity' of the 'technical schemas' immanent to technical objects (MECD 87) seem to determine their evolutionary tendency independently of psycho-social historicity.

Faced with the contemporary conflict between technicity and sacredness the goal is therefore reversed. It is not about bringing back the symbolic world to a dominant position in order to control a process of (technical) civilisation the huge acceleration of which would determine the crisis of the sacred and *therefore* of social stability. It is rather to rethink completely technicity and to promote its integration into culture *starting from* its programmatic assumption as a value. In short – as expressed by Simondon this time in quite Marxian terms – the goal would be to promote the liberation of the 'essential' evolutionary tendencies of technicity from the alienating historicity of both use value (determined by the relation to the *natural* milieu) and the historical-symbolic value (determined by the relation to the *social* milieu, in the sociological sense of the 'status symbol'). This liberation would inject technicity – as a destabilising and constitutive feature – within the interplay of cultural values (PST 320).

In short, the relation of structural isomorphism between sacredness and technicity, if projected on the contemporary opposition between culture and technology, serves Simondon's goal of authorising the research of their possible 'synergy in the psycho-social domain'. It is from this perspective that Simondon's discourse clearly adopts the pedagogical-political aim of producing a 'parallel demystification of sacredness and technicity', so to 'discover without prejudices the true structure and real nature of technicity, and verify whether the germs of value, the axiological lines it can provide, are in close concordance with sacredness' (PST 320). It is apparent enough that PST is based on the hypothesis that technicity might and should be the future basis for culture:

> Nothing proves – and this is precisely the hypothesis we shall advance – that technicity cannot constitute, as sacredness, the foundations of a culture. (PST 129)

10.4 *Culture et technique*: Acceleration and Conjuncture

As has been shown, Simondon's oeuvre displays a constant search for a scheme suitable to describe the original phase-shift from which the 'human field' would have emerged, the exasperation of which would have brought about the actual contrast between technology and culture. This is a kind of recurrent pattern he adopts in different fields and in relation to different problems. During his most productive period (which goes from the middle of the 1950s to the middle of the 1960s), he seems to consider many different hypotheses: the phase-shift between technics and religion in MEOT (1958), between technicity and sacredness in PST (1961–62), between technical action and symbolic production in IMIN (1965–66). Finally, in the essay on *Culture et technique* [Culture and Technics] (1965) Simondon confers to the term culture itself an ambiguous function: on the one hand 'culture' as opposed to (technological) 'civilisation' would derive from one of the two poles of the original phase-shift, on the other hand the expression 'act of culture' refers to a possible resolution of the contrast between technology and culture within a new 'Culture' (spelled with the capital letter).

According to Simondon the etymology of the term 'culture' carries an implicit axiology that would be worth reactivating in order to underline its specificity in relation to other reductive and dangerous ways of defining Culture. The term culture derives from a technique, agriculture, its method and effects. While breeding acts *directly* on its object (the animal organism) by adapting it to an artificial milieu established according to human timing and necessities, cultivation is a technique that acts *indirectly* on the vital milieu rather than on the living being (the vegetal organism) without producing an anthropocentric adaptation of its object, but rather preparing the ground for 'the genesis of a second nature' (CT 4).[27] On this functional analogy Simondon traces the distinction between the *values of culture* and the *values of technicity*:

> Wether or not he wills it, man is the technician of the human species; a form of feedback loop [*boucle fermée*] operates in human groups, alternately comparable to either the farmer or cultivator who prepares the soil, or to the gardener or breeder who deforms species and obtains new varieties. (CT 5)

And in fact *technical* activity itself can be conceived either as a direct or an indirect activity on the human group-milieu relationship: it is 'culture' when directly dealing with humans, and 'technics' when dealing with their milieu. In both cases human groups act on themselves. Hence Simondon concludes that it is absurd to oppose culture and technics, because they 'are both activities of manipulation, and thus techniques' (CT 5). We might also add along with Simondon that they are the techniques of a 'human technician' working on the human species itself.

It is clear that the Bergsonian dialectical closed/open schema, which operates each time that Simondon is theorising social dynamics (since the elaboration of the concept of transindividual in *Individuation*) in *Culture et technique* acquires the shape of a twofold nature of technics itself. That is why it would not be pertinent to suppose an opposition between phases of a different *nature*. On the contrary, the qualitative difference between culture and technique must be understood in terms of scale (CT 6). It depends in fact on the order of magnitude at which the feedback effect of the technical loop [*boucle technique*] is measured:

> When techniques outstrip human groups, the power of the feedback effect, through the modification of the milieu, is such that the technical gesture can no longer be just an isolated organisation of means. Every technical gesture engages the future, modifies both world and the human species, whose milieu that world is. The technical gesture does not

[27] It cannot be excluded, of course, that agriculture could adopt – as in fact it does – operative modalities comparable to those of breeding (Simondon himself provides such an example referring to grafted rosebushes, the hypertelic adaptation of which makes them fragile and entirely dependent on their artificial milieu, CT 3).

> exhaust itself in its utility as means; it leads to an immediate result, but also triggers a transformation in the milieu, which rebounds onto living species, human species included. (CT 7)

So the concept of technics can be assumed to fit and resume the entire closed/open dynamics by itself, and this allows Simondon to reformulate the opposition between culture and technology in terms of a distinction internal to technics itself. In *Culture et technique* 'culture' is in fact a set of techniques connected to the internal normativity of groups, while other techniques are endowed with values that outstrip the group itself:

> This is not a conflict between culture and technics, but between two forms of technique: between a state of technique that is intra-groupal [*intra-groupales*] and thus intra-cultural, and a state that exceeds the dimension of the group, and therefore exceeds any possible *cultural* dimension. (CT 6, italics added)

In fact, the centrality eventually attributed to technics forces Simondon to reformulate the notion of 'culture' itself. Thus in the second part of his essay, the notion of culture is split into two meanings paralleling the two different meanings he had previously reserved for technics. As 'education', cultures establish a self-entertaining milieu which does not develop, but is maintained and stabilised. As an '*act of culture* in the true sense of the term', culture is a gesture which grounds the relation of the group to its milieu (CT 8).

At the backdrop of *Culture et technique* it is therefore possible to reinterpret most of Simondon's theoretical efforts as a series of variations around the main theme of the relation between culture and technicity. In the third part of *Individuation* Simondon understood the domain of culture as a system of production and exchange of significations through the cybernetic concept of information in order to make explicit the common ground of culture and technics. In the third part of MEOT magic, conceived as the model of a primitive unity underlying the 'system of [collectively] created objects', the techno-symbolic 'envelop' of human groups, conveyed the hypothesis of a possible rehabilitation of technicity as the basis for a new technical culture. Relying on the concepts of the transindividual and technicity elaborated there, in the *Note complémentaire* Simondon advanced a theory of the processes crossing social systems based on the crucial metastabilising function of technical invention. In PST he scrutinised the original structural isomorphism of technicity and sacredness hidden under the contemporary conflict between *Kultur* and *Zivilisation*. In IMIN he enquired into the importance of techno-symbolic production by linking it to the dynamical relation between the organism and its milieu. And finally, in *Culture et technique* Simondon translated the contrast between culture and technology in a structural phase-shift within technics itself that opens up to the consideration of human evolutionary process as a political problem.

In effect, the evolutionary perspective adopted by Simondon in *Culture et technique* brings us back to his debt towards Leroi-Gourhan, and through him to Durkheim and Bergson: the two primary sources of Simondon's political questioning of technological progress. Although Leroi-Gourhan's oeuvre *Le geste et la parole* [Gesture and Speech] (1964, 1965), was published closer to the period of *Culture et technique*, it is rather the long narration of *Evolution et techniques* [Evolution and Technics] (1943, 1945) that captures Simondon's attention on the function of technics in the nature-culture shift.[28] In that early work technical progress, in distinction from any other 'evolutionary' process,[29] manifested an evident progressive tendency, as it is clearly expressed in the conclusion concerning *Le progrès technique* [Technical progress]:

> The technical acquisition cannot be lost, its transmission is secured from political circumstances [...] Moral, religious and social progress is perpetually called into question [while] technical progress imposes itself out of any possible debate. (Leroi-Gourhan 1945: 471)[30]

Among all the processes that cross or parallel the development of the human group or 'ethnic group', technical progress follows a line of development better defined than any other feature. And this is not only a perspective illusion due to the relative abundance of palaeontological documentation: technicity actually *mediates* the nature-culture shift as far as it is founded on the biology of *homo sapiens* that underlies the complex system of relations made of 'migrations, borrowings and spontaneous apparitions' of objects in which historical-cultural variables dominate. Technical 'inventions' or 'imports' – the appearance of an object within a group – can be fixed (and this is true for technical objects only, not ritual, artistic etc.) on the biological tendency of the species. This dynamic determines progress by selecting the most effective technical solutions (Leroi-Gourhan 1943).

In the general scheme Leroi-Gourhan derives from his researches in *Milieu et techniques*, technical progress is made of a slow trend made of minute assessments

---

[28] In a certain sense Simondon is more faithful to the first Leroi-Gourhan (*Evolution et techniques*), insofar as he eminently confers on techniques, rather then to the symbolic function, the status of the distinctive 'mark' of the human condition. In fact *Le geste et la parole* is structured around the relation-contraposition of two primordial activities of 'liberation' – technical and symbolic – of the human 'social organism' from the constraints of the natural milieu. The importance of the activity of symbolic production and exchange is particularly highlighted by Leroi-Gourhan, in particular the one providing a 'liberation of memory': 'the most striking material fact is certainly the "freeing" of tools, but the fundamental fact is really the "freeing" of the word and our unique ability to transfer our memory to a social organism outside ourselves' (Leroi-Gourhan 1965: 34). It is important here to refer to the key concept of 'prostheticity' recalled by Hyppolite during a discussion at the Royaumont Conference. After underlining how he appreciated the attempt to avoid any reference to 'consciousness', Hyppolite points at 'prostheticity' as the possible point of convergence of cybernetics and existentialism (RO 418). Also in Wiener's paper the question of the prosthesis emerged as crucial for the understanding of a 'human-technical system' (Wiener 1962: 103–12). The theme was crucial to the project of a 'general organology' carried on by Bergson and Canguilhem (see Chap. 9, notably n. 16), and more recently by Stiegler who – starting from Leroi-Gourhan and Simondon – made of 'prostheticity' a key concept for his philosophy by conceiving the 'prosthetic' object as the support of processes of (trans)individuation in order to ground symbolic production on the technical 'exteriorisation of memory' (*hypomnèmata*) (Stiegler 1994).

[29] On the problematic expression 'technical evolution', see Sect. 11.3.

[30] This allowed Leroi-Gourhan to classify human groups according to the degree of technological development: 'a less flexible language or a less developed religion can be borrowed; but one would not change the plough for the hoe' (Leroi-Gourhan 1945: 522).

and sudden mutations (Leroi-Gourhan 1945: 408), a scheme that admits decisive thresholds. In particular, the Neolithic acceleration is the crucial event through which, starting from the institution of a milieu of technical objects and an 'apparatus for social memory' (Leroi-Gourhan 1965: 24), the organism-milieu relationship is radically reconfigured; it amplifies and changes its order of magnitude, and human societies emerge. The Neolithic is, at the geological scale, the trigger of a circular causality between the biological and the cultural which measures the sudden emergence of human civilisations:

> In terms of geological time only an instant separates the last aurochs hunter from the first Mesopotamian scribe, and the emergence of the new economies is a sudden explosion. (Leroi-Gourhan 1964: 222, see also 204, 241)

Thus in the deployment of technicity at a worldwide scale, Leroi-Gourhan sees the irreversible exit from the state of nature (and the entering into history) as an event humanity has yet to deal with. The instantaneity of this 'original' acceleration is striking in the peculiar *memento mori* Leroi-Gourhan offers as a possible spectacle for future anthropologists retroactively studying industrial revolution:

> Let us imagine archaeologists of the CXX century provided with the same technical means we use today for studying the Neolithic era. They would have to ascertain that, over a thick layer of swords, guns and horse-drawn carriages, suddenly appears – instantaneously – a prodigious heap of airplanes wrecks, locomotives, radio stations and metal cans. (Leroi-Gourhan 1945: 406–407)

Focusing on the effects of the industrial rather than the Neolithic revolution, it is still at the large scale of humanity as a whole that in *Culture et technique* Simondon poses the question on technics. And it is precisely at this level that the 'conflict' between culture and technics eventually proves to be the outcome of a historical event:

> The apparent conflict between technics and culture is rather a conflict between two technical levels, the preindustrial level, for which technics are a concatenation of means in the service of intra-cultural ends within each human group, and the industrial level, which opens technology to a great autonormative gesture with an evolutionary meaning that modifies the human species' relation to its milieu. (CT 11)

Simondon poses thus in *Culture et technique* a question which crosses both the biological and the political domains: it is the question of the irreversible threshold beyond which human groups experience difficulties that depend on the contrast between their biological nature and the development of society. At that point a 'simple' political solution is not possible anymore.

Indeed, the early Leroi-Gourhan had posed the same question, explicitly assuming as his point of departure Bergson's 'extremely fruitful view' on closed and open societies in the *Deux sources* (Leroi-Gourhan 1945: 340).[31] There Bergson

---

[31] 'Henri Bergson, assuming a different point of view, has clearly defined in *Les deux sources de la morale et de la religion* a static condition [*état statique*] in which human groups would turn in a spiral, changing from generation to generation a limited number of concepts, of prescriptions progressively complicated, and a dynamic condition [*état dynamique*] in which the groups would take the straight line of development of their tendencies. We would like to resume this, this extremely fruitful view, by adapting it to our point of view' (Leroi-Gourhan 1945: 340).

conjoined the biological-evolutionary and the political sides of the problem, thus developing his argument: 'closed society' is 'human society fresh from the hands of nature', where nature 'left an opening for expansion' which eventually brought the process of civilisation; and yet in all societies – however 'open' they might be – 'the primitive instinct persists' bringing in a tendency towards closure. In short, for Bergson an event pertaining to natural history, the opening of originally closed human communities, instituted a major political problem the status of which is essentially aporetical and nevertheless cannot be escaped. The problem of government – Bergson argues – is a problem 'which the increased size of societies may well have rendered insoluble' (Bergson 1932: 292 ff.).

Reformulating again Bergson's problem in terms of the relation between the moral 'static' and the technical 'dynamic' development of society, Leroi-Gourhan radicalised his formula. He evidenced the disproportion between the complex and powerful techno-symbolic dispositive developed by human societies and the biological configuration of *homo sapiens* which, almost unchanged since its beginnings during the Upper Palaeolithic, continues 'in a disordered manner to satisfy predatory tendencies which hark back to times when humans were fighting the rhinoceros' (Leroi-Gourhan 1965: 25). It is clear that with this formulation the problem can hardly be accountable at the level of any political action. And in fact there can be no possible political solution to a problem posed at the (indeed 'geological') scale Leroi-Gourhan adopts in the conclusion of *Le geste et la parole*:

> The great problem of the world as it already exists summons up a solution: How shall this archaic mammal, with its archaic needs that have been the driving force of its ascent, continue […]? (Leroi-Gourhan 1965: 266)

It is sufficient to add a simple thermodynamic variable to this formula, and we reach Simondon's conclusion. Since a 'process of degradation' (CT 9) is present in all cultures as a closing regressive tendency, in order to adequately oppose its innate entropy society is compelled to evolve. This necessity to continue evolution is both biological and historical, and it can only work through the implementation of the structural openness of technicity in the social system, since techniques are, for the human species, 'the most concrete mode of the power to evolve; they express life' (CT 8–9).

References[32]

Barthélémy, J.-H. 2011. Quel mode d'unité pour l'œuvre de Simondon? *Cahiers Simondon* 3: 131–148.

Bergson, H. 1907. *L'évolution créatrice*. Paris: Alcan (English trans. *Creative Evolution*. London: Macmillan & Co., 1954).

Bergson, H. 1932. *Les deux sources de la morale et de la religion*. Paris: Alcan (English trans. *The Two Sources of Morality and Religion*. London: Macmillan & Co., 1935).

[32] *Simondon's complete bibliography and a list of abbreviations are provided in the* Appendix.

Canguilhem, G. 1955. Le problème des régulations dans l'organisme et dans la société. In *Ecrits sur la médecine*, 101–125. Paris: Seuil, 2002 (First edition: *Les Cahiers de l'Alliance Israélite Universelle* 92: 64–81).
Durkheim, É. 1893. *De la division du travail social*. Paris: Alcan (English trans. *The Division of Labour in Society*. London: Macmillan, 1984).
Durkheim, É. 1912. *Les formes élémentaires de la vie religieuse*. Paris: Alcan (English trans. *The Elementary Forms of the Religious Life*. London: Allen & Unwin, 1964).
Durkheim, É. 1924. Détermination du fait moral. In *Sociologie et philosophie*, 49–90. Paris: Alcan (English trans. The Determination of Moral Facts. In *Sociology and Philosophy*, 35–62. Cohen & West, 1965).
Eliade, M. 1952. *Images et symboles*. Paris: Gallimard.
Eliade, M. 1956a. *Forgerons et alchimistes*. Paris: Flammarion.
Eliade, M. 1956b. *Le sacré et le profane*, 1994. Paris: Gallimard.
Hottois, G. 1994. Gilbert Simondon entre les interfaces techniques et symboliques. In *Ordre biologique. Ordre technologique*, ed. F. Tinland. Seyssel: Champ Vallon.
Hubert, H., and M. Mauss. 1902–1903. Esquisse d'une théorie générale de la magie. In Mauss, M. 1950. *Sociologie et anthropologie*, 1–141. Paris: PUF (English trans. *A General Theory of Magic*. London/New York: Routledge, 2001).
Karsenti, B. 1997. *L'homme total. Sociologie, anthropologie et philosophie chez Marcel Mauss*. Paris: PUF.
Lacan, J. 1962–1963. *Le séminaire de Jacques Lacan. Livre X. L'angoisse*. Paris: Seuil, 2004.
Le Bon, G. 1895. *Psychologie des Foules*. Paris: PUF, 1988.
Le Cœur, C. 1939. *Le rite et l'outil*. Paris: PUF, 1969.
Leroi-Gourhan, A. 1943. *Evolution et techniques I. L'homme et la matière*. Paris: Albin Michel.
Leroi-Gourhan, A. 1945. *Evolution et techniques II. Milieu et techniques*. Paris: Albin Michel.
Leroi-Gourhan, A. 1950. L'*homo faber*: la main. In *A la recherche de la mentalité préhistorique*, ed. A. Laming, 75–98. Paris: Albin Michel.
Leroi-Gourhan, A. 1964. *Le geste et la parole. Technique et langage*. Paris: Albin Michel (English trans. *Gesture and Speech*, I. Cambridge-London: MIT Press, 1993).
Leroi-Gourhan, A. 1965. *Le geste et la parole. La mémoire et les rythmes*. Paris: Albin Michel (English trans. *Gesture and Speech*, II-III. Cambridge-London: MIT Press, 1993).
Lorenz, K. 1937. *Über Tierisches und Menschliches Veralten. Aus dem Werdegang der Verhaltenslehre*. München: Piper & Co, 1965.
Mauss, M. 1930. L'œuvre de Mauss par lui même. *Revue Française de Sociologie* 20(1): 209–220.
Mauss, M. 1948. Les techniques et la technologie. *Journal de Psychologie Normale et Pathologique* 41: 71–79.
Merleau-Ponty, M. 1956–1960. *La Nature. Notes. Cours du Collège de France*. Paris: Seuil, 1995 (English trans. *Nature: Course Notes from the Collège de France*. Illinois: Northwestern UP, 2003).
Petit, V. 2010. L'individuation du vivant (2) «Génétique» et «ontogenèse». *Cahiers Simondon* 2: 53–80.
Stiegler, B. 1994. *La technique et le temps. 1. La faute d'Épiméthée*. Paris: Galilée.
Von Uexküll, J. 1934. *Streifzüge durch die Umwellen von Tieren und Menschen – Bedeutungslehre*. Berlin: Springer (French. trans. *Mondes animaux et monde humain. Suivi de La théorie de la signification*. Paris: Denoël, 1965).
Wiener, N. 1962. L'homme et la machine. In *Collected Works* IV, ed. N. Wiener, 824–827 (First edition: In *Le concept d'information, 5ème colloque philosophique de Royaumont*. Cahiers de Royaumont. Philosophie 5. Paris: Minuit 1965, 99–100).

Chapter 11
The Mysticism of (Technical) Evolution

Simondon's research articulates different perspectives on the question of technics, far from both Heideggerian technophobia and the positivistic (and in part the cybernetic) technocratic dream.[1] The theme of technological progress never leads him to any nostalgic return to mythical origins, nor to a revival of the modern strive for a technical solution for political problems. Nonetheless his philosophy draws the line of an irreversible path, and sees the appropriation of technics as a binding necessity: technical 'openness' is not for Simondon a simple judgement of value, it is also a palaeoanthropological evidence that – as it is for Bergson in the *Deux sources* – 'calls for a bigger soul [*supplément d'âme*]', a goal which Simondon believed was intrinsic to technics itself.

As I shall explain, it is in close contact with such a vision – derived from both Durkheim and Bergson – that Simondon intends to develop a political pedagogy of the 'technical mentality'. But his philosophy of individuation entails a conception of technical progress and human historicity that contrasts with every 'mythical' quest for a foundation of social systems – natural or historical – on an alleged human *essence* or on the *élan vital* itself. On the contrary, Simondon advances a radically open and inventive conception of the evolution of social systems and of what he calls 'human progress'. It is from this perspective that Simondon's philosophy forces the political problem of connecting technological development and different cultures to be posed by rejecting, at the same time, the complementary simplifications of the Eurocentric faith in the civilising power of technological progress, *and* of the 'communitarian' regression inspired by Heidegger's anti-technological stance.

[1] On the relation between Heidegger's philosophy of technics and Simondon's, see at least Chateau (1994) and the volume edited by Vaisse (2006). In fact, Simondon's explicit references to Heidegger are quite rare and mainly concerned with a critique of his reduction of the essence of technicity, and consequently of the technical object, to *Gestell*, to a 'framing' and thus 'alienating objectivation of human experience': against this reduction of technical objects to mere 'utensils [*ustensiles*]', Simondon repeatedly affirms the essential 'historicity of the technical object' (MEOT 222, PST 128–29).

A. Bardin, *Epistemology and Political Philosophy in Gilbert Simondon*,
Philosophy of Engineering and Technology 19, DOI 10.1007/978-94-017-9831-0_11

11.1 Mechanics and Mysticism in Durkheim and Bergson

When Bergson concluded the final chapter of the *Deux sources* (*Mechanics and Mysticism*) referring to 'the essential function of the universe, which is a machine for the making of gods' (Bergson 1932: 338), he was not at all abandoning himself to some kind of teleological dream. He was in fact reviving a positivistic project: more precisely, he was reviving the way Émile Durkheim had questioned that project. Almost half a century before, Durkheim had reviewed the work of the Positivist sociologist Jean-Marie Guyau, *L'irréligion de l'avenir. Etude de sociologie* [The Non-Religion of the Future: a Sociological Study] (1887) by quoting an important remark of the latter:

> If nothing authorises us to suppose that evolution aims at a determined goal, yet nothing prevents us from 'conceiving it as resulting in beings capable of proposing to themselves a certain aim and of dragging nature after them towards it [...] It is not probable that we embody the highest achievement possible in life, thought, and love. Who knows, indeed, whether evolution might be able to bring forth – or has not already brought forth – what the ancients called "gods"?' (Durkheim 1887: 305–6, n. 23, quoting Guyau 1887: 439)

Guyau was radically assuming the positivistic hypothesis of the future disappearance of religion, progressively absorbed into scientific research. Durkheim replied arguing that 'belief results from practical causes', and therefore the degree of integration (and the eventual disappearance) of the dynamics of sacredness in scientific knowledge will depend on the degree of social development (Durkheim 1887: 310).[2] In effect, according to Durkheim and – later – to Bergson, the problem should be posed differently: given the universalising tendency of scientific and technological development, will it be possible to reconstruct a social bond at the level of the entirety of humanity? What role do science and religion play in view of this thoroughly political task? And finally, which is the role played by philosophical thought (or rather, *chez* Durkheim, by sociology)?

It is universally acknowledged that Durkheim's *Formes élémentaires de la vie religieuse* [The Elementary Forms of Religious Life] (1912) is one of the main polemical targets of Bergson's *Deux sources de la morale et de la religion* [The Two Sources of Morality and Religion] (1932). Since both works are concerned with the relation between religion and the origin of human society, a brief comparison on the themes of human evolution and technical progress shall clarify the standpoint from which Simondon questions what – following Leroi-Gourhan – he names 'technical evolution' (Sect. 11.3).

Durkheim and Bergson both assume that a structural contradiction inhabits society, crossed as it is by irremediably opposed tendencies that weaken the social bond. These tendencies also traverse single individuals, divided between the duties

[2] Durkheim (1912) will explicitly deny this possibility. In this text, however, he is already quite critical towards Guyau's hypothesis: 'In order to demonstrate that it [religion] has no future, one should demonstrate that the reasons that made it necessary have disappeared. And since these reasons are sociological, one should find out what change in the nature of societies has to take place, which would make religion useless and impossible' (Durkheim 1887: 310).

imposed by society and their personal needs (Durkheim), or between social obligation and the *élan vital* that inhabits each of them (Bergson). The internal tensions of the social system reach their apex in industrialised society, where the social bond is weaker and technology offers powerful instruments of destruction. Therefore, a philosophical and/or sociological science aiming at challenging this set of problems must first of all revisit the origins of the social system in order to display any possible solutions. In short, the two authors share the same fundamental interest in the problematic tension intertwining social systems and individuals, and the same methodological approach to the question concerning their ontological and historical nature – interests and approaches which also profoundly resonate in Simondon's research. Furthermore, the theme of religion is central for both, since they recognise in it a crucial feature of the social bond, which can be analysed by reflecting on the evidence gathered by ethnological field-research.

In fact, however, the differences between the ontological presuppositions of Durkheim and Bergson are strikingly evident. While for Durkheim the social 'organism' is part of a nature monistically conceived, according to Bergson social groups are the 'mixed' product of life's creative effort opposing the passive tendency of matter to mechanical repetition. Consequently, the very analysis of the 'source(s)' of social systems are radically opposed too. Durkheim speaks of society as the unique source of the categories of pure reason and of action, which structures both science and morality:

> Thus it is not at all true that between science on the one hand, and morals and religion on the other, there exists that sort of antinomy which has so frequently been admitted, for the two forms of human activity really come *from one and the same source*. (Durkheim 1912: 635, italics added)

On the contrary, for Bergson the sources of both morality and religion are split, characterised by the two different operative modalities typical of the *élan vital* and of inert matter: opening and closure. This clear divergence in the approach to social ontogenesis entails different answers to the actual problems concerning the social system, and explains the different conception of political action Durkheim and Bergson consistently derive from their respective premises.

In Durkheim the path towards universality resides in the new kind of international life that immediately entails the homogeneity of beliefs and the consequent growing expansion of the collective horizon. This way towards religious internationalism is, in Durkheim's argument, *a priori* homogeneous and implicit in the universality of scientific and technical development, which also follows the same path since it comes from the same source as religion. The religious and the techno-scientific features of the social body display a common tendency, and although scientific thought is probably destined to replace religious thought in the theoretical dimension, at the practical level religion will probably remain an irreplaceable guide for social praxis (Durkheim 1912: 614–15). This grounds Durkheim's basic faith in institutions – considered rational inasmuch collective – which can be extended not only to political institutions, but also to the whole of the symbolic universe, religious institutions included. A rational faith is derived from the evolutionary postulate assumed by sociology:

> In fact, it is an essential postulate of sociology that a human institution cannot rest upon an error and a lie, without which it could not exist. If it were not founded in the nature of things, it would have encountered in the facts a resistance over which it could never have triumphed. (Durkheim 1912: 3)[3]

All 'categories', those grounding scientific and religious thought included, are for Durkheim the result of a symbolic accumulation which took place in human history along with the material accumulation of technical instruments, and which *de facto* entails 'a close relationship between the three ideas of tool, category and institution' (27, n. 1). This progressive accumulation determines a tendency towards universalisation according the ideal of a 'truly and properly human thought' that 'is not a primitive fact' but 'the product of history'. In short, a process of universalisation towards a 'new kind of social life' is inborn in human historical societies. These societies that cease to appear to themselves 'as the only whole' are necessarily pushed to extend the collective horizon and therefore require an ecumenical political project founded on the formation and diffusion of 'logical thought' and universal morality (634–35).

According to Bergson, on the contrary, nature provided human societies with a biological 'simple schema', non-evolutive and immutable, which predisposed human beings to a certain social form: that of small societies. And nevertheless, the fact that this 'simple schema' is 'vague and uncompleted', entails the structural openness of these same societies: in Bergson's words, nature 'left them an opening for expansion' (Bergson 1932: 293–94). Thus, once societies have reached an order of magnitude superior to their original institution, the tendency towards openness starts operating as a disintegrating force. Bergson never ceases to note that in the actual conjuncture technology is the continuation of a tendency to openness that breaks with the moral and institutional 'instruments' originally elaborated by humans in their 'natural' milieu. Consequently from this perspective he poses the political problem:

> If our organs are natural instruments, our instruments must then be artificial organs. The workman's tool is the continuation of his arm, the tool-equipment of humanity is therefore a continuation of its body. Nature, in endowing us with an essentially tool-making intelligence, prepared for us in this way a certain expansion. But machines which run on oil or coal or 'white coal', and which convert into motion a potential energy stored up for millions of years, have actually imparted to our organism an extension so vast, have endowed it with a power so mighty, so out of proportion to the size and strength of that organism, that surely none of all this was foreseen in this structural plan of our species: here was a unique stroke of luck, the greatest material success of man on the planet. A spiritual impulsion had been given, perhaps, at the beginning: the extension took place automatically, helped as it were

[3] 'Thus it is seen that whatever has been done in the name of religion cannot have been done in vain: for it is necessarily the society that did it, and it is humanity that has reaped the fruits' (Durkheim 1912: 600). This clearly recalls Comte's confidence in a society that 'cannot be completely wrong concerning its real needs' because of its 'founding aphorism: there is neither society without government nor government without society' (Comte 1852, II, V).

> by a chance blow of the pick-axe which struck against a miraculous treasure underground. Now, in this body, distended out of all proportion, the soul remains what it was, too small to fill it, too weak to guide it. Hence the gap between the two. Hence the tremendous social, political and international problems which are just so many definitions of this gap, and which provoke so many chaotic and ineffectual efforts to fill it. What we need are new reserves of potential energy: moral energy this time. (Bergson 1932: 330)

At the basis of the contemporary 'excess' of mechanics is an only partially accomplished process, and nothing seems to announce that a second 'movement' is going to take place. According to Bergson mechanics operates in effect as a process of universalisation, but there is no destiny in it: this progressive enlargement of human societies is only one side of the '*twofold frenzy*' characterising any 'vital tendency' (316). According to the '*law of dichotomy*' Bergson formulates at the end of his work, mechanics and mysticism seem to alternate in human history in a kind of progress through 'oscillation' between ascetic simplification and technical complication of human life (311 ff.). The 'twofold frenzy' reaches its extreme in both the 'industrial' drive towards the multiplication of needs and in the renunciation to invention. The apex of this separation of mechanics and mysticism respectively results in a society of commodities on the one hand, and its 'ascetical' refusal, on the other.

Despite this apparent symmetry, in which technological development represents one aspect of a twofold process, clearly the outcome is not virtuous. If it is true that 'the mystical summons up the mechanical' (329), this is only true in the sense that the material conditions of humanity must be brought to an adequate level for the purpose of producing an adequate mysticism.[4] In short, Bergson's diagnosis is the following: with industrial society and occidental mysticism human evolution has reached a crucial and irreversible threshold. The gap between the two 'frenzies' is so wide today that it poses an unprecedented problem of scale to political practice (and to political thought): mechanics summons up a mysticism which cannot be limited to ascetic closure into a regressive simplicity, but primarily a diffusion of – rather than education to – 'true mysticism'.

In conclusion, while for Durkheim religion was simply to be accepted as the necessary practical nourishment of the social bond and science was alone the motor of social progress, for Bergson the *élan* of progress was mainly a matter of mystical invention. What science was able to accomplish only at the intellectual level, mysticism could do at the level of the 'invisible emotion' that adheres to the 'huge inorganic body' that is 'the seat of our potential or theoretically possible actions'. Only great mystics may be able to carry on a political pedagogy, by 'blazing a trail along which other men may pass' (Bergson 1932: 273–74).

These questions and formulas describe the horizon within which Simondon develops a political thought and a pedagogy of technicity. His acceptance of the problematisation of the positivistic project is carried on through the inversion – and apparently not the refusal – of Bergson's hierarchy. In Simondon's philosophy

---

[4] 'How could it spread, even diluted and enfeebled as it must necessarily be, in a humanity obsessed by the fear of hunger?' (Bergson 1932: 329).

the priority of technics over mysticism could be said to ground his own pedagogical-political project, based on the initiation-to, and the diffusion-of what he called 'technical mentality'.

11.2 Simondon's Pedagogical-Political Project

A pedagogical program crosses the whole of Simondon's work. This project is based on the *postulate* of a common source for technicity and sacredness, and on the *evidence* of their possible convergence in a global technological network. It relies on the discovery of some essential features of 'technical mentality' and culminates in the pedagogical commitment to 'reflexive thought' and, notably, to 'philosophical thought', which Simondon considers 'political' insofar as it is pedagogical.

Since *Place d'une initiation technique dans une formation humaine complète* [Place of Technical Initiation for a Complete Human Formation] (1953–54),[5] Simondon offered a series of reformulations of his project to introduce a 'technical culture'. This remained a primary concern also in his last interview, *Sauver l'objet technique* [Saving the Technical Object] (1983), where he did not fail to reaffirm that 'techniques are never completely overcome. They hide an unalienable schematic force which is worth preserving' (SOT 152). There is however a moment, during the 1970s, when Simondon clearly defines the schemas that techniques carry on *across* cultures and explains their pedagogical value. *Mentalité technique* [Technical Mentality] (1970) is a programmatic text which openly aims at extending the positive values implicit in technical normativity in the psycho-social field:

> This paper is not concerned with ontology but with axiology. It aims to show that there exists a technical mentality, and that this mentality is developing, and is therefore incomplete and at risk of being prematurely considered as monstrous and unbalanced. (MT 343)

The 'technical mentality' brings about operational schemas which prove to have epistemological, aesthetical, ethical and political implications.[6] Simondon argues that such implications are not yet transparent within the domains 'of the affective categories' and 'of the will', but in the domain of 'cognitive schemas' technical

---

[5] This is the first text Simondon published when he was still teaching at the *Lycée Descartes* of Tours, where he carried on didactical experimentation in a small technology lab (PI 115).

[6] On 'technical mentality' as a prism through which one can glimpse Simondon's overcoming of phenomenology (in relation not only to Merleau-Ponty, but also to Mikel Dufrenne), see Carrozzini (2011). On this ground, the author accurately analyses Simondon's technological paradigmatism in his courses of general psychology, reads his conception of 'technical culture' against the background of Jacques Lafitte's '*science des machines*', and develops a critique of Simondon's techno-aesthetics. On the normative function and social effects of (the relationship with) technological artefacts, one can widely draw from Bruno Latour's work (see for instance Latour 2002, where he hints at Simondon's theoretical contribution), such as Simondon's claim that technologists have 'to be the representative for technical objects' (MEOT 151); on the regulatory function of Simondon's philosophy of technology see also Schmidgen (2012).

mentality is already 'coherent, positive, productive'. It is therefore possible to provide its formulas, that is the two 'postulates' of technical mentality: (a) '*The subsets are relatively detachable from the whole of which they are a part*'; (b) '*If one wants to understand a being completely, one must study it by considering it in its entelechy, and not in its inactivity or its static state*' (MT 346–47).

The first postulate entails an heuristics, but also an ethics and a politics that would counter the 'holistic postulate' that, 'often presented as an attitude of respect for life, a person or the integrity of a tradition', is in fact the more modest and conservative act 'to accept or reject a being wholesale' avoiding the more generous attitude of a 'careful examination' (MT 346). The second entails the necessity of differentiating the regimes of functioning and the threshold conditions of a system's functioning. In this sense, the whole set of relations between an individual being and its milieu is in fact conceived as part of the being itself, and must be therefore integrated in its definition (MT 347). Once the two postulates have been assumed Simondon tries to extend the cognitive schemas connected to them – the intertwining of the subsets (first postulate) and the existence of the thresholds of functioning (second postulate) – onto the affective and the ethical-political domains (respectively, the domain of the modes of production and labour relations and of individual and social normativity). Simondon's basic assumption is that the process of rationalisation of production should be also a process of concretisation and integration of technical objects in a technological 'reticular structure' capable of freeing production from the alienating constraints of inessential exigencies induced by the market.[7]

With the introduction of the 'cognitive schemas' implicit in technical mentality in the 'tense incoherence of the affective modalities' characterising the social system, social normativity would finally fit the 'optimum' functioning for the development of social systems. This development would take place according to a 'single criterion' uniting 'the manifestation of cognitive schemas, affective modalities and norms of action: that of the *opening*' (MT 354–57). The cognitive schemas dependent on the postulates of technical mentality and concretised (or 'crystallised') into the '*open* object', would thus be integrated in the functioning of the social system. And the pedagogical diffusion of technical mentality would be inscribed in a virtuous circle as both the condition of possibility for a correct understanding of technicity and the major effect of its actual integration in social normativity.

In a similar fashion Simondon had already argued in the *Note complémentaire* that technical objects are in fact 'germs of thought, endowed with a normativity' which carries on a 'function of civilisation' by producing relations in the field of culture that would maintain the openness of community (NC 514). And this is also what the pedagogical neo-encyclopaedic mission of MEOT is based on: the assumption that the technical object is '*a set of sensorimotor schemas* rationally intertwined and organised, as an organism' the pedagogical reactivation of which is

[7] His emancipatory promise only deals here with the pedagogical-political function of the *open* 'post-industrial technical object': because of its double-layered structure, it would both maintain a relation with its actual functioning and the opening to future invention.

an act of 'liberation' for the individual and therefore a crucial factor of social transformation (Van Caneghem 1989: 824). This liberation through acquisition of a technical mentality tends towards universality. Through the generalisation of the fundamental 'technical schemas', a 'technics of all techniques' could be developed: a 'general technology' capable of abstracting from the actual functioning of technical objects a schema shared by different objects, technical and also biological. The categorisation of these schemas of general processes would allow for the construction of a 'general theory of causality and conditioning' which would contribute to the valorisation of technology, and through which 'the normative value [of techniques] would penetrate culture' (MEOT 218, 221).

As an ideological ground for this pedagogical-political project, a fundamental faith in the universalising convergence of technics and culture – although never conceived as a destiny – is derived by Simondon from the postulate of a common origin to technicity and sacredness (PST), or of technicity and culture (CT).[8] On this basis Simondon maintains throughout his whole intellectual production an optimistic attitude towards the on-going acceleration of technology, and he identifies the problem of the contrast between technology and culture in contemporaneity as the crucial political problem, the origin of which is the genesis of industrial 'civilisation'. This broadly corresponds to what in MEOT Simondon names the second phase-shift. As explained in Sect. 10.2, according to Simondon something crucially irreversible took place in the relation between technics and religion at the moment of the second phase-shift, opposing social and political thought to industrial technology. The structural metastability of the new couple of phases contrasts the rigid contraposition between traditional religion and the parcelled operational attitude typical of artisanal techniques. Thus 'the exigency of totality and unconditional unity' which in the first phase-shift opposed religion and technics, is now reconfigured by a new technical milieu which tends to a new kind of reticular unity that pushes the universality of religion – formerly assumed as 'given' once forever – to become the 'virtual universality' of social and political thought (MEOT 208).

For this reason it is precisely at the level of the historical development of industrial techniques and of social and political thought, that a compatibility between the phases of technicity and religion can finally take place. As Simondon claims, the genesis of industrial civilisation entailed in the long term the construction of technical networks and the development of a correspondent form of social and political thought: 'both born out of becoming', they build compatible structures which tend to coincide 'from the perspective of a permanent changing of the technical and socio-political structures' (MEOT 230). On this basis Simondon will be able to affirm that 'the technical mentality can be developed into schemas of action and into values to the point of yielding a morality in human environments that are entirely dedicated to industrial production' (MT 351–52). It is clear that once this perspective has been assumed, the actual technological development of 'technical networks' might appear as an evidence of the possible actualisation of the technical schemas implicit in technical mentality.

[8] See above Chap. 10, in particular Sects. 10.3 and 10.4.

In PST in particular, Simondon draws the lines of the *Les coïncidences actuelles du sacré et du technique* [The Contemporary Convergence of Technology and the Sacred], as the virtuous result of the original tendency towards the *Rencontre possible de la sacralité et de la technicité dans l'avenir: l'unité de la Culture* [The Possible Encounter between Sacredness and Technicity in the Future: the Unity of Culture] (PST 329–34). According to Simondon, technicity and sacredness display two homologous forms of reticulation, but also two asymmetrical tendencies towards ecumenism. As far as the technical gesture 'fulfils the equivalent function for bigger groups', this difference grows as the order of magnitude of the social system changes and technicity tends more and more to substitute 'the ritualisation and solemnity typical of the manifestations of sacredness' (PST 344).

The different scale of the two forms of ecumenism derives from the different logic they depend on. While technicity always refers to an *analogical* extension of the system in view of the integration of new 'elements', sacredness recalls the *binary* and exclusive logic of the 'biological community' (PST 340–41). As Simondon argued in the *Note complémentaire*,

> On the ground of the primitive categories of inclusion and exclusion, corresponding to [biological] actions of assimilation and de-assimilation, the annexed categories of purity and impurity, good and harmful, develop as the social roots of good and bad. (NC 509)

In short, being essentially *bio*logical, sacredness tends to communitarian closure, while, being *analogical*, technicity follows the transductive logic of sovra-communitarian collective individuation: 'technicity implies, on the contrary, that norms are not given, they have to be discovered' (PST 345).[9] Thus the separated logics of sacredness and technicity call into question two realities the status of which radically differ: we have in the first case an accomplished reality, the elements of which can be classified according to common and pre-constituted categories derived from the basic sacred/profane opposition, while in the second case we have a reality to be accomplished, not primarily dependent on the norms assuring the stability of the system, but rather on processes of normative invention.

Consequently, although the two reticular structures are isomorphic and subject to the same regressive risks (PST 324–27), and although in both cases the process of reticulation needs to be triggered (which in sacredness is perceived as a supernatural event, PST 340), the relative impact of invention in the two domains substantively differs. In sacredness invention does not go too far: because of the presupposition of a 'unique' relationship with the divine, any invention is in fact the positive 'counterpart' of 'a negative disposition towards other networks of sacredness' (PST 341). On the contrary, technicity is endowed with a power 'of actual ecumenism, displayed by the international exchanges between technicians and scientists, which saves it from the danger of reproducing the *uniqueness* of the categories of sacredness'

---

[9] In this sense an ethical decision compatible with the status of technicity depends on both culture as the milieu which mediates different norms, and the immediate decision according to an 'already given intellectual schema or vital attitude' (NC 506–7) (Sect. 8.3). The two limit cases can be formalised as two different relations between values and decision displayed by Dumont 1983: 290–98.

(PST 345). Ultimately, the reticulation of the 'technical sets' that has taken place at the level of the 'great human groups' has developed 'a ground of relativity and universality that the pre-technical universe lacked'. And this reticulation is endowed with an intrinsic political efficacy (PST 343, see also PST 234–36).

As Simondon had already clarified in MEOT, 'the structures of this reticulation become social and political' (MEOT 220). The technological structures, being 'more stable than the economical ones', entail 'a modification of what one might name the political constellation of the universe' and they reinstall 'key-points' through which 'social and political thought enter the world', in analogy with what has taken place with religion after the magic 'phase' (Sect. 10.1). It is according to this new reticular structure that social and political thought is compelled to measure its own efficacy: the original tendency of religion to 'present itself as absolute' is in fact conserved in social and political thought only on the condition of 'posing problems' at the scale of the technicity of global networks (MEOT 223). It is in this sense that Simondon speaks of a political 'effect' of the network of technicity [*réseau de technicité*]. Celebrated during the 1930s and early 1940s for their effects at the national scale, the radio networks soon exceeded the original identitarian function of manifesting power. After the second world war, this network function was supplemented by a new natural, technical and human 'polytechnic universe', which brought about the launch of missiles and satellites (true modern 'hierophanies') which did not point towards the past, i.e. a closed identity, but towards the 'open' group, i.e. humanity itself. According to Simondon, the manifestation of technology has thus become capable of reversing a given system of values (PST 333), since it

> Presupposes nothing, it refers neither to a previous tradition nor to a revelation: it is self-justifying, and becomes the most adequate symbol of a group discovering its dynamism and power of expansion. (PST 344)

A strong confidence in the spontaneous capacity of institutional self-regulation of the social organism crosses the whole of Simondon's oeuvre until *Trois Perspectives sur l'Ethique et la technique* [Three Perspectives on Ethics and Technics] (1983) where – studying the 'conditions for the establishment of a thorough technology' – Simondon basically confirms what he previously stated in MEOT, i.e. that the reactivation of systems and individuals carrying on technical schematism must be 'stabilised by institutions that can fix and continue them, by installing them' (MEOT 76).[10] But Simondon's dream eventually appears with the disquieting traits of a mysticism of technical invention, accompanying a disarming

[10] In this late essay Simondon displays some of his alternatives to nuclear development and the integration of alternative energetic sources. Furthermore, he refers to the loss of elders and their social function, in order to explain how 'closed social groups' are the outcome of an only apparently open and progressive society, in fact closed to the operative schemas of the past (TP 117). The active 'recovery' of those schemas is a political priority both for technical and social functions (TP 108 ff.). In fact, in Simondon's words, 'technical devices have a fundamental schema which can be at times untimely […] this schema can come back into existence, be reactivated and integrated into a new, more complex device. There is something eternal in a technical schema. And this is what is always present and can be conserved in things' (MECD 87).

confidence in its virtuous impact on the functioning of social systems, such as in the passages that bear witness to an esoteric endorsement of technicity in *Sauver l'objet technique* [Saving the Technical Object] (1983):

> If Adam and Eve had never left the Garden of Eden, they would have never become human persons or inventors, and their children would have never been a shepherd and a farmer from whom techniques came into existence. Finally techniques and transgression appear to me as the same thing. Long ago blacksmiths were considered cursed. (SOT 149–50)

This dream is sometimes converted into a technocratic approach, of which Simondon's article on the *Aspect psychologique du machinisme agricole* [Psychological Aspect of Agricultural Machinism] (1959) seems quite emblematic. There he displays the results of an '*action research*' devoted to the aim of 'establishing a "human engineering" as complete as possible' (APM 13): a research 'in the field' he carried out by himself for the possible solution of a problem of cultural integration between agriculture and industry through the introduction of 'transfer-machines' in the agricultural milieu in order to induce 'transcultural' effects.[11]

It is in fact through a 'thorough techno-*logy*' [*technologie approfondie*] that technological progress appears to Simondon to offer an unforeseen possibility for the regulation of contemporary society. Technical schemas would finally acquire, within contemporary technological networks, a normative force and a transductive power which makes of them the possible germs for a new modality of the regulation of the human being's cultural milieu, a new synthesis between culture and technics that Simondon sometimes identifies with the 'new magic' represented by cybernetics (MEOT 103). And yet despite these strikingly triumphal tones, Simondon repeatedly underlines that the hypothesis of a structural determinism of the technical structure cannot be assumed. The technical 'structure' can only determine the conditions of possibility of an efficient insertion of technology in the milieu of social groups. But the main condition of this 'insertion' is a partial renunciation to universality that would allow the different cultures to become compatible with the technical mentality developed within technical networks.

The mediator of this operation should be what Simondon calls 'reflexive thought'.[12] Now, what can this expression mean within the framework of a

---

[11] Simondon's proposed solution is the introduction of the tractor insofar as an 'indefinitely utilisable concrete open machine' (APM 16). The project of a 'human engineering' is always present in Simondon's research between 1958–1962, i.e. in the majority of his writings, plus his summary of the *Entretiens de Mysore* (1959) and his paper at the *Colloque de Royaumont* on the concept of information. See also his reference to the 'human engineering' developed by Myrdal (1944) (PST 132). See also Friedmann (1956), which inspired his criticism of Durkheim's optimistic views on the relationship between division of labour and the development of an organic solidarity. This book was in general for Simondon a major reference to the problem of alienated labour (PST 333, MT 350).

[12] It has not been possible to establish a straightforward connection between the philosophies of Simondon and Jean Hyppolite, who was his *directeur de thèse* for *Individuation*, although the concept of 'reflexive thought' seems to invite such an attempt in relation, for instance, to the latter's (Hegelian) overcoming of the (Kantian) opposition between reflection and being (Hyppolite 1953, Chap. 2). Yet perhaps a more meaningful link has been traced through the concept of 'prostheticity' in Chap. 10, n. 28.

philosophy in which the subject has no privileged place, let alone consciousness? The answer resides in Simondon's theory of symbolic function, and in particular in the concept of 'modes of thought' he displayed in the third part of MEOT, which makes the distinction between thinking and acting ineffective (Sect. 10.2). 'Reflexion' thus defines all the concrete activities (among which also intellectual activities are included, of course) emerging out of the mixed milieu of technical and symbolic objects (which Simondon often calls 'culture') and affecting the functioning itself of the social systems from which they have originated. In short, any transindividual activity, thought included, is 'reflexive' insofar as it can change the cultural milieu from which it emerges.

It is in this precise sense that 'reflexive thought' can be said to operate *politically* on social systems precisely when it makes the second phase-shift of technics and religion converge into a new, 'third' 'cultural reality' (MEOT 217). This is the aim of what in the last part of MEOT, devoted to *Pensée technique et pensée philosophique* [Technical Thought and Philosophical Thought] Simondon calls 'philosophical thought'. With this expression he designates the activity that continues and revives at the level of the second phase-shift the same function of the 'invention of compatibility' formerly accomplished by aesthetics (MEOT 216).[13] Only philosophical thought finally allows for the tendency towards totalisation proper to religious thought to be transformed into a 'plurality of political-social insertions' that Simondon explicitly defines as 'ecumenical':

> It would have been difficult to build ecumenism in the past, since it is not possible to build it out of a reflexive thought concerned with the foundation of a culture. This is essentially a philosophical enterprise [...] Until today, only limited ecumenisms have come into existence (such as within Christianity), but philosophical reflection has to develop a universal ecumenism in order to integrate religious reality into culture. (MEOT 232)[14]

Simondon's final hypothesis in MEOT is that the institution of 'a techno-*logy* [*une technologie*]' – i.e. a *logos* of technics – coincides with the institution of 'ecumenism' (MEOT 232).[15] The task of philosophy is therefore to follow technics in its development from 'primitive technicity' into the technology of networks. Philosophical thought thus reveals its double face, ecumenical and technological – depending on whether it is considered in relation to the development of the

---

[13] In other writings Simondon admits that 'technical culture' and 'technical taste' are both preconditions of the integration of technicity into culture (NC 520–22). This function is shared by the technical and the artistic object, with the difference that the latter is in general 'accepted only if it reflects an already existing vital dynamism' (NC 515), while the technical object carries on a normativity essentially antagonistic to the communitarian one, since it 'modifies the code of values of a closed society' (NC 513), at least as long as it is not 'captured' by communitarian symbolism.

[14] It is interesting to note that Simondon apparently shapes his thesis *against* Bergson *through* a Bergsonian argument: 'it cannot be granted that open religions actually exist, nor that the opposition between closed and open religions is as sharp as Bergson claims; but the opening is a function common to different religions, each of them being also partially closed' (MEOT 232).

[15] 'The ordinary meaning of the word "techno-*logy*" refers to modern technics in so far as it would be the application of the logos to science. Simondon reinterprets this word as the study (logos) of technics' (Barthélémy 2012: 229).

religious or technical phases –, but also its peculiar function of convergence towards the establishment of a culture capable of making the two tendencies compatible. This finally explains the philosophical finality of his book *Du mode d'existence des objets techniques* [MEOT] concerned with the incorporation of technics into culture, through the same pedagogical effort to which the entire oeuvre of Simondon seems sometimes to commit its own justification: the institution of a technical culture to proliferate the technical mentality.

What is still far from being understood is how institutions would organise such pedagogy out of the ecumenically spontaneous tendency of technological development and the innate power of 'reflexive thought' assumed as paradigmatic of the political tasks of philosophy. It is clear that such a project can make us doubt that Simondon's philosophy, only partially aware of its own Eurocentric stance, presents more than a mark of a humanism-mediated technocracy. The point is that Simondon is strongly affected by a strange mix of the Durkheimian conception of institutionalised culture as the motor of progress and by the Bergsonian faith in the power of the *élan vital*. But a political pedagogy based on the cultural integration of the biological schemas of technicity cannot be esteemed as the last word of his philosophical enterprise.

It is true that, despite all his terminological innovations, the core of Simondon's philosophy is not a plain break from the philosophical and sociological tradition it emerges from. Since his earliest writing in 1953–54, Simondon aimed at a 'constituting enterprise [...] founded on sociology' (PI 117) which should allow for the construction of a 'thorough technology connected to the history of thought and to social consciousness' (PI 120); and in doing this he did not fail to refer to Comte's theorisation of the basic value of 'technical understanding' as far as it 'contains germs of necessary positivity' (PI 119). This heritage cannot be overlooked, because it is the actual ground on which his epistemology and ontology emerge with a shape that marks the originality of his philosophy. All his attempts at reforming the concepts of information, society, technics, and the human being, indeed consumed the conceptual framework he inherited from the inside. As I will explain in Chap. 12, this allowed him to give a new shape to the political significance of 'philosophical thought'.[16]

Yet firstly, in order to display Simondon's way out of the apparently suffocating alternative between Bergson and Durkheim, I shall challenge one of the most 'untimely' concepts of his philosophy, an expression clearly compromised with an Eurocentric, technocratic and colonialist stance, in order to show how Simondon's philosophy of individuation allows for its undermining from within: the concept of 'human progress'.

[16] What I imply here is that there is much more for political thought in Simondon's epistemology and ontology than we could imagine on the basis of his more strictly political claims. I owe a debt here: 'we will leave to others the task to evaluate the value and success of Simondon's pedagogical reform. What is worth noting is that this cultural perspective does not allow him to develop the problem his oeuvre nevertheless poses' (Aspe and Combes 2004). For a critical interpretation of the limitations implicit in Simondon's 'political' thought and the opportunities it offers, see Stiegler (2006a, b). Mine will be displayed in Chap. 12.

11.3 Technical Evolution and *Les limites du progrès humain*

With *Étude critique*: *Les limites du progrès humain* [The Limits of Human Progress: A critical Study] (1959) Simondon intends to respond to an essay published one year in advance in the same Journal, the *Revue de métaphysique e de morale*, by Raymond Ruyer on *Les limites du progrès humain* (1958). Assuming that the concept of organisation extends from matter to life, Ruyer also challenged the distinction between life and technics, refusing any 'romantic' differentiation between 'a natural living community and a mechanised society': human being itself – he wrote – is a 'hybrid' of life and technics (Ruyer 1958: 413–15). Despite these premises that seem highly 'compatible' with Simondon's stance, in what follows Ruyer provided the essential lines of a philosophy of history based on a clearly Bergsonian trust in the inventive power of life. What is more interesting for my purpose in the present section is that Ruyer theorised there an 'end of history' (414): his formula for this widely discussed philosophical problem (or philosophical 'myth') will allow me to display the peculiar stance Simondon adopts in response to Ruyer's essay.

In his essay Ruyer assumes that we are in the brief transition between a past 'ethnological phase' and a future 'phase of civilisation and rational administration' (414). The premise of the whole argument is that we are now living the 'most formidable evolutive explosion that ever took place on earth', and after this 'explosion' 'it is rigorously certain that the accelerated march of technological progress will slowdown', until the final inertia of the system will be reached 'once the industrial system will be one and the same with the social system'(416–22). The schema is three phased – acceleration-slowdown-stabilisation – and represented by a 'sigmoid curve' at the end of which 'the organic life of cultures [...] will regain importance', i.e. biology will eventually dominate the functioning of human societies (422). But this will take place at a different level thanks to technological improvements and will result in what Ruyer draws as a kind of utopian life almost entirely devoted to '*jeux divers*' (such as 'cinema, radio and television') offering 'psychic nourishment' to a childish humanity strongly 'independent from its milieu':

> A civilisation at a high technical level gives more possibilities of relaxation and more money in the pocket, and therefore more possibilities of life in the true sense, unselfish. An industrial progress in development always represents a hard and brutal period [...] once the technical structure has stabilised, however, life can again resume its diversions and phantasies. (Ruyer 1958: 423)

It is precisely against Ruyer's utopianism that in his essay Simondon denies the absolute function of the technical phase and advances his definition of 'human progress'. According to Simondon 'human progress' is a tendency towards universality that results from the durable overlapping of different domains or 'phases': language, religion and technicity (LPH 269).[17] This general tendency

[17] Simondon hardly resists the temptation to provide his own 'grand narrative'. Each phase of 'human progress' would follow the same pattern: growth, saturation, hypertrophy of automatism, and the opening of a new modality of concretisation.

defines a series of processes the progressive continuity of which displays a more and more complex systemic integration of 'what the human being *produces* and what it *is*' (LPH 268). Yet the human tendency towards universalisation is not equally distributed on all phases: it is more strongly connected to the technical phase, the 'primitivism and materiality' of which is a 'condition of universality' (LPH 271). Simondon's remark does not refer to the first steps of the process of hominisation here, but rather to the unprecedented opportunity technology represents today. In fact, the two aspects are strictly connected: it is precisely because it is rooted in some fundamental human needs that technics, developed as technology, can be extended – at least in principle – to the whole of humanity:

> Religion, in effect, concerns a more primitive reality, less localised, somehow more natural for human beings than that to which language addresses itself [...] Technics is even more primitive than religion: it connects with the elaboration and satisfaction of biological desires themselves. It can therefore intervene as a link between the people of different groups or between people and the world [...] The impression of a relapse into primitivism, into vulgarity, which we feel at the passage from religion to technology, the Ancients felt watching the most perfect monuments of language abandoned in favour of a religious upsurge which they judged vulgar, destructive and filled with the seeds of barbarism. Yet this step-by-step descent towards primitivism and materiality is the condition of universality: a language is perfect when it is congruent with the *polis* that is reflected in it; a religion is perfect when it achieves the dimensions of a continent whose diverse ethnicities are at the same level of civilisation. Technics alone is absolutely universalisable, because that part of the human being that resonates with it is so primitive, so close to the conditions for life, that every human possesses it in him/herself. (LPH 271–72)

Simondon thus welcomes the hypothesis of a tendency of progressive universalisation of biological normativity through civilisation, assuming a clear stance in relation to the role played by technics along this process. A previous clarification of what he names 'technical evolution' is therefore needed in order to understand how he conceives 'human progress'. The schema Simondon applies to all evolutionary processes, biological, technical and 'human', is unique: it is a tendency made of breaks and provisional compatibilities, where, although not necessary, once the overcoming of a threshold has taken place, it becomes irreversible and entails systemic consequences. To this general schema Simondon gives different configurations and quite different names during the convoluted path of his research. Better than the concept of transduction, the concept of 'relaxation' seems to be formulated precisely to explain technical evolution.

In MEOT, in the paragraph *Enchainements évolutifs et conservation de la technicité. Loi de relaxation* [Evolutive Chains and Technicity Conservation. The Law of Relaxation] Simondon illustrates the 'seesaw' [*en dents de scie*] process through which technical evolution takes place, according to a 'rhythm of relaxation' which 'finds no equal in the geographical nor in the human world', since it is the proper 'technical temporality' (MEOT 66–67).[18] The 'law of relaxation' is a paradoxically cyclical evolution in which we have 'the conservation of technicity as information

[18] The macroscopic technical model is an underground spring cyclically emitting water displayed by Ruyer (1954) and schematically drawn by Simondon in MEC 140.

through subsequent cycles'. And yet this 'law of conservation of technicity through the sequence of elements, individuals and sets' conveys a 'schema of relaxation' that compels the social system to open (PR 265–66). Hence in Simondon's writings the schema tends to extend beyond the original boundaries. In *Individuation* the 'law of relaxation' is the way in which the quantic passes into the biological, and therefore it can explain at the same time physical and physiological processes, suddenly amorced after a regular quantitative growth that did not manifest evident effects (I 204). In a similar fashion in *Le relais amplificateur* [The Amplifying Relay] (1976) Simondon recurs to the concept of 'relaxation' to explain a series of mechanical and electrical phenomena, not omitting to question whether it would be the case to 'push the research of models further' attempting to interpret the metabolic phenomena of growth as processes of amplification conceived according to the same model (MEC 139).

Simondon often declares he derives this model from technical reality, but the original field seems rather to be biological, at least if we recall Bergson's statement, according to which life is essentially 'everywhere the same, a slow accumulation of potential energy to be spent suddenly in free action' (Bergson 1932: 271). And this is clearly the same 'model of any vital process' (I 209) Simondon explicitly derived from Gesell and possibly from Piaget, and extended in *Individuation* to the physical-chemical and biological fields.[19] Thus for Simondon 'life deploys itself through transfer and neoteny: evolution is rather a transduction than a continuous or dialectical process' (I 171).[20] In short, through his systemic approach, Simondon elaborates

[19] On Simondon's debt towards Gesell (1946), see Sect. 5.1. On the debt towards Piaget see Petit (2010) and the way it is highlighted in Barthélémy's simondonian 'dictionary': 'Like Jean Piaget before him, Simondon uses this term [transduction], which is at the same time technological and biological, in order to give it a new meaning, one that will become absolutely central in the thought of individuation' (De Boever et al. 2012: 230). In effect, Piaget's description of the passage of the child from the egocentric pre-operational to the operational stages might be the model of the process of individuation as a 'transduction' itself (Sect. 1.3). As Guchet explains, Simondon's analysis of 'technical evolution' is strictly connected to Leroi-Gourhan's description of the process of objectivation that makes the human being-nature relationship less and less anthropocentric (Guchet 2008: 23). I will just add that in Leroi-Gourhan's view 'technical evolution' seems to acquire a meaning at the scale of 'geological evolution', as Deleuze and Guattari did not fail to notice in the third chapter of their *Mille Plateaux* (1980), which widely draws from Leroi-Gourhan and Simondon.

[20] It is worth recalling here the brief text *Pour une notion de situation dialectique*: some posthumously published working-notes originally drafted – as Carrozzini explains – in view of the *Colloque de Royaumont* (18–23 September 1960) *La dialectique* to which Simondon eventually did not take part (Carrozzini 2005: 107). As it often happens, Simondon seems to attempt a 'reform' of the concept arguing that 'dialectics only exists in the form of a situation' (SD 114). In this sense from this truly 'ontological-political' text one can derive that there is no actual historical process that is not 'dialectical' in the sense of a discovery-invention of a new compatibility. The same strategy is also carried on in *Individuation*, where the term 'dialectics' is opposed to 'transduction' (I 111) and 'phase' (I 322–23), but at the same time 'redeemed' by conceiving 'dialectical stages' as 'phases of being' (I 323). As Guchet noted, despite Simondon moves from an explicit denial of dialectics to the dialectical drawing of 'technical evolution', 'an exam of the texts demonstrates that the break is less deep than it appeared to be' (Guchet 2005: 251).

a quite discontinuous conception of evolutionary processes, which entails a similar conception of technical evolution, a model possibly derived from Leroi-Gourhan, according to whom 'there is room for a real "biology" of technics', precisely because

> Analysis of techniques shows that their behaviour over time resembles that of living species, as though driven by an apparently inherent evolutionary force that places them outside human control. (Leroi-Gourhan 1964: 206–207)

Along the same path, in MEOT Simondon looked for a 'more primitive genetic schema' which, overcoming the dichotomy between *élan vital* and adaptation, could integrate technical and biological evolution in one and the same model of 'subsequent stages of individuating structuration, going from a metastable state to another through subsequent inventions of structures' (MEOT 155–56). This schema becomes in fact the ground for Simondon's conception of 'human progress' since it takes into account the incidence of the cultural and technological progress over social systems. 'Technical evolution' is crucial to progress precisely because it provides the emergence of a technical milieu functioning according to a regime of information exchange which retroacts on the biological 'phase' always present in social groups. This determines the emergence of culture: the (symbolic) milieu that provides a 'compatibilisation' of biological and technical normativities (Sect. 8.3). Simondon understands 'technical evolution' as a cumulative propagation in which homeostatic and continual processes are as crucial as the aleatory emergence of invention, and therefore entail the impossibility of a definitive synthesis. This is also the schema of what he calls 'human progress'.

As far as Ruyer is concerned, he had already presented his own formula a few years before in *La cybernétique et l'origine de l'information* (1954), where he displayed his 'metaphysics' of information ultimately based on the 'absolute overview' of consciousness (Sect. 2.3). The required 'bigger soul [*supplément d'âme*]' was pictured there as the result of the previous establishment of a 'bigger brain [*supplément de cerveau*]' (i.e. a virtuous relationship between 'human brain *plus* information machines') as a condition and outset of wisdom (Ruyer 1954: 18–19).[21] In this light the astonishing picture Ruyer provided a few years later in his *Les limites du progrès humain* of a civilisation finally achieved – in which, after an 'hard and brutal period', 'life can again resume its diversions and phantasies' (Ruyer 1958: 423) – reveals his profound trust in the inner force of cultures as something that remains intact, even latently, ready to sprout, and immune from the destructive influence of technological development. In effect, although admitting the over-communitarian tendency of technology, Ruyer optimistically conceives it as a continuation of the same evolutive path life had began far before the appearance of human society and will continue far beyond the final structuration of a new 'indefinite phase' of humanity: a 'post-historical', neither 'mineral' nor 'rational' phase that 'will be as much organic as the ethnographic phase' (423).

[21] This would have occurred thanks to the accomplishment of the process of automation and the consequent liberation of labour prospected by Georges Friedmann (1946).

The naiveté of Ruyer's perspective culminates in his rhetorical question concerning why 'the variety and organic life of cultures' should ever be menaced by contemporary 'technological unity', given that the 'technological unity at the level of the chipped stone' did not at all prevent 'the variety of primitive cultures that are today dying out in front of us' (415–16). In fact, I argue that what makes Ruyer oversee the actual catastrophic risk run by the diversity of cultures as they face the planetary deployment of the capitalistic mode of production, is precisely his Bergsonian 'quasi-mystical' trust in the all-encompassing and irreducible inventive power of the *élan vital*. According to Bergson, the long term process we are concerned with calls for an adequate effort, and this 'destiny' can be accomplished only if posed by a thought and implemented by a policy relying on the vital *élan* which generated our species: an 'extra effort required [for humankind] to fulfil, even on a refractory planet, the essential function of the universe, which is a machine for the making of gods' (Bergson 1932: 330). The whole process relies, once again, on the fundamental faith in the fact that technology is basically an outcome of life, and therefore there can be no obstacle in it 'which cannot be broken down by wills sufficiently keyed up' (312–313).

Partially following Ruyer, Simondon draws from Bergson's question on technology the same need for 'a bigger soul'.[22] And, still following Ruyer, he also concludes that the progressive tendency 'implicit' in technicity will not be a solution if it does not become 'of an organic type, and part of the specific evolution of human beings' (LPH 274). And nevertheless, Simondon is in fact posing the problem from a completely different perspective here, far from both Bergson's and Ruyer's. Although in Simondon's text 'human being' [*l'homme*] is the grammatical subject of 'human progress', this is not to be intended as an indefinite aspiration grounded on a kind of invariant human nature. Also when he speaks of 'human thought' as a specific issue of homo sapiens, he does not simply refer to the brain or, worse, to a metaphysical 'faculty', he refers to 'thought':

> The questions of the limits of human progress cannot be posed without also posing the question of the limits of thought, because it is thought that appears as the principal repository of evolutionary potential in the human species. (LPH 275)

Again, 'thought' is to be intended here as a 'reflexive' activity, part of the cultural milieu, and therefore provided with a power of retroaction on the social system it is originated from. Thus the 'evolutionary potential' of the human species grows in this milieu by developing it at the same time. Indeed, according to Simondon there

[22] Simondon's question on technology ('Will technology become industry as language became grammar and religion theology?' LPH 271) clearly relies on Bergson's understanding of history as a risk: 'We do not believe in the fatality of history' (Bergson 1932: 312). And yet Bergson's mystical optimism often seems to exorcise the risk: 'Let us not merely say, as we did above, that the mystical summons up the mechanical. We must add that the body, now larger, calls for a bigger soul, and that mechanicism should mean mysticism. The origins of the process of mechanization are indeed more mystical than we might imagine. Machinery will find its true vocation again, it will render services in proportion to its power, only if mankind, which it has bowed still lower to the earth, can succeed, through it, in standing erect and looking heavenwards' (Bergson 1932: 330–31).

is human progress '*only if*, when passing from one self-limiting cycle to the next, the human being increases the part of itself which is involved in the system it forms with the objective concretisation' (LPH 270). Therefore technical evolution cannot be simply reduced to an aspect of the general evolution of our species, because the shift its explosion triggered has made human social systems irreversibly cross the threshold of the biological, and enter the accelerated field of transindividual individuation and its 'technical loop'.

In this sense Simondon could be said to be a kind of 'Lamarckian' at the epoch of the triumph of the Modern Darwinian Synthesis. In effect, his philosophy explains the impact of the technological milieu on human evolution in a way very similar to the active role played by the milieu on embryogenesis: something completely ignored by Neo-Darwinism. Indeed, Simondon seems sometimes to play Lamarck against a deterministic reading of Darwin, in order to oppose a discontinuist theory better compatible with his general model of ontogenesis: 'it is one of Lamarck's major merits to have considered evolution an incorporation of aleatory effects coming from the milieu and going into the individual' (I 213, n. 23).[23]

In effect, it would be difficult to deny today that – once a certain irreversible threshold has been crossed – technological development has become crucial also at the evolutionary level, and this is the level at which Simondon is looking for a possible answer to the Bergsonian question. For Simondon 'human progress' takes place, out of any contraposition of the social system and the individual, and, most notably, out of any contraposition of humans to other 'beings', since in a philosophy concerned with processes of individuation there are no predetermined boundaries prescribing what 'human' means. And first of all, posing the Bergsonian question within the domain of the transindividual, means for Simondon to pose it out of any faith in the unexhausted *élan vital* a 'heroic' individual might 'mystically' embody. In fact the domain of the transindividual is incompatible with Bergson's metaphysics of life as much as with Ruyer's neo-finalistic metaphysics of information.

11.4 The Transindividual Historicity of Cultures

Simondon's conjoint refusal of mechanicism and teleology on the ground of a fundamental duality of tendencies might be assimilated – at least in principle – to Bergson's, but it clearly does not flow from the same source. For Simondon technics actually raises high hopes because the normativity it carries offers an unprecedented

[23] Simondon's stance should be probably considered still quite far from Gould's theory of 'punctuated equilibrium' (on this topic see LaMarre 2013: 101 ff.), but it is certainly close to Gould's claim – against the 'ultra-Darwinian' Dennet (1995) – for the fecundity of the Lamarckian paradigm for the explanation of cultural change: 'Human cultural change operates fundamentally in the Lamarckian mode, while genetic evolution remains firmly Darwinian. Lamarckian processes are so labile, so directional, and so rapid that they overwhelm Darwinian rates of change. Since Lamarckian and Darwinian systems work so differently, cultural change will receive only limited (and metaphorical) illumination from Darwinism' (Gould 1997: 52).

evolutionary opportunity for universalisation. This hope calls for a political response that positivism erroneously committed to technocracy and that, from Simondon's perspective, cannot be answered by Bergson's metaphysics of life. As already demonstrated, the heroic stance that Canguilhem also derived from Bergson deeply contrasts Simondon's epistemology (Sects. 7.2 and 8.4). The latter transfers the Bergsonian opposition onto a different level: situating the source of all processes of collective individuation not in the *élan vital* of the individual creator, but rather in the 'field' of the transindividual. The whole theoretical operation can be easily grasped through the lens of the concept of historicity.

In effect, the form assumed at the level of transindividual individuation by 'transductive' processes, so crucial to Simondon's philosophy, brings about his peculiar use of the concept of 'historicity'. What is most interesting for present purposes is to discern the link between the concepts of 'singularity' and 'historicity' which traverses the whole text of *Individuation* where processes are concerned, triggered as they are by singularities which are 'historical and local' (I 81). The coupling of the terms 'singularity' and 'historicity' functions at all levels (or 'regimes') of individuation, from human societies to matter, i.e. all the fields of 'being' Simondon defines: at a physical level there are 'historical singularities brought about [*apportées*] by matter' (I 57); in crystallisation 'there is therefore a historical issue in the occurrence of a structure in a substance: the structural germ has to appear' (I 79), and 'the individuation of an allotropic form starts from a singularity of historical nature' (I 80); at the biological level 'the individualisation of the living being is its real historicity' (I 268); at the psychic level 'we believe any thought, precisely as far as it is real, is a *relation*, i.e. it entails a historical aspect in its genesis' (I 84).[24]

In brief, processes of individuation are always 'historical' as far as transduction, which operates in a singular and progressive mode, is involved:

> We understand by transduction an operation – physical, biological, mental, social – through which an activity propagates gradually within a domain, by founding this propagation on a structuration of the domain that is realized from one place to the next: each region of the

[24] Given the importance of the concept of 'singularity' in Deleuze's philosophy and the way he draws on Simondon's terms since his review to Simondon's IGPB (Deleuze 1966), a clarification is needed concerning the difference in usage. For both philosophers the concept of 'singularity' points to a discontinuity concerning processes, but they conceive the relation between 'singularity' and 'individual' differently. Deleuze situates the individual on a different scale (molar) in relation to the pre-individual regime of singularity-events (molecular) his 'transcendental-empiricism' is concerned with. On the contrary, for Simondon's philosophy of individuation, the individual is to be understood as a part of a discontinuous process *without reducing it* to a kind of epiphenomenon of molecular features. Indeed, Simondon's use of the terms 'singular' or 'singularity' is a very restricted one, which refers to a structured individual *when* it is the result or the trigger of a process of individuation. Transductive processes are therefore aleatory precisely due to this 'historical' aspect in their genesis. In this sense Toscano's 'idea of the individual as a "theatre" rather than an "agent" of individuation' (Toscano 2006: 150) probably fits Deleuze's philosophy better than Simondon's: according to the latter the individual *as a system* is both a 'theatre' and an 'agent' of individuation such as 'the living is both the agent and the theatre of individuation' (I 29) (see above Chap. 1).

> constituted structure serves as the principle for the constitution of the following region, in such a way that a modification is thus progressively extended at the same time as this structuring operation. (I 32)

In this sense, in the domain of psychic and collective individuation, only the 'true' collective is historical, and it is such only when it operates transductively, i.e. as transindividual. Therefore – although he does not depart from speaking of biological and technical 'evolution' or progress – it would be an error to connect Simondon's concepts of 'historicity' and of 'evolution' to either vitalistic or mechanistic patterns of development. Thanks to epistemological enquiry, his philosophy of individuation overcomes the postulate of a 'natural society' implicit in both Bergson's metaphysics of 'life' and in Durkheim's positivistic sociology, that in fact imply the same optimistic faith in institutions and technology inspiring their philosophy of history.

It is true that Simondon is always strongly optimistic concerning the human capability of organising the process of cultural transformation that results from technical progress. The main problem presented by the process of technical evolution is due, according to Simondon, to the fact that, since technics cannot *in itself* provide the social bond, it must necessarily rely on the homeostatic processes of closure that are exerted on its open functioning. Now, this is only possible if such processes do not reach the point of shutting down the social system. This is true at any scale, local, national, or global, but once technics has developed as technology on a planetary scale, an unprecedented event takes place: 'technical networks [...] enlarge the dimensions of their mesh, and interfere with the order of magnitude of national or continental groups'. This change in the capacity of technicity is first of all an actual problem for the values of social groups the scale of which still corresponds to artisanal techniques. But this is also the opening of a completely new scenario for a possible ecumenical culture corresponding to the size of societies that previously would have been unimaginable in dimension and configuration; they respond neither to territorial nor class-based logics (PST 343–45).

Simondon seems thus to point to the same universalising perspective that Merleau-Ponty displayed in his course on *L'institution* (1954–55). Although explicitly distancing himself from Bergson's *Deux sources*, Merleau-Ponty in fact adopted the same stance when asking institutions to provide an 'unlimited historical effort' pointing 'to the *Miteinander* [one with the other] or to the *Füreinander* [one for the other], the universal embrace'. In a way, posing the political problem at the level of humanity, Merleau-Ponty assumed the risk of supplying a myth of a powerful and ineluctable progress that celebrated the opening of 'true' historical societies, the only ones 'faithful to the spirit of institution' and therefore capable of 'playing the mysterious game which consists in taking all humanity into account' (Merleau-Ponty 1954–55: 122).

The whole philosophical and ideological discourse concerning the connection between technical normativity and human progress is grounded in the implicit identification of industrial civilisation and historicity. As Bergson had taught, 'civilised man differs, above all, from primitive man by the enormous mass of knowledge and habits which he has absorbed' from a rich 'social milieu' (Bergson 1932: 24). Although it is worth refuting the ethnocentrism that is today quite clear in such a formulation, it is also necessary to assume the *entire* problematic concerning the inner historicity of social systems, in fact hidden by the distinction between 'cold' and 'warm' societies.[25] On the one hand we witness, of course, the massive epochal destruction of all the different historicities of 'cold' societies (allegedly 'without history') under the expansion of 'warm' industrial civilisations. It also is worth highlighting that, on the other hand, and at a different level, we witness industrial societies *becoming* without history. In a certain sense the entire problem points to a global disappearance of historicity in a kind of globalised advanced industrial society, the model of which seems to be a mechanical pendulum 'without memory'.[26]

Also in Simondon's philosophy of individuation it is only at this scale, the same scale chosen by Durkheim and Bergson, that politics can fit a humanity actually facing the problem of its own survival. And such a politics cannot escape taking care of the 'regulation' of the evolutive tendency carried on by technology and of its effects on the natural and social milieu. But if this were all we can say of Simondon's political philosophy, we would be confronting yet another version of the modern faith in progress, slightly tempered by the acknowledged necessity of its political governance, and possibly touched by the ineluctable disappearance of some 'exotic' cultures. In fact, to the Eurocentric stance implied by the concept of 'human progress' two crucial remarks have been added by Simondon's philosophy of individuation and of technics. Firstly, although human groups play a decisive role in triggering the development of technology, 'human progress' is a process of complex systems made of human, animal, symbolic and mechanical operations which all shape the composite milieu in which social systems evolve. Secondly, since all processes are singular and historical, no metaphysics of nature or of history can support the concept of progress: it must be understood out of both a mechanistic and an eschatological perspective.

[25] With these expressions Lévi-Strauss differentiates 'cold societies' 'the internal climate of which is close to the zero degree of historical temperature' (i.e. with low and constant number of components and mechanical functioning) from 'warm societies, which appeared in different places on earth following the Neolithic revolution' (i.e. with a growing number of components and ever-expanding functions) (Lévi-Strauss 1962: 309–310; see also Lévi-Strauss 1960). As highlighted by Clastres (1974), ethnological ethnocentrism prevalently focuses on the distinction between a-historical ('primitive') societies and societies the historicity of which is primarily linked to the process of industrialisation and therefore to the concept of progress. To this criticism, however, it would be perhaps interesting to add Levi-Strauss's magical-religious characterisation of 'savage thought' as inherently related to the affective force of 'pure historicity' and yet not affected with the concern for continuity typical of 'domesticated thought': savage thought is essentially 'discontinuous and analogical' (Lévi-Strauss 1962: 320–21, 348–49).

[26] On the image of society as a pendulum 'endowed with memory', which Canguilhem (1955) derives from Bergson's *Deux sources*, see Sect. 7.3.

The concept of transindividual individuation thus makes any attempt to challenge the relation between bio-technical evolution and historical-cultural universalisation irremediably partial, inasmuch as both perspectives are insufficient to grasp the complexity of the system they are part of. This is why both a philosophy of the living and a philosophy of culture are epistemologically limited when they attempt to grasp the complex process of individuation called 'human progress'. The effects of Simondon's shift is patently displayed by Merleau-Ponty's resistance to the cancellation of the nature/history dichotomy:

> What there is in common between history and nature is that they are individuations – but they are irreducible *precisely for that reason* – Historical individuation is irreducible –. Simondon conceives of it as a borrowing from the pre-individual – from the same pre-individual whence comes the physical or living individual – Is it *the same*? Must it not be that the being to which the collective or even the psychic opens up be other than that from which physical individuals come? (Merleau-Ponty 1959b: 42)

According to my reading, the absorption of 'nature' – as the pre-individual –, and of 'history' – as the transindividual – in the dynamics of social systems, is the mark of Simondon's abandonment of that very dichotomy. In fact for Simondon 'pre-individual' and 'transindividual' name the simultaneity of phases the relation of which is partially regulated in the psychic and collective regime of individuation. This 'pre-individual' nature is neither undifferentiated nor, on the contrary, entirely determined. In transindividual individuation the 'physical-biological' pre-individual potential (and not simply 'vital') neither remains as is nor is it completely integrated. Once 'enveloped' by significations, the pre-individual cannot be exhausted: it enters an evolutive tendency that is neither natural nor historical, and is not even exclusively technological. Seen through the concept of the transindividual, the whole system of subsequent symbolic individuations of the original biology and technicity of homo sapiens appears the evolutive tendency Simondon calls 'human progress'. This is why all questioning of technical evolution must necessarily connect it to the process of cultural universalisation out of any possible explicit or hidden philosophy of history.

In conclusion, it can be said that to this set of problems Simondon's writings display a twofold approach. On the one hand his oeuvre patently affirms the necessity of cultivating the 'myth' of technical evolution, conceived as a transmission of 'technical schemas' with archetypal functions that would breed a new social bond at a global level. From MEOT up until *Culture et technique*, Simondon does not abandon the belief that to the structural risk of technological evolution it is sufficient to oppose the pedagogical-political effective institutionalisation of philosophical thought. He also claims that the cybernetic aspiration to a new encyclopaedism represents the political perspective proper to twentieth century: the connection of technology and 'technical culture' at the scale of 'technical systems' (MEOT 116–17).

On the other hand, the concept of transindividual individuation makes any possible institutional theory of the regulation of human progress insufficient. This entails that the solution to the problem that emerges when the notion of the transindividual is introduced *cannot be simply administrative, but must be political.*[27] Which also means that it cannot be planned from the standpoint of a singular Culture, but rather arises from the intertwining of the main trends in technological evolution with the different paths to universalisation displayed by heterogeneous cultures. It is in this sense that the concept of 'progress' is to be reformulated in order to include the repeated invention of new compatibilities of community and society:

> Progress is a kind of development that integrates in a whole the sense of a sequence of discoveries and of the stable unity of a community. It is through the mediation of technical progress that community and society can become synergic. (NC 514–15)

It should be clear how this perspective radically questions the nature of 'technological evolution' itself, which cannot be directly received within a culture as the supposed necessity of technological development, because as such it would result in the dominion of a singular culture (namely Western culture). What Simondon tries to imagine is a process of universalisation that would not be simply reduced to a process of cultural colonisation. As Simondon explains in *Culture and Technics*, society necessarily tends to evolve, both historically and technically, but this very process changes at the same time its scale and its nature:

> [Animal] functions are internalisations or incorporations of physical effects that had been, more or less fortuitously, achieved by the external milieu, incorporations corresponding to needs and stabilised by the appearance of progressively differentiated organs. Now, thanks to the technical gesture human evolution takes place along the same functional line. A certain physical effect is incorporated into what amounts to the internal milieu of the human group, this effect becomes available and reproducible through the deployment of a technical dispositive, and that availability is equivalent to the incorporation of the effect into the collective organism: it is a supplementary function. Everything takes place as if the corporeal schema of the human species had been modified, dilated, as if it had received new dimensions: the order of magnitude changes, the perceptual grid is broadened and differentiated, and new cognitive schemas are developed, as when a child leaves his village and takes stock of his country's extent. It is not a matter of *conquest*: that notion is the fruit of a closed culture. It is a matter of *incorporation*, which, on the collective level, is functionally equivalent to the appearance of a new species. (CT 12)

If this process of 'incorporation' entails neither adaptation nor, on the other hand, technological conquest, it cannot be conceived as the progressive absorption of all cultural heterogeneity by a pre-constituted 'western' subject. In fact, it triggers new transindividual individuations the result of which is the emergence of a new 'culture' fitting the state of technological and historical 'evolution', i.e. the singular historicity of what Simondon calls 'human progress'. This view opens up an interpretation of Simondon's political thought that leads far beyond his explicit pedagogical-political claims, without abandoning the very perspective that his philosophy of individuation and technics contributed to open.

---

[27] This is the problem I will deal with in Sects. 12.3 and 12.4.

References[28]

Aspe, B., and Combes, M. 2004. L'acte fou. *Multitudes. Politiques de l' individuation. Penser avec Simondon.* 2004/18. http://www.multitudes.net/category/l-edition-papier-en-ligne/Multitudes-18-Automne-2004/. Accessed 30 Dec 2014.

Barthélémy, J.-H. 2012. Fifty key terms in the works of Gilbert Simondon. In *De Boever* 2012, 203–231 (French edition: Glossaire Simondon. Les 50 grandes entrées dans l'œuvre. *Cahiers Simondon* 5: 105–137).

Bergson, H. 1932. *Les deux sources de la morale et de la religion.* Paris: Alcan (English trans. *The Two Sources of Morality and Religion.* London: Macmillan & Co., 1935).

Canguilhem, G. 1955. Le problème des régulations dans l'organisme et dans la société. In *Ecrits sur la médecine*, 101–125. Paris: Seuil, 2002 (First edition: *Les Cahiers de l'Alliance Israélite Universelle* 92: 64–81).

Carrozzini, G. 2005. Nota a 'Per una nozione di situazione dialettica'. *Il Protagora* 33(5): 105–109.

Carrozzini, G. 2011. *Gilbert Simondon filosofo della Mentalité technique.* Milano: Mimesis.

Chateau, J.-Y. 1994. Technophobie et optimisme technologique modernes et contemporains, suivi de La question de l'évolution de la technique. In *Gilbert Simondon, Une pensée de l'individuation et de la technique*, Bibliothèque du Collège international de philosophie, 115–172. Paris: Albin Michel.

Clastres, P. 1974. *La société contre l'État.* Paris: Minuit.

Comte, A. 1852. *Système de politique positive ou Traité de sociologie, instituant la religion de l'humanité. Tome deuxième, Contenant La statique sociale ou Le traité abstrait de l'ordre humain.* Paris: Carilian-Gœury et Dalmont, Mathia, Ladrange.

De Boever, A., et al. (eds.). 2012. *Gilbert Simondon. Being and technology.* Edinburgh: Edinburgh University Press.

Deleuze, G. 1966. Gilbert Simondon, L'individu et sa genèse physico-biologique. In *L'île déserte et autres textes*, 120–124. Paris: Minuit.

Deleuze, G., and F. Guattari. 1980. *Milleplateaux.* Paris: Minuit.

Dennet, D. 1995. *Darwin's dangerous idea: evolution and the meanings of life.* London: Penguin.

Dumont, L. 1983. *Essais sur l'individualisme. Une perspective anthropologique sur l'idéologie moderne.* Paris: Le Seuil.

Durkheim, É. 1887. De l'irréligion de l'avenir. (Critical Review) *Revue philosophique* 23: 299–311. http://classiques.uqac.ca/classiques/Durkheim_emile/textes_2/textes_2_03/irreligion_avenir.html. Accessed 30 Dec 2014.

Durkheim, É. 1912. *Les formes élémentaires de la vie religieuse.* Paris: Alcan (English trans. *The Elementary Forms of the Religious Life.* London: Allen & Unwin, 1964).

Friedmann, G. 1946. *Problèmes Humains Du Machinisme Industriel.* Paris: Gallimard.

Friedmann, G. 1956. *Le travail en miettes.* Paris: Gallimard.

Gesell, A. 1946. The ontogenesis of infant behavior. In *Manual of child psychology*, chap. 6, ed. Carmichael, L., 295–331. New York/London: Wiley (French transl. *L'ontogénèse du comportement de l'enfant.* In *Manuel de psychologie de l'enfant*, vol. I, chap. 6, 470–527).

Gould, S. J. 1997. Evolution: The Pleasures of Pluralism. *New York Review of Books*, June 26 44/11:47–52.

Guchet, X. 2005. *Les Sens de l'évolution technique.* Paris: Léo Scheer.

Guchet, X. 2008. Evolution technique et objectivité technique chez Leroi-Gourhan et Simondon. *Appareil 2.* http://appareil.revues.org/580. Accessed 30 Dec 2014.

Guyau, J.-M. 1887. *L'irréligion de l'avenir, étude de sociologie.* Paris (English trans. *The Non-Religion of the Future: a Sociological Study.* London: Heinemann).

[28] *Simondon's complete bibliography and a list of abbreviations are provided in the* Appendix.

Hyppolite, J. 1953. *Logique et existence*. Paris: PUF (English trans. *Logic and Existence*. Albany: The SUNY Press, 1997).

LaMarre, T. 2013. Afterword: Humans and machines. In *Gilbert Simondon and the philosophy of the transindividual*, ed. M. Combes, 79–108. Cambridge MA: MIT Press.

Latour, B. 2002. Morality and technology: the end of the means. *Theory, Culture & Society* 19/5-6: 247–260. http://www.bruno-latour.fr/sites/default/files/downloads/80-MORAL-TECHNOLOGY-GB.pdf. Accessed 30 Dec 2014.

Leroi-Gourhan, A. 1964. *Le geste et la parole. Technique et langage*. Paris: Albin Michel (English trans. *Gesture and Speech*, I. Cambridge/London: MIT Press, 1993).

Lévi-Strauss, C. 1960. Éloge de l'anthropologie. (Leçon inaugurale au Collège de France, 5 Janvier 1960). In *Anthropologie structurale deux*. Paris: Plon, 1973 (English trans. *Structural Anthropology. Vol 2*. Harmondsworth: Penguin, 1968).

Lévi-Strauss, C. 1962. *La pensée sauvage*. Paris: Plon (English trans. *The Savage Mind*. London: Weidenfeld & Nicolson, 1966).

Merleau-Ponty, M. 1954–1955. *L'institution. La passivité. Notes de cours au Collège de France (1954–55)*. Tours: Belin, 2003.

Merleau-Ponty, M. 1959b. Notes de travail inédites. Cours 1959. *Chiasmi International 7: Merleau-Ponty. Vie et Individuation*: 39–42.

Myrdal, G. 1944. *An American dilemma. The Negro problem and modern democracy*. New York: Harper & Bros.

Petit, V. 2010. L'individuation du vivant (2) «Génétique» et «ontogenèse». *Cahiers Simondon* 2: 53–80.

Ruyer, R. 1954. *La cybernétique et l'origine de l'information*. Paris: Flammarion.

Ruyer, R. 1958. Les limites du progrès humain. *Revue de Métaphysique et de Morale* 4: 412–427.

Schmidgen, H. 2012. Inside the black box: Simondon's politics of technology. *SubStance* 129 41/3: 16–31.

Stiegler, B. 2006a. Chute et élévation. L'apolitique de Simondon. *Revue philosophique de la France et de l'étranger* "Gilbert Simondon" 2006/3: 325–341.

Stiegler, B. 2006b. Le théâtre de l'individuation. Déphasage et résolution chez Simondon et Heidegger. In Vaysse, J.-M. 2006, 57–73 (English trans. The theatre of individuation; phase-shift and resolution in Simondon and Heidegger. In De Boever, A. et al. 2012, 185–202).

Toscano, A. 2006. *The theatre of production: philosophy and individuation between Kant and Deleuze*. Basingstoke: Palgrave Macmillan.

Vaisse, J.-M. (ed.). 2006. *Technique, monde, individuation: Heidegger, Simondon, Deleuze*. Hildesheim/Zurich/New York: Georg Olms Verlag.

Van Caneghem, D. 1989. Hommage à Gilbert Simondon. *Bulletin de psychologie* 42(392): 815–836.

Chapter 12
Regulation and Invention: Simondon's Political Philosophy

In Simondon's oeuvre there are no texts directly devoted to politics and almost no bibliographical references to political theory or political philosophy; furthermore, his rare political reflections are always mediated (or elicited) by reflections on technology. Yet nevertheless, a conception of human nature as a 'work in progress' is implicit in his epistemology of the social system, which has drawn the interest of many political thinkers to the concept of the transindividual. Since Balibar (1993), Stiegler (1994) and Combes (1999), a whole series of attempts flourished, which have aimed to discover in Simondon's 'transindividual' a supposedly latent political philosophy.[1]

Drawing on Stiegler, who argues that 'there is no Simondonian politics, while the question of individuation is entirely political' (Stiegler 2006: 339), I have in fact maintained the very 'existence' of a Simondonian political project. This is quite problematic, and indeed scattered – particularly in the third part of MEOT and in his minor writings –, but I believe Simondon's philosophy of individuation is better understood as having something actually political *at stake*. On this basis, an

[1] Although the complex debate on the 'political' Simondon can be said to begin with Deleuze (1966) it was, in fact, initiated by Balibar (1993), Stiegler (1994) and Combes (1999). Also Hottois (1993) attempted a 'political' reading of the issue of 'technical culture', although less concerned with the theme of the transindividual. Other interesting interpretations, more or less critical of the 'political' Simondon, can be found in Stengers (2004) and in Guchet (2010). The interest for the 'political' Simondon has also extended beyond the Francophone milieu. In Italy Simondon has directly been received as a political philosopher since Virno's translation (2001), occasionally raising the interest of radical left thinkers, such as Virno himself, Agamben, Negri, Esposito, and Morfino (2008): the latter is particularly concerned with the connection between ontology and politics, a line of research which seems to also orient South American researchers (see for instance Rodriguez 2009). In English, Toscano (2002) had from early on challenged the topic, while more recently Del Lucchese (2009) and some collective works and journal issues have appeared. Following the translations of Stiegler's and Combes's works interest is spreading, while Simondon's oeuvres are in course of translation in many languages worldwide, English included. I discussed the earlier stage of this debate on the 'political' Simondon in Bardin (2010), in particular *Intermezzo* I and II.

A. Bardin, *Epistemology and Political Philosophy in Gilbert Simondon*,
Philosophy of Engineering and Technology 19, DOI 10.1007/978-94-017-9831-0_12

explanation of the 'archetypal' role played by 'philosophical thought' in his texts will show how it carries on a political function, by endorsing the destabilising power of collective invention and allowing for its integration in the regulatory apparatus of social systems. From this perspective Simondon's philosophy can be a useful tool for questioning the ideological efficacy of the models provided by the traditional epistemology of social systems within the global political *milieu* emerged through the planetary development of technology.

12.1 Is There a Theory of the Political in Gilbert Simondon?

The basic closed/open schema presented by Simondon in *Individuation* and in the *Note complémentaire* for a theory of social systems shares similar concerns that also appear in MEOT and in PST, where he draws the lines of his ecumenical project, in which technicity is the source of innovation and philosophical thought the mediator of a political pedagogy aiming at the integration of technical normativity into the regulatory apparatuses of social systems.

But this schema that sees in technicity the 'open' counterpart of the closing tendency of religious and socio-political thought cannot be integrally applied to the second phase-shift Simondon presents in MEOT, where politics emerges as a phase of 'religious thought', the counterpart of the 'techniques on the human being' (Sect. 6.2). The basic schema contrasting closed = religion-politics to open = artisanal-industrial techniques fails here. Firstly because the new kind of techniques are exemplified through operations of 'closure' as well, such as the 'integration to social groups [and] the cohesion of groups' (MEOT 215). Secondly because politics is conceived as operating at a scale in which the unity and stability of social groups is called into question:

> Political thoughts integrate [the human world] into a superior unity: the unity of the becoming totality of humanity. There it looses its actual unity, such as the individual within the group. (MEOT 215)[2]

Would then politics be in some way an exception to the homeostatic function of religion, although situated by Simondon in the same genealogical line? To answer this question it is worth going back once again to Simondon's source for the open/close paradigm. This will provide an explanation of the twofold nature of politics in Simondon's thought.

Bergson in the *Deux sources* attempts to explain religious phenomena on the basis of the structural discontinuity of the *élan vital*. Religion is either static or dynamic, it either tightens the social bond or, taking on the original *élan*, makes new forms of (social) life possible. The two sources of social practice are in tension. The 'myth-making function [*fonction fabulatrice*]' provides images and symbols that

---

[2] Concerning Simondon's use of the expression 'mode of thought' as referring both to theory and practice, see Sect. 10.2.

regulate society through obligation, such as animal societies are regulated by instinct, while 'mystical experience' allows for a few individuals to introduce social innovation. In fact, the two are strictly connected, and 'dynamic religion is only propagated through images and symbols supplied by the myth-making function' (Bergson 1932: 285). In short, the 'open society' is for Bergson an inner tendency of all 'closed societies' that exceeds their actual configuration but necessarily relies on it. Thus 'static' and 'dynamic' are the two operating modes of the same *élan vital* in religion, and they operate on human society as a double kind of regulation. In this sense religion has a direct political function.

When reading the *Deux sources* as a philosophical-political work, Soulez (1989) noticed therein the juxtaposition of the domains of closed morality, religion and 'politics' (Soulez 1989: 268–69), and he introduced the concept of 'the political', in order to understand the internal discontinuity of each of the three domains. In his final remarks, Soulez argued that 'Bergson passes from a reflection on the political [*le politique*] to some "considerations", in his words, on politics [*la politique*]' (278). Through this basic scheme Soulez reads a text in which the two functions – exemplified in the disjunction between open and closed society – are indeed closely related and yet irreducible to one another in all social domains.

The terminological opposition of 'politics' and 'the political' translates in English language a conceptual distinction that emerged in continental political theory as a critical tool to contrast the modern conception of state policy.[3] Following Soulez, I will refer here to 'the political' as what the practice of politics – i.e. of ordinary and institutional production and distribution of power – simultaneously derives its force from and cannot definitely neutralise. According to Soulez, in Bergson's text this duality relies on the way he conceives the nature/culture threshold: *the political* describes the genesis of societies out of the process of social self-organisation of life ('human being is a "political animal" precisely because it is an animal', 280), while *politics* describes the typical modality in which a particular form of life – homo sapiens – can organise and regulate its social life. The opening tendency carried on by *the political* cannot thus be entirely exhausted in the modes in which politics organises and governs human societies: it persists in the 'mystical' exception of the legislator and hero, who cannot be said to pertain to the human species only, but rather to life itself in its never exhausted *élan* (277–279). From this

[3] The terms 'politics' and 'the political' respectively translate 'das Politische' and 'die Politik' in German, 'la politique' and 'le politique' in French, and 'la politica' and 'il politico' in Italian. But this terminological distinction, and notably the expression 'the political', is in fact 'the name of a problem which traces the conceptual and empirical incompletion of politics' (Valentine 2006: 506). The opposition is typical of a tradition of political thought concerned with the critique of the modern conception of state sovereignty, well exemplified by Carl Schmitt's (1932) conception of the political as a 'transhistorical category' pointing to the original formation and defence of the community (Bates 2012: 22). In radical political thought the concept traditionally served to oppose extra-institutional action to the institutional structure and regulation of the state. In this sense, the true space of 'the political' would be, in short, what a politics of state governance submitted to the laws of global capitalistic economy tends to ideologically and physically cancel. I am unaware of whether Soulez was or was not interested in this kind of operation. I am just attempting to transfer his close reading of Bergson's *Deux sources* to what Simondon mainly derived from this same text.

perspective, the identification of 'politics' and 'static religion' in Bergson's *Deux sources* would allow us to read the reference to mysticism as 'the political' activity of innovation carried on by life itself within human societies, in continuity with the *élan* that brought about their original genesis.

According to Simondon, religion only carries on the regulative function, while the exceedingly inventive function is that of technicity. But, as my analysis of the in-group processes evidenced in Sect. 6.2, social ontogenesis cannot be reduced to one of the two dynamics, since the closure of 'community' is the precondition for any opening of 'society'. In *Individuation* Simondon attempted to keep this duality together – and in fact he hid it – in the paradoxical concept of transindividual, while later, in particular in PST, he tried to resolve it into the opposition between sacredness and technicity. But, as I said, in the third part of MEOT the contradiction explodes: at the level of the second phase-shift between technicity and religion, the essential ambiguity of the latter becomes evident when the dynamical function usually attributed to techniques is quite surprisingly situated by Simondon on the side of what he calls 'social and political thought' (Sects. 10.2 and 11.2). While religions express an 'absolute' and 'actual' totality as it is given to the social system, on the contrary social and political thought functions as the 'announcement of new structures', an opening towards a 'virtual' and 'relative' totality of which the existing social system is the condition but not at all the actual realisation (MEOT 229).

Is it thus possible to understand Simondon's political thought by opposing an absolute totalising tendency of 'politics' and a virtual totalising function of 'the political'? Can we conceive the (transindividual) political as the risky invention that constantly pushes the metastability of social systems to a critical point, but can *also* interrupt its innate entropic tendency? Are we allowed to read the second phase-shift of religion in MEOT as the emergence of 'the political' within the phase of religious thought? Simondon's texts only allow us to sketch such a hypothesis. Neither in MEOT nor elsewhere does Simondon clearly refer to social and political thought as internally exceeding religious thought. In his conception of 'human progress' the potentials for social invention consistently remain on the side of the technical genealogical chain, where an open technical normativity constantly opposes the closed normativity of religion. Since the original tension between the 'universal religion' of the absolute subject and the 'artisanal techniques', technicity is the active pole of the phase-shift. Similarly, in the relation between 'technical systems' and 'social and political thought', the former appear to be the determining factor:

> The introduction of technicity in the systems that entail the human being as an organiser or an element makes techniques evolve. In the same way and at the same time this evolutionary feature of human groups becomes conscious, and this consciousness creates social-political thought. (MEOT 231)

In short, since social and political thought is in MEOT the form assumed by religion at the level of contemporary technical networks, it maintains both the structural tendency to closure and the ambiguity Simondon's concept of religion inherited from Bergson's. This explains the structural duality, if not the aporetic nature of the concept of 'political thought' in MEOT: on the one hand it carries on the homeostatic function of religion within the second phase-shift,[4] on the other hand it is an exception to it, since it can organise the convergence of religion with the technics on human beings. This twofold nature expresses the need for regulation *and* invention: the two movements are simultaneous and there is place neither for a definitive solution to the problem of social order nor for any emancipatory formula of its virtuous destruction. As static religion cannot grant community from collapsing into its own entropic tendency, dynamic religion cannot be *the solution* to these problems.

From this background, it is now possible to advance the thesis that through the concept of transindividual Simondon attempts to name 'the political' as a process, by differentiating it from 'politics' as the structure of social systems. The political would be in this sense what exceeds any possible organisation of social systems because it cannot be reduced to homeostasis, and yet it cannot be neutralised insofar it is the ontogenetic process which allows social systems to emerge an develop. This process crosses what in the *Note complémentaire* Simondon calls community and society, and therefore it exceeds both the instinctual organisation of groups at the biological level and the technical administration of groups at the rational level. As I have shown in Chaps. 6 and 7, Simondon's attempt to name the transindividual in *Individuation* crosses different domains, each time undermining the reduction of transindividual individuation to an individuated structure: language is the always partial result of a process of production of signification/information; labour derives from the economical integration of the processes of technical invention; myth is the crystallisation of a shared belief grounded on the collective individuation of emotions. In all these cases Simondon is trying to grasp the process of collective individuation that is hidden by its own structural effects. And all these different attempts to name the transindividual are carried on by the same challenging reference to the nature/culture threshold, which Simondon identifies as a bio-technical threshold crossed at the moment of the emergence of a techno-symbolic milieu. This rejects any presupposed anthropology, advancing the idea that social systems are always mixed systems, made of men, animals and technical objects within (and part-of) a common techno-symbolic milieu.

Now, the reason why – through the concept of the transindividual – Simondon ceaselessly attempts to *name*, *define*, and finally *keep in motion* the concept of 'the political' is, in my hypothesis, a clear mark of his iconoclastic struggle against any mythical filling of the image of 'Man', as it had taken place in his times in the political field, through the 'easy ontology' carried on by 'the great collective mythologies' (HU 53).[5] Since early modernity, the attacks of scientific research to the classical image of the human being's place in the theological cosmos opened a gap for ideological struggle, both religious and political. If Simondon constantly forces the sciences to fill in that gap, it is not because he believes that a definitive knowledge

---

[4] 'Political and social thought is considered here of the same order as religion, and can be treated in the same way' (MEOT 217).

[5] Simondon refers to pragmatism, communism and national socialism, all of which he pictures as characterised by a mythology of technology (Bardin 2013: 27 ff.).

of such processes can be provided by science, but rather because he knows that the gap has to be kept open beyond any saturation by mythical contents. This is what his philosophy of 'processes of individuation' always does, an aim that 'philosophical thought' cannot achieve without maintaining its exercise in close contact with the open attitude of scientific and technological research.

With the expression 'philosophical thought', in the final part of MEOT Simondon brings about the necessity of a permanent regulation of the processes traversing social systems, i.e. of the collective production/invention of new 'compatibilities'. This is an operation he denominates 'reflexive' and which – precisely for this reason – can be said to be 'political' only in an indirect and quite problematic sense, as I have shown in Sect. 11.2. But it would not be unfair to maintain that this political aim is – in Simondon's view – what 'philosophical thought' emerged and continues to exist for.

12.2 Philosophical Thought as an Archetype

Providing a basic explanation of the 'archetypal' role played by 'philosophical thought' in Simondon's texts is now necessary. This will explain how what he calls philosophical thought carries on a regulatory and therefore political function in social systems, and will allow us to test the consistency – or even the existence – of Simondon's political philosophy.

In his lecture at the *Société française de philosophie* (1960), Simondon linked the notions of 'schema' and 'archetype'. Attempting to develop them as the essential elements of his project of 'axiomatisation' of the human sciences, he aimed at extending the biological concept of schema he had openly derived from Gesell's studies on the development of children and probably from Piaget.[6] To make a long story short, Simondon showed how the growing organism builds '*systems of potentials* starting from the metastable field made of liquefied elementary schemas [… that] will be able to structure themselves' (FIP 546). These schemas would remain latent in the organism and capable of being reactivated in case of encounter with 'structural germs connected to external circumstances' orienting further processes of structuration thanks to the triggering of new processes of adaptation (FIP 546). Also the 'technical gesture' expresses a defined disposition to the actualisation of bodily energy. Furthermore, schemas and ethological *patterns* also have a transindividual function in social groups (FIP 550): transductively propagated through symbolic and technical invention, they become the source of collective symbolic and technical activity and contribute to structure social systems.

[6] It is perhaps possible to push this hypothesis up to the point of stating that what Simondon calls 'archetype' is quite close to Piaget's notion of schema. It is not by chance that, highlighting Bergson's influence on Simondon, Van Caneghem refers at the same time to Piaget, since both would 'conceive the living being not as a *thing*, but rather as superposition of dynamical schemas' (Van Caneghem 1989: 818). Also along the way Gesell (1946) and Piaget influenced Simondon's concept of 'evolution', see Sect. 11.3, n. 19.

In general, the importance of the concept of schema in Simondon's thought can be measured by the variety of domains it crosses and the different forms it assumes. Its usage is split on a broad-spectrum that goes from the biological to the gnoseological, and from the technical to the psycho-social, where it finally becomes an 'archetypal form' capable of explaining the 'possibility of a transductive propagation' of a process in a (social) field (FIP 549). And it is on this ground that the relation of Simondon's notion of archetype with Carl Gustav Jung's must be (although carefully) taken into account by confronting the respective concepts of individuation.[7]

In 1960 – when Jung's *Psychologische Typen* [*Psychological Types*] was being republished in the sixth volume of his *Gesammelte Werke* – Simondon thus concluded his lecture at the *Société Française de Philosophie*:

> Differently put, we will consider the process of dedifferentiation within a social body, or within an individual entering a period of crisis [...] After this crisis and this sacrifice, a new differentiation takes place: it is the *Albefactio*, or *Cauda pavonis*, that sends out objects from the obscure night, as the dawn shapes their colours. Jung discovers, in the aspiration of Alchemists, the translation of the *operation of individuation* and of all kinds of sacrifices that entail a return to a birth-like state, that is a return to a state rich in potentials, as yet undetermined: the domain for a new propagation of Life. (FIP 551)

In *Individuation*, in the paragraph *Signification de la subconscience affective* [Signification of the Affective Subconscious] Simondon expresses his view on the subconscious with an explicit reference to quantum physics and to Jung. He understands the 'centre of individuality' as made of affective and emotive processes 'subject to quantic reorganisations' as far as they 'go through brusque leaps according to degrees, and they obey threshold laws'. This stratified centre – he argues – permits a partial compatibility between his concept of '(group) personality'[8] and psychoanalytic clinics, in particular Jung's: 'what Jung discovers in his analysis of the unconscious (or the subconscious) are affective-emotive themes' (I 248). This same connection Simondon restates elsewhere and not strictly in *Individuation*, and it can be considered a constant of his thought that brought him to adopting the hypothesis of a 'structure of personality made of layers and levels (depth psychology)' (IMIN 74).

In effect, in Jung's theory the 'psyche' rather shows a certain affinity with a 'thermodynamic' approach: it is a system in which contrasting issues provide a continuous energetic supply that destabilises but also stimulates its dynamic self-regulation. This is according to Jung the 'process of individuation' i.e. the progressive integration of the different parts of a system in a totality the structural conflict of which marks the creative nature of psyche. In this light some consonances with the themes of *Individuation* evidently appear, beginning from the centrality both authors reserve for the concept of individuation in the understanding of psychic-collective

---

[7] Unfortunately, only Chabot (2003) and Carrozzini (2005) have taken Simondon's reference to Jung seriously.

[8] See above, Sect. 6.2.

processes.[9] But the distance between Jung's and Simondon's models is remarkable in at least two senses, both connected to their different conception of the individual: one concerns the meaning of the concept of archetype and the other the relationship between the individual and the collective.

Firstly, Simondon's reformulation of the concept of archetype takes its distance from the Jungian as far as the archetype is neither an *a priori* 'datum' in the individual, nor a collective construction starting from individual norms. It rather operates transductively in a regime of transindividual individuation in which the role played by the biological 'phase' is as important as the role played by the sedimentation of tradition and its resumption by invention. Although initially recognising that 'Jung had already established the overdetermined [*surdéterminé*] characterisation of archetypes' (PST 319), Simondon later notes that the 'seducing' interpretation of some images as the permanence of anterior evolutive phases must be explained through the influence of the schemas anticipating motion onto symbolic production (IMIN 34–35): an influence always mediated by cultural issues. According to Simondon archetypes are in fact 'schemas of imagination', a 'mould [*moule*] of images pertaining to the past of humanity (possibly to pre-human phases of evolution)' which only act – differently from Eliade's hypothesis, he claims – in a mixed individual *and* collective regime (IMIN 129).[10]

Rather than a given structure, the archetype is to be understood as 'an originary archetypal source' (IMIN 123). The 'archetypal function' is thus entirely resolved into its operationality and therefore – to be rigorous – without origin. It is displayed in the production of objects and symbols carrying on a normativity that is either biological, technical and social. As it should be clear, in Simondon this phase-shift of the different normativities does not rely on any essence to be realised, and therefore individuation cannot be understood as a process of adaptation – even though creative – relying on an alleged 'original state of identity' (639) of the individual: a non phase-shift 'nature' is in fact impossible to conceive for Simondon, as it is impossible to conceive a 'complete' individual whose 'normativity' would pre-exist collective

[9] 'The concept of individuation plays a large role in our psychology. In general, it is the process by which individual beings are formed and differentiated; in particular, it is the development of the psychological *individual* as a being distinct from the general, collective psychology. Individuation, therefore, is a process of *differentiation* having for its goal the development of the individual personality [...] As the individual is not just a single, separate being, but by his very existence presupposes a collective relationship, it follows that the process of individuation must lead to more intense and broader collective relationship and not to isolation [...] Individuation is practically the same as the development of consciousness out of the original state of *identity*. It is thus an extension of the sphere of consciousness, an enriching of conscious psychological life' (Jung 1921: 637–39).

[10] The reference to Eliade is important here, because he was part of the *Eranos* circle (whose members used to meet annually in Switzerland). Since the beginning in the 1930s until the 1970s the circle reunited scholars and intellectuals of different specialisation (biology, psychology, anthropology religious studies), some of whom – namely Jung, Eliade and Portmann – Simondon often made reference to.

individuation.[11] On the contrary, according to Jung, the 'psychic individuality is given a priori as a correlate of the physical individuality' (Jung 1921: 639) because the 'individuum' is an 'a priori psychological and physiological datum' (637).

Secondly, and consequently, Simondon's distinction between the individual and the collective is far from Jung's. For Jung the individual is 'everything that is not collective' (636), and collective normativity is the result of the interaction of already constituted individuals: 'a norm is the product of the totality of individual ways'. Therefore, on the basis of their a-priori 'grounded' 'psychic constitution', individuals contract, so to speak, new configurations of collective normativity against norms formerly deposited by other individuals (638–39).

On the contrary, within Simondon's conceptual framework it is impossible to conceive social normativity as simply resulting from the interaction of individual normativities, as it is also impossible to suppose that the intertwining of the different normativities crossing the social system through individuals, objects, symbols, elements, might ever be composed or integrated in a whole which would not be purely imaginary. Consequently, what Simondon calls the 'subconscious of the living beings' is conceived as a network of relations. Although this network is 'the basis of myths', any identification of it – whether with an individual nature or with a collective unconscious – would equally be a misplaced myth itself. In effect, according to Simondon, at the level of the affective-emotive themes 'it is possible in a way to speak of the individuality of a group or of a people' (I 248). But this identity is in fact only the 'communitarian' side of social processes, namely it is what contrasts the transductive process in which individuals play a systemic role as 'germs of consciousness and action'.[12]

[11] To this regard it is worth recalling Adolf Portmann, another 'adept' to the *Eranos* group (see the previous note), whose text on zoology *Animal forms and patterns* (1948) Simondon included in the bibliography of *Individuation*. In *Les bases biologiques d'un nouvel humanisme* (1951), Portmann notes that the peculiarity of homo sapiens is grounded on the anomalous length and articulation of what the author names 'gestation period', which would last for the 266 days from conception to birth plus at least another thirteen extra-uterine months (Portmann 1951: 221). Biology makes the human being thus a *'form of life'* dependent on the 'particularity of each existence' and – so to speak – ontologically linked to an ethics of responsibility: 'this responsibility summons up a table of values that biology can only contribute to establish, but that must essentially derive from a wider vision of our existence, a vision that will provide principles apt to orient human behaviour' (79). Also Virno's interpretation of Simondon's concept of the 'pre-individual' points to a political anthropology, and he sees in 'neoteny' a typical human feature that would characterise political praxis (Virno 2003: 158). If the contingency of political praxis is well underlined by Virno, it seems thus to be definitively located in the biological phase, at the outset of the process, and not in the process itself. For this reason Virno's reading appears on this point surprisingly closer to Portmann than to Simondon.

[12] 'To postulate that [...] there are no lost islands of becoming, no domains eternally closed in themselves, and there is no absolute autarchy of the instant, is to claim that each gesture has a meaning as information and it is symbolic in relation to life or to the totality of lives' (I 333). Hence the importance, for Simondon, of the Romanian figure of the inheritor (I 250), and his conception of the individual as 'a "temporal parallax" in the organisation of the intermediate zone between our present action and the cultural horizon, far, stable and collective' (IPM 1451). This is also the meaning of the "perpetual *nekuia*" Simondon evokes in I 250 probably relying on Cumont (1949). An interesting artistic development of this theme is offered by Duhem 2013b.

In this sense myths, as much as opinions, are part of the continuous process through which a community structures and organises through the 'interindividual' collective memory a shared normative system, at times triggering also 'powerful collective movements' thanks to the force of identification carried out by the 'symbol-images' (IMIN 133). But this is still an imaginary and homeostatic, magic-like efficacy of symbols, in which a regressive tendency persists in a latent manner. In the terms used by Simondon in IMIN, any the fixation of a phase of the 'cycle of the image' entails the degradation of the symbol to the instrument of an operation that completely fails the properly historical – i.e. transindividual – dimension of the transductive processes. On the contrary, the irreducible opening of the archetypal 'source' allows for 'the heterogeneity of the footprints connected to the same source which provides the symbol with its internal tension' (IMIN 125), a symbol the transindividual-historical power of which allows us to constantly cross and overcome any once-and-for-all imagined identity.[13]

Against this backdrop it is also possible to understand how Simondon reformulates the contraposition between 'imaginary' and 'historical' through the conceptual couple interindividual (intersubjective)/collective (transindividual): 'interindividuality is an exchange between individual realities [...] that look for an image of their own existence in other individuals', while 'the collective does not really exist out of an individuation that institutes it, it is historical' (I 167). In this sense the theme of cultural heritage can be said to shape Simondon's peculiar elaboration of the concept of 'archetype': it is the form schemas assume at the level of the peculiar historicity of social systems. It is again Merleau-Ponty who provides the key for understanding the way Simondon enters the debate concerning the epistemological status of psychoanalysis at the threshold between the natural and the 'human' sciences, trying to define the historicity that characterises social systems out of any contraposition of nature and culture:

> From the point of view of modern biology, on the contrary, there is the idea (Simondon p. 231) of heredity as a prolongation of ontogenesis, of individuation, themselves understood as *vital processes* and not as phenotypal adventures – The constitution of a tradition, of a memory, of a past, of a history, of an order of "choice" do not therefore indicate a creation *ex nihilo* – Its ambivalence or "unconsciousness" [*inconscience*] does not mean that there exists of it a clear (unconscious) text whose appearance would be its *masking* [...] Freedom is always the {care-taking} [*gestion*] of an inheritance. (Merleau-Ponty 1959b: 41)[14]

Simondon's political stance can thus be drawn back to the way he very classically posed the question of humanism as a crucial political commitment to technology in

[13] It is in this sense that – following Bergson's idea that, in order to expand, open religion requires a myth-making function – Simondon conceives of belief and myth also as 'amplifying projections' (IMIN 44).

[14] As already noted in Sect. 9.4, Merleau-Ponty's seminar on *L'institution* offers one of the possible perspectives for the understanding of Simondon's conception of institution at the threshold between the biological and the social. According to Merleau-Ponty, organised series of 'symbolic matrices' 'appear where men and the givens of nature or of the past meet' and can 'leave their footprint of the course of things' and then disappear due to 'internal disintegration' or because they change their nature. (Merleau-Ponty 1955: 28).

his early essay on *Humanisme culturel, humanisme négatif, humanisme nouveau* [Cultural Humanism, Negative Humanism, New Humanism] (1953). In this essay Simondon still acknowledged Sartrian humanism as capable of contrasting 'forces of "mystification" such as pragmatism, national-socialism and communism' (HU 53, see also MEOT 223) on the condition of assuming the demystifying and anti-ideological force of science and technology, that carry on an actual '[human] heritage full of implicit meaning that reflection can make explicit' (HU 56).[15] The same connection between the demystifying power of technicity and the philosophical effort of integrating it into culture in order to make it politically effective still resonates in *Culture et technique*, where philosophy was the name Simondon attributed to the specific task of conferring to the 'demystifying' power of the schemas of technicity the political effect of operating a dynamical convergence of culture and technics (CT 16).

Thus conceived as a critical 'effort' to integrate technicity as a tendency within culture as a tradition, philosophical thought reveals its intrinsically political nature as a part of the regulatory apparatus of social systems. Culture is in fact the milieu through which social systems act on themselves, structuring and reconfiguring their own regulatory mechanisms:

> Culture is the part of human reality that can be modified [...] the active mediator between subsequent generations, between simultaneous human groups, and between subsequent or simultaneous individuals. (MEOT 227)[16]

The concrete activity which emerges from culture as a mixed milieu of technical and symbolic objects is – in Simondon's words – 'reflexive': an activity in which the biological, technical and social normativities that constitute the social system confront and modify the conditions of their own exercise.[17] And it is precisely the part of culture we call philosophy that, by retroacting on the cultural milieu it emerges from, can provide the social system with a further regulatory mechanism: philosophy 'is constructive and regulatory of culture' (MEOT 212).

So, what definition of philosophical thought and of its political efficacy can we derive from Simondon's writings? In IMIN Simondon refers to a tradition in which philosophy would be conceived as the invention of schemas, archetypes, '*a priori* images' which function as germs of political (reforming) action:

> A priori images are fertile, also and first of all when they are integrated into the world as anticipations in the long term, after the extensive passage – *tèn makran hodon* – of philosophical thought [...Plato's] philosophical doctrine, rich in *a priori* images, has been capable of naturally inspiring the highest philosophical-political school of the ancient world, and becoming the most audacious model for reformists. (IMIN 61–62)

---

[15]As many others of his generation, Simondon saw in Sartre the champion and the model of a 'fighting wing of humanism', to whom one should address a philosophical request of emancipation. And this is what Simondon does, in this early writing, when requesting the integration of 'cultural humanism' with the traditionally excluded issue of technics.

[16]In other words it is the milieu 'through which the human being regulates its relation to the world and to itself' (MEOT 227).

[17]I partially owe this formulation of the concept of 'reflexivity' to Guchet, *Ontologie sociale et technologie*, a paper he delivered at the conference *Between Deleuze and Simondon*, organised by *Warwick University* in Venice the 18/09/2009.

If collective invention (technical, scientific, sacred or linguistic) is for philosophy the source of constantly novel developments, this is possible because philosophical thought is characterised by a specific modality of functioning, a tendency towards 'amplifying' the schemas implicit in other processes. But the postulate of Simondon's enterprise is that philosophy can be *itself* effective through analogical invention. More precisely, the actual efficacy of philosophical thought within the milieu of culture is to be conceived of as a paradoxically 'conservative' activity of *invention*:

> The philosophical effort can conserve technicity and religiosity by discovering their possible convergence through a genesis that would not have been accomplished out of the genetic intention of the philosophical effort. Philosophy would aim not only at the discovery, but also at the production of genetic essences. (MEOT 213)

Thus conceived philosophy cannot be an essential feature of human nature, nor the destiny or the mark of any superior achievement, but rather a tradition capable of endorsing the destabilising power of collective invention and allowing for its integration in the ordinary functioning of social systems. This is something Simondon seems to implicitly recognise when he claims that 'Thales, Anaximander, Anaximenes were first of all technicians' partially detached from the community and therefore capable of 'reflexive thought', through which they could both invent and mediate the relationship between the community and 'the world' (NC 511–12).

This tradition emerged itself from the invention and transduction (i.e. the always singular historicity) of a certain number of operational schemas, of 'archetypal' techniques, of scientific paradigms, which can continue, as structural germs, their own history, under the inescapable condition of a renovated collective activation and integration within different social systems. It is not a tradition primarily made of contents, but rather of operative modalities, along with the schemas and examples for their assumption, reactivation, for their renovated invention and transposition on the different domains of culture, first of all derived from 'technical operation', where philosophical thought can find 'both a ground for reflection and a paradigm' (MEOT 256).

If thought is a transductive process that can only continue by propagating into new domains and determining their radical reconfiguration, its functioning and efficacy cannot be granted only once and for-ever. And also the continuation of philosophical thought's transductive history ultimately depends on the vicissitudes of the milieu through which it was and still is propagated. Hence the philosophical-political meaning of Simondon's two major works – *Individuation* and MEOT – can be finally clarified. Simondon's masterpieces appear in this light as two quite different outcomes of the same effort to *continue* the political action of philosophical thought on culture, the regulatory apparatus of social systems. On the one hand MEOT aims at a normative pedagogy of 'technical culture' against the mirroring risks of a collapse of the social system in the ideological closure of a communitarian mythology and of its dissolution in the indiscriminate opening induced by accelerated technological expansion. On the other hand, the intellectual adventure of *Individuation* is a ground breaking example of the twofold nature of philosophical research, both experimental and 'reflexive': capable of discovering 'genetic essences' within the very processes it emerges from. In this sense, the 'study of individuation' – writes Simondon – can be itself 'a source of paradigms' precisely insofar as it grasps the actual processes from which it derives its own schemas (I 324).

12.3 Imagining the Collective Without Human Nature

If there is an undisputable merit in Simondon's philosophy, it is the capacity of integrating scientific and technical concepts, images, and operative modalities into the philosophical imaginary. It is from this perspective that Simondon's philosophy questions the mechanistic image of nature we inherited from early modernity by reviving the intertwining of biological, technological and social models in the light of the discoveries that took place in the natural sciences at the beginning of the twentieth century. Out of this mixed scientific and philosophical background Simondon outlined a new epistemology of the social sciences aimed at providing a critique of substantialism and determinism and at reforming the cybernetic concept of information as he conceived it.

Drawing on the Bergsonian metaphysics of the *élan vital*, and on Canguilhem's critique of the biological and social (self)regulatory mechanisms, Simondon provided a critique of 'automatism' which had and still can have remarkable consequences also outside of the technological domain.[18] In fact, this allowed him to dispel the cybernetic myth of a self-regulating automaton also from the biological and social domain. Simondon refused to consider society 'the domain of unconditioned homeostasis' and therefore rejected Wiener's postulate that 'a good homeostatic regulation is the ultimate end of societies, the ideal which should guide any act of government' (MEOT 151).[19] Simondon's 'machine' is in fact an 'open' one, connecting the metastability of the system (in accord with life sciences) to the partial indeterminacy of processes (in accord with quantum physics). This 'open machine' is the basic schema from which Simondon developed his epistemology of the collective: the model of a social system opened by the recurrent emergence of invention through the reactivation of the schemas of technicity.

Simondon's model for the understanding of social systems is far from both the biological and the mechanical models: the model inspired by the ancient concept of an ordered cosmos dominated by final causality, and the juridical model inspired by the image of a soulless universe of precision sketched by seventeenth century mechanical physics. Simondon's model plays thus a demystifying role against this apparent alternative, by demounting, first of all, the very image of human nature that

[18] Also Baudrillard explicitly associates Simondon's work with the critique of the model of the automaton, when writing that 'AUTOMATISM [...] is the major concept of the modern object's mechanistic triumphalism, the ideal of its mythology' (Baudrillard 1968: 153–54).

[19] For Simondon's critique of the cybernetic concept of the automaton as a model for social systems, see above Chap. 7.

all philosophical political imagination has ever been based on. Simondon's ground breaking contribution is neither a restoration of the classical role played by human beings between divinity and nature, nor the discovery of a new '*place de l'homme dans la nature*' (De Chardin 1956). It is rather the dissolution of the very myth of a human nature grounding both sides of this false alternative.

As explained, Simondon's perspective entails the full acceptance of the achievements of the empirical sciences and the integration of evolutionism in the philosophical worldview. This means not only the acceptance, of course, that homo sapiens are an animal species, but also the clarification that political problems do not strictly pertain to a species, because societies are complex systems made of so many differently evolving processes taking place at so many different levels, that they cannot be reduced to any ultimate 'model'. Finally, such processes can only very approximately be qualified as 'human progress' (Sect. 11.3). And, more importantly from a philosophical point of view, this allows for a rereading of all that has been traditionally referred to as 'human nature' in terms of a complex intertwining of processes, that it makes no sense anymore to reduce it to any supposed stable identity, whether individual or collective.

From this perspective it is possible to frame Simondon's invention of the concept of the transindividual not as the alleged solution to a political problem, but rather as a problem that *Individuation* poses without actually expecting *a solution.*[20] In fact, the concept of the transindividual marks the boundaries of the political problem concerning the decision on what human nature is, and denies that a solution can be derived out of any given anthropology. On the contrary, the research path opened by Simondon entails the study, the explanation, the imagination and the actual structuration – theoretical and practical – of a collective that 'exists *physikòs* and not *loghikòs*' (I 314). Simondon's anti-substantialistic stance allows for no 'vital' or 'cultural' characterisation of the human that can serve such a purpose, since it offers no means to trace a definitive threshold between nature and culture. It would be meaningless to have to get rid of a false anthropology in order to substitute it with another.

As I have tried to demonstrate, the transindividual entails a kind of 'intermittence' of the human field and therefore of 'the political'. Different approaches are therefore needed in order to enquire into the different threshold conditions of the human, such as Simondon does in *Individuation* (Sect. 6.3). But the first paradigmatic cut, the matrix of all the others, is marked by the recurrent theme in Simondon of the human/animal difference. In *Deux leçons sur l'animal et l'homme* [Two Lessons on Animal and Man] (1963–1964) Simondon provides a brief and insightful historical picture of the philosophical gesture that built the concept of 'animal' in differential relation with the concept of 'man', showing the key role played in this operation by the concept of 'instinct'. What resulted was an anthropology entirely

[20] 'The whole spectre of communitarianism and of essentialist identitarian demands simultaneously deflate when one draws the consequences of Simondon's transindividualism. Each identity (personal, collective) is a *problem*, and not a given: a response provisional and *in progress* of one's effort to persevere in being, in constitutive interaction with a certain milieu, and not a stable solution' (Citton 2004).

functional to the neutralisation of all that is aleatory, singular, unpredictable in human beings, in animals and – finally – in being itself. In brief, according to Simondon's argument, when the study of the human being has been eventually absorbed into the field of the natural sciences the problem of failing to grasp its concrete existence still has not been resolved, rather it has been aggravated. In fact, while believing to grasp the (biological) 'essence' of the human being out of any metaphysical presupposition, what is actually studied is a completely 'abstract' animal, whose behaviour is entirely determined by the genetic and environmentally given conditions (Bardin 2008, in particular on DL 60–63). This is precisely an animal conceived as a 'fictive being' that would have with nature 'relations [entirely] regulated by specific characteristics' (I 302). Such an animal does not exist: what actually exist are the individuals here dissolved in an articulated analysis of the specific patterns and needs characterising their species. The ideological move enacted by modern anthropology appears to Simondon complementary to the progressive reduction of the human to the measure of governmentality policies, as Foucault denounced soon thereafter, still following in Canguilhem's footsteps.

To this approach Simondon opposes a biology capable of redeeming some concepts from their ideological use and to give back to living systems their complexity. Therefore he not only attacks mechanism, but also 'a vitalism founded on a partial analysis of life, that valorises the forms more close to human species, thus instituting a de facto anthropomorphism' (I 171). In fact, the concept of the transindividual allows Simondon to enquire and define the human field – as any other regime of individuation – without presupposing a common 'essence' shared by a defined set of individuals: rather describing and formalising all the relational activities and processes that are called 'human' (from perception to affectivity, from language to technical activity, spirituality and the construction of the collective). This will allow for them to be made into the objects of a science by definition *in fieri*:

> Anthropology cannot be the principle for studying the Human Being: on the contrary, human relational activities [...] can be taken as principles on which a possible anthropology can be built. (I 297)

The perceived necessity to 'build' an anthropology on which social and political thought might possibly be grounded, is probably what still encloses Simondon's philosophy within the limits of modern 'humanism'. Yet Simondon's humanism always remained anti-essentialist and immune to doctrinal formulas that would result in their dysfunctional relation to the struggle that, alone, justifies their evocation: 'humanism can never be a doctrine or an attitude that might be defined once and forever: each epoch must discover its own humanism, by orienting it towards the major danger of alienation' (MEOT 102). No 'human reality', in sum, can either ground the emergence of the collective or, once it has emerged, guard its stabilisation.[21] The transindividual *does not define human nature* because, to be rigorous,

[21] This is one of the earliest claims resonating in Simondon's research: 'it is too easy to rely on a permanent and universal human nature [...] no universal human nature can be defined, since all events and singularities are part of humanity' (HU 52).

there is no human nature: neither is an anthropology based on the biological concept of species possible, nor a psychology that would maintain the primacy of perception or consciousness, nor a sociology of the human field, given that strictly 'there is no "humanity" in sociology' (FIP 533). If it is true that political theory always entailed an anthropology, a political theory at the level of Simondon's philosophy of individuation would entail an anthropogenesis in a double sense: a discovery *and* an invention of the human field. And this is precisely the ontological and epistemological status of the transindividual.

What surfaces again through this concept is the very aleatory 'nature' (of humans and, of course, for Simondon, of any other being) that both ancient and modern thought had attempted to neutralise: the first situating human nature in the harmonious and organic order of the *cosmos*, the second securing it in the cold cage of mechanical determinism. Cancelled in the theories of the nature-machine and of the state-machine, or – in a complementary fashion – hypostatised outside of nature as a sovereign *res cogitans*, 'human nature' is the philosophical myth that Simondon helps challenging through the concept of the transindividual, without seeking refuge in the *ur*-myth of the *zoon politikon*. An apparently epistemological problem demonstrates thus to be a political one. The epistemology of the human sciences reveals to have always been a battlefield between different models of social regulation, concerning the nature and opportunities, obligations and limitations of political planning and intervention. To this battle in the field of the collective imaginary all 'reflexive' activities necessarily take part: sciences, religions, arts and, finally, philosophy.

A strong faith in this kind of 'reflexive' power of 'philosophical thought' is precisely what drives Simondon's philosophical enquiry and inspires his pedagogy of technicity. Through an activity of invention and pedagogy, philosophical thought establishes the conditions of possibility of its own existence and takes part in the collective process of the regulation of the social system. But the system concerned is a peculiar one, in which Simondon isolates and analyses in particular the key role of technicity, not without detecting in it some ambiguities intrinsically connected to the very nature of the transindividual. This ambiguity is evidenced by the fact that – both in MEOT and in *Individuation* – technicity appears at times as the primary mode of 'initiation' to individual and collective life (NC 511), and at times as one of the modalities of transindividual individuation, or even as *the* first propulsive force of social mutation through invention.[22] In general, Simondon's reader experiences a relay between the problem of the contemporary opposition of technology and culture, and the attribution of a crucial role to technicity in the ontogenesis of human societies. This attitude culminates in the apparent identification *tout court* of technicity with the transindividual in some passages of the *Note complémentaire*, and in the conclusion of MEOT, where technical activity is explicitly assumed as 'the model of *transindividuality*' (MEOT 248), while, on the other hand, this

[22] In some passages, to the technical 'choices' of a society, Simondon seems to attribute a dominant structuring impulse (MEOT 86–87).

strongly contrasts with many other passages, not to mention the direct claim, in *Individuation*, that 'religion is the domain of the transindividual' (I 250).

Indeed, Simondon's oscillations depend on the way he is each time analysing the phase-shift between homeostatic and exceeding tendencies in the social system. As an evidence of this we can take the extension covered by the set of concepts he variously assimilated to the sacred along his oeuvre, such as 'religion', 'sacredness', 'political thought': all tending to fulfil the institutional, normalising function of social regulation against the inventive, normative one attributed to technicity. In general, it is possible to claim that in the domain of technicity Simondon looks for the opening tendency proper to all social dynamics, each time opposing it to a closing tendency differently conceived. And yet this contraposition is neither an ontological assumption, nor the dialectical key for a philosophy of history, and least of all for a political philosophy celebrating the promise of technological progress. It is rather a conceptual framework Simondon adopts for a diagnosis in the conjuncture. Therefore, instead of looking for a supposed systematic consistence in Simondon's 'political' philosophy of technicity, I have tried to establish the theoretical constellation of *the terms* defining the way he poses the political problem. From this standpoint, it is now possible to appreciate what theory of social regulation can be derived from his epistemology.

If there is in general – at all levels: physical, biological, psychic and collective – an excess, a surplus of the 'operation' over the 'structure' (the key terms of what Simondon names 'Allagmatics'); and if this surplus cannot be eliminated precisely because it constitutes the structure itself, Simondon's philosophy authorises the reformulation of the relation between the social structure and political action in terms of an 'internal excess' characterising social systems. In this sense, as explained in Sect. 12.1, politics can be read either as the surplus of the collective operation of normative invention of the social bond, or as the whole set of existing apparatuses of homeostatic regulation in the social system: beliefs, myths, norms, jurisprudence, institutions. This is probably the ambiguous meaning of 'political thought', situated in the third part of MEOT on the side of religious thought, whose 'exigency of totality' it represents, as said, at the level of the second phase-shift (MEOT 224). It is starting from this 'phase-shift' of the concept of 'the political' – not explicitly authorised, indeed, by Simondon's text – that I have tried to read Simondon's invention of the concept of transindividual individuation.

From this perspective transindividual individuation can be intended as what 'opens' the domain of the collective, which is more and less than a society: it is the factor providing energetic support to the social system, and it is also, precisely for the same reason, the factor permanently menacing its structural crisis. Transindividual individuation cannot be normalised without being neutralised; it cannot be entirely liberated without becoming ungovernable: its neutralisation brings about regressive processes; its excess triggers the revolutionary crash of the actual configuration of the system. These processes overcome the biological continuity of communitarian life, but also depend on it, since the transindividual relies on a former structure in order to take place: this is why 'societies cannot exist without communities' (NC 508), and 'communitarian life is communitarian and social', although – following

Bergson's lesson – 'these two features are antagonistic' (NC 512) (see above, Chap. 6). Now, if the social system is both the origin of norms and the effect of normative invention (i.e. through norms the social system structures itself), then the 'central obscure zone' of politics is situated where this circularity cannot be closed. It is where the established normativity of the social system does not hold, that the inevitable emergence of the political opens to the 'domain of the collective or transindividual' (I 261) that – from within – repeatedly calls into question the actual configuration of the system.

As I have widely demonstrated, Simondon's major references for his theory of social systems are Bergson's *Deux Sources*, the French sociological tradition mediated by Leroi-Gourhan, and Norbert Wiener's cybernetics. But the outset of his reflection – such as the entirety of the references he adopts – can be said to be the heritage of his master Georges Canguilhem. Although refusing the Positivistic idea of an *ineluctable* progress towards the solution of the political problem, Canguilhem never abandoned the perspective of a possible *definitive* synthesis of culture and technics:

> The whole problem of social regulation consists of knowing *if* these remains of ideological divergence can be eliminated in order to allow, one day, the functioning of the cybernetic social machine, through self-regulation, as an organism. (Canguilhem 1972)

The same question also grounds Simondon's political agenda. As I am going to explain in the concluding section of this book, Simondon embraced Canguilhem's challenge by providing a new model of the body politic from which he drew a novel conception of social regulation, in which technicity played a peculiar but not exclusive role. Simondon's approach was essentially differentiated not by his refusal of the Positivistic hypothesis of a possible *definitive* resolution of the political problem, but rather by what he radically renovated as the interpretive *model* of the social system, overcoming the organism/machine alternative, and *therefore* the perspective they both entailed of a possible physiological or technological solution – in both cases a normalising one – to *the* political problem. Through his philosophy of individuation Simondon succeeded in keeping at a distance the reassuring image of a 'human nature', an essence to which political philosophy had for a long time secured its promise of a 'normal' functioning of the 'body politic'.[23]

[23] As Citton highlights, the concept of the transindividual functions in Simondon's philosophy as an alternative to the 'naive analogy' of the body politic in which individual 'members' would play a fixed role, and therefore any reductionist reading of Simondon's philosophy towards sociobiology should be avoided. Yet Citton's reading should be complicated if we do not want to contradict Simondon's struggle against a supposed 'anthropological difference' (Citton 2004). Relying on Simondon's critique of the 'anthropological prejudice' shared by social positivism and cybernetics (both reducing the social system to homeostasis, Guchet 2011: 74–75) Guchet concludes his essay on the 'social body' in Simondon by proposing an interpretation that is quite close to mine concerning the relation between the themes of the transindividual and of technics Simondon developed in the two theses, and nevertheless he relies on a different conception of the anthropological difference: 'humans are the only living beings for whom the forms of social organization can be transformed through the eruption of new modes of engaging with materiality' (Guchet 2011: 92).

12.4 For a Political Regulation of Collective Invention

A consistent picture of Simondon's political philosophy has been provided in Chap. 11 by reassessing his philosophical sources and collecting different issues scattered in his work. The resulting picture was that of a pedagogical-political project of integration of the social processes of invention – notably technical – into the functioning of social systems. Although Simondon's project might seem inspired by a technocratic stance, it is in fact profoundly critical towards the relation of dominion that technocracy carries with it since its early-modern Cartesian roots, and towards the entirety of the conceptual constellation that philosophical thought can still provide – thus becoming itself ideology – to the metaphysical settlement of these relations of power:

> One might call autocratic a philosophy of technics that takes the technical set as a domain where machines are used to gain power. The machine is only a means, while the end is the conquest of nature, the domestication of natural forces through an initial enslavement. The machine is a slave that serves to make other slaves. Such an inspiration to domination and enslavement might converge with a demand of liberty for human beings. But it is difficult to liberate oneself by transferring slavery to other beings, humans, animals or machines. Reigning over a people of machines enslaving the whole world is still to reign, and any reign entails the acceptance of schemas of slavery. (MEOT 126–27)[24]

A philosophy according to which any anthropology essentially remains – as Simondon claims – 'to be built [*à édifier*]' (I 297), can neither *deduce* political organisation from human nature, as in classical political theory, nor *prescribe* a policy regulating the life of individuals and of social groups according to the values proper to one of the different 'communitarian' traditions. The 'reflexive' efficacy of philosophical thought is not normative in the sense that it has to apply a theoretical principle or an actual truth to an imperfect reality. And in fact, it requires the same conditions of efficacy of any other 'reflexive' action, and all 'reflexive' activities

[24] Quoting this text Marcuse underlines how – on the basis of modern science's 'pure and neutral' theoretical reason – the capitalistic control of technology developed a process going from the 'instrumentalization of things' to the 'instrumentalization of men' (Marcuse 1964: 159). Along a similar path Feenbeerg explains that Simondon's strive for a 'concrete technology' integrated to the development of social systems cannot rely on technics itself, it has to be ultimately linked to a non-capitalistic political choice (Feenberg 1991: 194–95). However, in his criticism to 'technological rationality' Marcuse does not consider that Simondon sees precisely in the liberation of machines the *precondition* for the liberation of human beings, as far as they carry on schemas of liberty opposed to the 'schemas of slavery' typical of their usage without knowledge. In this sense Simondon's perspective cannot comply with any Marxist hypothesis of liberation of human beings through the 'enslavement' of machines: as Simondon clearly states, philosophical thought has to play a role towards technical objects 'analogous to the one it played in the abolition of slavery and in the affirmation of the value of the human person' (MEOT 9). As Toscano notes Marcuse, being unaware that Simondon's nature 'has a very different relation to technical reality than Hegelian nature', fails to grasp that to Simondon the necessary condition for the emergence of a new finality in the collective is the freeing of the machines as 'intercessors, mediators and converters' that relate social and political collectivity to the 'disparate' becoming of nature (Toscano 2012: 114–15).

depend on a series of factors: the scale of the intervention, or the compatibility with the existing regime of information exchange, that is the peculiar metastability of the system.

That is why the hypothesis of an exclusively philosophical-political *solution* to the problem of social regulation can only be a dream and a dangerous simplification, through which politics – and this has been for a long time the dream of political philosophy – would 'apply an elementary thought to global realities' taking for real what in fact was the projection of a communitarian mythology on a global scale. Borrowing Simondon's words we could maintain that the resulting 'combination of a set of procedures and of a mythology' would not represent the actual 'encounter of technicity and of the respect of totality'. In fact, such an identity would just be 'the mythology of a group' erected as a 'universal doctrine' (MEOT 225). This imagined identity forcibly depends on 'techniques of manipulation of the human beings' such as publicity and propaganda; otherwise it would not be effective.[25] And there is no other way to imagine the collective on the ground of an established anthropology.

The very outset of Simondon's philosophy of individuation is rather the living being defined by the two equally necessary and contradictory aspects of its normative activity: one of opening, mutation, invention of new opportunities, the other of closure, ordering, regulation. It is on this ground that Simondon opposes in the *Note complémentaire* a 'closed' biological-binary axiological system, in which established ends determine the suitable means, to a technical-analogical axiology aimed at an 'open' system capable of inventing new finalities (NC 508–09). The two tendencies appear in human groups as desire: the simultaneous and diverging demand for liberty and security. This polarity between normativity and normality is the impassable horizon of the dynamics through which human groups organise and risk their structural configurations. In effect, the technical gesture responds to a logic that does not follow a binary biological operation of inclusion or exclusion. On the contrary it aims, so to speak, at a maximum of implementation of the system with the values conveyed by technicity:

> To seek a means to limit the technical gesture according to cultural norms is to wilfully bring a potential evolution to a halt, and view the status quo as something that already enables us to define a kingdom of ends, a final code of values. It is to consider the notion of end as final, as the most high, though it may itself be nothing more than a provisional concept allowing certain vital processes to be grasped while neglecting others. (CT 11)

In technical invention there is something that overcomes the community and institutes a transindividual relationship going from individual to individual without crossing the communitarian integration granted by a shared mythology. Therefore there is no *political solution* to a constitutive dynamic that is impossible to give up. There is, instead, a *politics of possible solutions* thanks to the direct access – apparently with no social mediation – to technical normativity, which makes 'the penetration of a new normativity in a closed community' (NC 513–14) possible. The

[25] The same notion allows Simondon to conceive publicity in relation to a 'zone of technicity' productive of cognitive structures with an archetypal function of 'dispersion' (EH 14–16).

political wager of Simondon consists, thus, in the hypothesis of a reconversion of the apparatus of social values through the development of a technical mentality: the transformation of a system grounded on the evaluation of the means appropriated to the ends posed by a given community into a system grounded on the value 'in itself' of *metastable* functioning. It is in this direction that Simondon arrives at a plan that might be named an 'economy of functioning' opposed to an apparently libertarian 'ethics of productivity [*morale du rendement*]', which in fact would be a new kind of 'communitarian ethics' presupposing a finality to evaluate *a priori* the performativity of the system (NC 526). The expanding process of this ethics entails a 'communitarian resurgence' that is simply disastrous and rapid: 'this notion [...] affects every educational system, every effort, every job':

> A civilisation of productivity, despite the civil liberties it apparently allows to individuals, is in fact extremely constraining and prevents their development since, simultaneously enslaving the human being and the machine, it realises through the machine a constraining communitarian integration. (NC 527)

Simondon repeatedly affirms that the concept of 'functioning' is related to the opening of a system, beyond all the dynamics depending on pre-established correlations between means and ends. From this perspective the problem of providing a technical-political regulation of the human is the problem of regulating a field essentially lacking a pre-established nature and finality:

> The human being is set free from her/his condition of being subjugated to the finality of the whole by learning how to produce finality, how to organise a finalised whole that she/he judges and appreciates, in order not to passively suffer a de facto integration. (MEOT 103)

In effect, the theoretical gesture through which Simondon cancels the possibility of assuming a substantialist anthropology as the ground for political decision and action, shifts politics from the level of the organisation of the relations between interests, means and ends, to the level of the functioning of social systems. And in the light of the concept of functioning, the political problem is twofold: it is the problem of collective invention, and it is also the problem of the compatibility of invention with the actual configuration of the social system.[26] It is, in short, the problem of 'regulation' that ultimately inspired Simondon's project of an axiomatisation of the social sciences.

Through the lens of the notion of regulation it is thus possible to reconstruct Simondon's steps for a reform of the human sciences relying on a larger epistemological reformism. First, he elaborates a general theory for defining systems on the basis of the different typologies of processes by which they are crossed and constituted. Second, he builds a theory of the social system that is conceived of as a system internally discontinuous, the functioning of which exceeds all homeostatic dynamics. The paradoxical exteriority of the regulatory apparatus characterising

[26] In this sense I perfectly agree with Toscano when he argues that 'Simondon tries to think together nature and excess, technology and revolution, in a manner that might at least dislocate some of the common-places of the contemporary debate on ontology and politics' (Toscano 2012: 107–108).

Simondon's model, 'added on [*surajoutée*]' to society – i.e. out of the domain of homeostatic processes – defines precisely what Canguilhem names 'justice':

> Although there are institutions of justice in society, justice, the supreme regulation, does not appear in the form of an apparatus produced by society itself – it can only come, writes Canguilhem – from elsewhere [*d'ailleurs*]. (Canguilhem 1955: 122)

That is why society would be characterised by periods of – so to speak – the tendency of the rate of wisdom to fall, and exceptional moments of heroic invention, in full accordance with the hypothesis of the 'twofold frenzy' presented by Bergson in *Les deux sources*.

But Simondon criticises altogether the technocratic universalism of cybernetics – which reduces justice to a social mechanism of internal self-regulation – and the heroism implicit in the Bergsonian postulate of a structural exteriority of justice – which would definitively situate the question of social regulation out of the boundaries of any possible knowledge of social systems. As explained in Sect. 7.2, Simondon plays the two positions one against the other and demonstrates – through the concept of the transindividual – the necessity of the internal energetic supply provided by non-homeostatic processes, thus opening his theory to the possible emergence of justice from *within* the social system. Hence we might finally ask what kind of regulatory apparatus we are concerned with, when dealing with a social system. But this question would be ill posed. In fact, the living, the machine and society are systems that, in Simondon's philosophy, share the same basic schema, and therefore to try to identify a specific regulatory *apparatus* characterising society, possibly assimilating it to the living or to the machine, would be a step backwards from Simondon's philosophy of the *processes* of individuation.

In short, Simondon inherits from French sociological positivism, through Bergson and Canguilhem, a way to pose the political problem which forces him to *provide* a formula for the question of justice. And yet, by crossing his philosophy of individuation the whole question radically changes its form. Once it is understood that homeostatic stability tends to entropy, an intervention aimed at interrupting the processes of structural stabilisation, and more precisely the entropic inertia of the social system, reveals itself as inevitable. Such an intervention is always risky because it breaks with an instituted order, and yet it is necessary for that order not to result in its own slow dissolution. It is an intervention emerging from within the 'metastable situation' (SD 116), capable of building relations between different homeostatic processes, phases that would not – by themselves – cross or merge, since they function at different orders of magnitude. This intervention is precisely the act of government:

> Such as the living being relies on homeostases for developing and becoming, instead of perpetually remaining in the same state – in the act of government there is a power of absolute event [*une force d'avènement absolu*], which, although resting on homeostases, uses and exceeds them. (MEOT 151)

Government is thus not the regulation of the existing situation, but is conceived as the aleatory invention of new 'compatibilities': recognising the origin of the act of government out of the *given* set of social homeostases, means in fact accepting the impossibility of granting its outcomes. And yet, this does not mean Simondon conceived government as some kind of revolutionary event. In his view, by starting from a complex science of the social system it would be possible to establish not only the conditions of *exertion*, but also the conditions of *emergence* of the act of government. There are always, in effect, both systemic and aleatory conditions defining the possibility of an act of invention: the different homeostases structuring the social must have a sufficient degree of 'disparation' so that they can generate a metastable tension; and an encounter of the system with a compatible functional schema is also necessary for a new process of structuration to emerge.

It is evident that Simondon was not a political revolutionary. Even when he provides the model of'a pre-revolutionary state' as the model of a metastable state from which a process of social individuation takes place (FIP 549–50), this cannot be read in the light of the operating of 'the political' within and *against* the instituted order of 'politics'.[27] In fact, what Simondon's philosophy clearly demonstrates is that an act of government is intrinsically political as far as it always takes place – so to speak – within an endless phase of transition, i.e. in a phase-shift system always developing *between* the opposite risks of the dissolution *and* of the fatal stabilisation of its communitarian base on a mythical identity. Justice as an 'institutional category' cannot fill this open space: it rather poses a problem at the same time of structural stability and of operative innovation, because it always exceeds the 'normal' oscillation of institutions between the complementary attitudes of a 'mythological and sacrificial sacredness' and of a purely 'operational [*opératoire*]' technicity. Justice in fact entails 'both technicity and sacredness', and therefore it essentially 'lacks internal coherence' (PST 348–49); but this 'lack of coherence' marks also the only possibility for political invention to take place.

Simondon believed he could contribute to the emergence of a collective space where processes of social transformation emerge, through his project of formation and diffusion of a 'technical culture'. Far beyond this project, his integration of the study of techniques in the epistemology of the social sciences can contribute today to shift political philosophy from the assumption of a 'structural' anthropology to

[27] When collective movements are the subject of Simondon's reflection, he manifestly denies any unconditional *a-priori* approval of processes of innovation or social re-structuration. On the contrary, he clearly shows the essential ambiguity of any resurgence of the original energetic potentials of the magic phase, kind of attempts to revive 'the political' as an emergency measure in a state of crisis. Although on the one hand he highlights the efficacy of the belief in symbolic objects, capable of producing 'powerful collective movements' (IMIN 133), on the other hand he does not cease to underline the obscure double represented by the persisting actuality of the magic phase, which comes back in 'the history of groups and cultures', according to modalities that risk to be regressive (IMIN 137–38). And nevertheless, it is worth underlining that the 'reformist' approach to philosophical concepts typical of Simondon's research appears to be the necessary basis for a 'revolutionary' philosophical invention of terms such as 'transduction', 'metastability', and 'transindividual' that have proved and still can give proof of philosophical and political fecundity. As De Boever notes by contrasting Simondon to Agamben, 'to approach politics starting from a technical mentality brings us to consider the relation between human beings and the State apparatus as a metastable relation [...] This brings about not the opposition of two poles, but rather the exploration, through accurate examination, of their common becoming' (De Boever 2010: 127–28).

that of an 'operational' technology. According to Simondon, a philosophy rooted *itself* in technical schemas should be capable of accompanying the never accomplished operation of synthesis in which social invention always consists. This requires the technical and 'clinical' exercise of thought that in *Culture et technique* Simondon defined as a regime of functioning in which 'the ability to grasp each problem is perhaps the highest task philosophy [*l'effort de la philosophie*] can assume' (CT 16). Such an exercise can actually bend the inventive activity of political decision towards opening society to the metastable tension peculiar to a 'political field' irreducible either to the jurisprudential and technocratic domain or to the supposed naturalness of political power and spontaneous rebellion: the two complementary facets of the same philosophical-political myth of an *essence* of the 'body politic', whether conceived as a machine or as a living being.

References[28]

Balibar, É. 1993. Individualité et transindividualité chez Spinoza. (conference held on May the 15th 1993). In *Architectures de la raison. Mélanges offerts à Alexandre Matheron,* ed. Moreau, P.-F., 35–46. Fontenay-aux-Roses: ENS Editions (English trans. "Spinoza: from individuality to transindividuality." *Mededelingen vanwege het Spinozahuis* 71. Delft, 1997).

Bardin, A. 2008. Gilbert Simondon. Studio della nozione di animalità: note per un'antropologia politica. *Kath'autón* 2: 40–45.

Bardin, A. 2010. *Epistemologia e politica in Gilbert Simondon. Individuazione, tecnica e sistemi sociali*. Valdagno: FuoriRegistro.

Bardin, A. 2013. De l'homme à la matière, pour une *ontologie difficile*. Marx avec Simondon. *Cahiers Simondon* 5: 25–43.

Bates, D.W. 2012. *States of war: enlightenment origins of the political*. New York: Columbia University Press.

Baudrillard, J. 1968. *Le Système des objets*. Paris: Gallimard (English trans. *The System of Objects*. London: Verso, 1996).

Bergson, H. 1932. Les deux sources de la morale et de la religion. Paris: Alcan (English trans. *The Two Sources of Morality and Religion*. London: Macmillan & Co., 1935).

Canguilhem, G. 1955. Le problème des régulations dans l'organisme et dans la société. In *Ecrits sur la médecine*, 101–125. Paris: Seuil, 2002. (First edition: *Les Cahiers de l'Alliance Israélite Universelle* 92:64–81).

Canguilhem, G. 1972. Régulation. In *Encyclopaedia Universalis*, vol. 14. Paris: Encyclopaedia Universalis.

Carrozzini, G. 2005. Nota a 'Per una nozione di situazione dialettica'. *Il Protagora* 33(5): 105–109.

Chabot, P., ed. 2003. La philosophie de Simondon. Paris: Vrin (English trans. *The Philosophy of Simondon: Between Technology and Individuation*. London: Bloomsbury, 2013).

Citton, Y. 2004. Sept résonances de Simondon. Multitudes. "Politiques de l' individuation. Penser avec Simondon" 2004/18. http://www.multitudes.net/category/l-edition-papier-en-ligne/Multitudes-18-Automne-2004/. Accessed 30 Dec 2014.

Combes, M. 1999. *Gilbert simondon and the philosophy of the transindividual.* Cambridge, MA: MIT Press, 2013 (First edition: *Simondon. Individu et collectivité, pour une philosophie du transindividuel*. Paris: PUF).

Cumont, F. 1949. *Lux perpetua*. Paris: Librarie Orientaliste Paul Gauthier.

[28] *Simondon's complete bibliography and a list of abbreviations are provided in the* Appendix

De Boever, A. 2010. Agamben et Simondon: ontologie, technologie et politique. *Cahiers Simondo* 2: 117–128.

De Chardin, T. 1956. *La place de l'homme dans la nature*. Paris: Albin Michel.

Del Lucchese, F. 2009. Monstrous individuations: Deleuze, Simondon, and relational ontology. *Differences* 20(2–3): 179–193.

Deleuze, G. 1966. Gilbert Simondon, L'individu et sa genèse physico-biologique. In *L'île déserte et autres textes*, 120–124. Paris: Minuit.

Duhem, L. 2013b. *Nekuia*. Évocations de Simondon. http://www.ludovicduhem.com/#!nekuia/c17e2. Accessed 30 Dec 2014.

Feenberg, A. 1991. *Critical theory of technology*. Oxford: Oxford University Press.

Gesell, A. 1946. The ontogenesis of infant behavior. In *Manual of child psychology*, chap. 6, ed. Carmichael, L., 295–331. New York/London: Wiley (French transl. *L'ontogénèse du comportement de l'enfant*. In *Manuel de psychologie de l'enfant*, vol. I, chap. 6, 470–527).

Guchet, X. 2010. *Pour un humanisme technologique. Culture, technique et société dans la philosophie de Gilbert Simondon*. Paris: PUF.

Guchet, X. 2011. Technology, sociology, humanism: Simondon and the problem of the human sciences. *SubStance* 129(41/3): 76–92 (First edition: "Le corps social du sujet." *Cahiers Simondon* 3: 71–92).

Hottois, G. 1993. *Simondon et la philosophie de la "culture technique"*. Bruxelles: De Boeck.

Jung, C.G. 1921. *Psychologische typen*. Zürich und Stuttgart: Rascher Verlag (English trans. *Psychological Types*. Princeton: Princeton University Press, 1971).

Marcuse, H. 1964. *One-dimensional man*. London: Routledge and Keagan.

Merleau-Ponty, M. 1955. *Les aventures de la dialectique*. Paris: Gallimard (English trans. *Adventures of the Dialectic*. Illinois: Northwestern UP, 1973).

Merleau-Ponty, M. 1959b. Notes de travail inédites. Cours 1959. *Chiasmi International 7: Merleau-Ponty. Vie et Individuation*: 39–42.

Morfino, V. 2008. Simondon e il transindividuale. *Il Protagora* 36(12): 395–400.

Portmann, A. 1948. *Die Tiergestalt. Studien über die Bedeutung der tierischen Erscheinung*. Basel: Rheinhardt Verlag (English trans. *Animal Forms and Patterns*. London: Faber and Faber, 1952).

Portmann, A. 1951. Les bases biologiques d'un nouvel humanisme. In *Cahiers de philosophie: le problème de la vie*, vol. 32, 59–79. Neuchatel: Éditions de la Baconnière.

Rodriguez, P.E. 2009. Prólogo. Individuar. De cristales, esponjas y afectos. In G. Simondon, *La individuación a la luz de las nociones de forma y información*, 11–19. Buenos Aires: Editorial Cactus y La Cebra Ediciones.

Schmitt, C. 1932. *Der Begriff des Politischen*. Berlin: Duncker & Humblot. (English trans. *The Concept of the Political*. Chicago: University of Chicago Press, 1996).

Soulez, P. 1989. *Bergson politique*. Paris: PUF.

Stengers, I. 2004. Résister à Simondon? *Multitudes. Politiques de l' individuation. Penser avec Simondon* 2004/18: 55–62. http://www.multitudes.net/category/l-edition-papier-en-ligne/Multitudes-18-Automne-2004/. Accessed 30 Dec 2014.

Stiegler, B. 1994. *La technique et le temps. 1. La faute d'Épiméthée*. Paris: Galilée.

Stiegler, B. 2006. Chute et élévation. L'apolitique de Simondon. *Revue Philosophique de la France et de l'étranger "Gilbert Simondon"* 196(3): 325–341.

Toscano, A. 2002. L'essere interattivo. Appunti 'simondoniani' su informazione e intelligenza sociale. *DeriveApprodi* 21(2002): 88–91.

Toscano, A. 2012. The disparate: ontology and politics in Simondon. *The Warwick Journal of Philosophy* Special Volume. Deleuze and Simondon: 107–117.

Valentine, J. 2006. The political. *Theory Culture and Society* 23(2–3): 505–511.

Van Caneghem, D. 1989. Hommage à Gilbert Simondon. *Bulletin de Psychologie* 42(392): 815–836.

Virno, P. 2001. Multitude et principe d'individuation. *Multitudes. "Politiques de l'individuation. Penser avec Simondon"* 2004/18: 103–17. http://www.multitudes.net/category/l-edition-papier-en-ligne/Multitudes-18-Automne-2004/. Accessed 30 Dec 2014.

Virno, P. 2003. *Quando il verbo si fa carne. Linguaggio e natura umana*. Torino: Bollati Boringhieri.

Appendix

List of Abbreviations

A	1957–58a*. Allagmatique
AI	1957–58b*. Analyse des critères de l'individualité
API	1962a. L'amplification dans les processus d'information
APM	1959b. Aspect psychologique du machinisme agricole
CSP	1964–65. Cours sur la perception
CT	1965. Culture et technique
DL	1963–64. Deux leçons sur l'animal et l'homme
EH	1960d. L'effet de halo en matière technique: vers une stratégie de la publicité
ENC	1957. La psychologie moderne
ET	1966. Interview on 'La technologie'
FIP	1960a. Forme, information, potentiels
FIPD	1960b. Discussion following FIP
HNI	1958d*. Histoire de la notion d'individu
HO	1974–75. L'homme et l'objet
HU	1953a. Humanisme culturel, humanisme négatif, humanisme nouveau
I	1958a. L'individuation à la lumière des notions de forme et d'information
IMIN	1965–66. Imagination et invention
IGPB	1964a. L'individu et sa genèse physico-biologique
IPC	1989. L'individuation psychique et collective
IPM	1966–67a. Initiation à la psychologie moderne

**All references are provided with the date of the first edition in the original language or, in case of posthumous works, with the date of the presumed draft (my hypotheses are followed by the symbol *). Conferences and interviews are ordered by the date of release. I refer to unpublished texts by quoting the source. Only primary sources have been used. Existing translations (provided in bibliography) have been sometimes changed according to my interpretation and on my responsibility.*

A. Bardin, *Epistemology and Political Philosophy in Gilbert Simondon*,
Philosophy of Engineering and Technology 19, DOI 10.1007/978-94-017-9831-0

IT	1971a. L'invention dans les techniques
LPH	1959a. Les limites du progrès humain
MEC	1976a. Le relais amplificateur
MECD	1976b. Discussion *in* Jones Colloque sur la Mécanologie
MEOT	1958b. Du mode d'existence des objets techniques
MT	1970b*. Mentalité technique
NC	1958c*. Note complémentaire sur les conséquences de la notion d'individuation
PI	1953–54. Place d'une initiation technique dans une formation humaine complète
PR	1958e*. 'Prospectus' of Du mode d'existence des objets techniques
PST	1960–61. Psycho-sociologie de la technicité
RO	1962b. Colloque de Royaumont
RPE	1954a. Réflexions préalables à une refonte de l'enseignement
SD	1960e. Pour une notion de situation dialectique
SOT	1983b. Sauver l'objet technique
TP	1983a. Trois perspectives pour une réflexion sur l'éthique et la technique

Works of Gilbert Simondon

1953a. "Humanisme culturel, humanisme négatif, humanisme nouveau." In *Actes du Congres de Tours et Poitiers, 3–9 Sept. 1953*, 51–54. Paris: Les Belles Lettres, 1954.

1953b*. "Cybernétique et Philosophie (unpublished)." Quotations in Guchet, X. 2005b.

1953c*. "Epistémologie de la Cybernétique." (unpublished) Quotations in Guchet, X. 2005b.

1953–54. "Place d'une initiation technique dans une formation humaine complète." *Cahiers pédagogiques pour l'enseignement du second degré* 1953 9/2: 115–120; (with Zadou-Naïsky) 1954 9/7: 429–34; 1954 9/8: 581. Lyon: ANECNES.

1954a. "Réflexions préalables à une refonte de l'enseignement." *Les Papiers du Collège International de Philosophie* 1992/12 (First edition: *Cahiers pédagogiques pour l'enseignement du second degré* 1954 20/2: 83–89. Lyon: ANECNES).

1954b. "Note sur l'objet technique." *Les Papiers du Collège International de Philosophie* 1992/12 (First edition: *Cahiers pédagogiques pour l'enseignement du second degré* 1954 20/2: 89–90. Lyon: ANECNES).

1957. "La psychologie moderne." (with F. Le Terrier) In *Encyclopédie de la Pléiade - Histoire de la science*, 1668–703. Paris: Gallimard.

1957–58a*. "Allagmatique." In I, 559–66.

1957–58b*. "Analyse des critères de l'individualité." In I, 553–58.

1958a. *L'individuation à la lumière des notions de forme et d'information*. Grenoble: Millon, 2005. (First edition: PhD thesis, *Archives de Georges Canguilhem*. Paris ENS, CAPHES: GC. 40.2.1).

1958b. *Du mode d'existence des objets techniques*. Paris: Aubier, 1989.

1958c*. "*Note complémentaire sur les conséquences de la notion d'individuation*." In I, 503–27.

1958d*. "Histoire de la notion d'individu." In I, 339–502.

1958e*. "Prospectus" of *Du mode d'existence des objets techniques*. In *Gilbert Simondon, Une pensée de l'individuation et de la technique*, Bibliothèque du Collège international de philosophie, 265–67. Paris: Albin Michel, 1994.

1959a. "Étude critique. Les limites du progrès humain." In *Gilbert Simondon, Une pensée de l'individuation et de la technique*, Bibliothèque du Collège international de philosophie, 268–75. Paris: Albin Michel, 1994. (First edition: *Revue de métaphysique et de morale* 1959/3: 370–76; English translation: Cubitt, S. "The Limits of Human Progress: A Critical Study." *Cultural Politics* 6/2: 229–236).

1959b. "Aspect psychologique du machinisme agricole." In *Premier symposium national de médecine agricole. Colloque de psycho-sociologie agricole. Tours 14 juin 1959*. Institut national de médicine agricole, II: 13–17.

1959c. "Les entretiens de Mysore (Inde). 28 Aout – 1er Septembre 1959." In Nikam, A. N. – Simondon, G. – Prasad, R., *Traditional Cultural Values, East and West. Entretiens: International Institute of Philosophy and Indian Philosophical Congress, Science and Philosophy, Individual Freedom and Community, Traditional Values*, 133–36. Mysore: Wesley Press.

1960a. "Forme, information, potentiels." In I, 531–51 (First edition: *Bulletin de la Société française de Philosophie* 54/5, Séance du 27 février 1960: 143–74).

1960b. "Débat" following "Forme, information, potentiels." *Bulletin de la Société française de Philosophie* 54/5: 751–65. http://www.sofrphilo.fr/telecharger.php?id=82 Accessed 30 December 2014 (First edition: *Bulletin de la Société française de Philosophie* 54/5, Séance du 27 février 1960: 174–88).

1960c. "Attitudes et motivations." In *Communication et information. Cours et conférences*, 375–98. Chatou: La Transparence, 2010.

1960d. "L'effet de halo en matière technique: vers une stratégie de la publicité." *Cahiers philosophiques* 1990/43: 7–22 (First edition: *Cahiers de l'Institut de Science Economique et Appliquée I.S.E.A. Collection Recherches et dialogues philosophiques et économiques* 1960/99: 55–69).

1960e. "Pour une notion de situation dialectique." *Il Protagora* 33/5: 110–119, 2005.

1960 f. "Psychosociologie du cinéma." In *Sur la technique*, 355–61. Paris: PUF, 2014.

1960–61. "Psycho-sociologie de la technicité. Aspects psycho-sociaux de la genèse de l'objet d'usage." *Bulletin de l'école pratique de Psychologie et de Pédagogie, Université de Lyon* 1960/2: 127–40; 1961/3: 227–38; 1961/3: 263; 1961/4-5: 319–50.

1961a. "Anthropo-technologie." In *Sur la technique*, 367–71. Paris: PUF, 2014.

1961b. “Objet technique et conscience moderne.” In *Sur la technique*, 363–65. Paris: PUF, 2014.
1962a. “L’amplification dans les processus d’information.” In *Communication et information. Cours et conférences*, 157–76. Chatou: La Transparence, 2010 (paper originally delivered at the *Colloque de Royaumont* in 1962. Only the abstract was published in Simondon 1962b).
1962b. *Le concept d’information, 5ème colloque philosophique de Royaumont*. Cahiers de Royaumont, Philosophie n° 5. Paris: Minuit, 1965. Présentation of Wiener, N. “L’homme et la machine”, 99–100 (Also in N. Wiener, *Collected Works* IV: 824); Discussion following De Possel, R. “Transformation de l’information sous forme de texte imprimé en information codée sur ruban magnétique”, 153–172. “Résumé de la séance de travail sur l’amplification dans les processus d’information”, 417 (abstract of Simondon’s paper, posthumously published in Simondon 1962a).
1962c. “Objet économique et objet technique.” In *Sur la technique*, 373–77. Paris: PUF, 2014.
1963–64. *Deux leçons sur l’animal et l’homme*. Paris: Ellipses, 2004 (English translation: Burk, D.S. *Two Lessons on Animal and Man*. Minneapolis: Univocal Publishing).
1964a. *L’individu et sa genèse physico-biologique. L’individuation à la lumière des notions de forme et d’information*. Paris: PUF.
1964b. “Cours sur l’instinct.” In *Communication et information. Cours et conférences*, 247–373. Chatou: La Transparence, 2010. (“*L’Instinct. Plan du cours. Première partie: chapitres a, b, c, d*.” *Bulletin de Psychologie* 1-2/235: 75–104).
1964c*. “Introduction au projet d’encyclopédie génétique.” (unpublished) Quotations in Bontems, V. 2006, 311–24.
1964–65. *Cours sur la perception (1964–65)*. Chatou: La Transparence, 2005.
1965. “Culture et technique.” *Bulletin de l’institut de philosophie morale et enseignement. Université libre de Bruxelles* 55–56: 3–16 (English translation: Fraser O. L. and Menegalle G. “Culture and Technics.” *Radical Philosophy* 2015/189: 17–23).
1965–66. *Imagination et invention (1965–66)*. Chatou: La Transparence, 2008.
1966. “Entretien avec Monsieur Gilbert Simondon” (Radio Interview by Y. Deforge). In *Dix entretiens sur la technologie. Ministère de l’éducation nationale ‘Institut pédagogique nationale’ radio télévision scolaire*, 33–35. Paris: Nouveaux Chantiers.
1966–67a. “Initiation à la psychologie moderne. Première partie.” *Bulletin de Psychologie* 1966 5/254: 288–98; 1967 8-9/256: 479–90; 1967 23-24/262: 1449–459.
1966-67b. “La sensibilité. Plan du cours.” *Bulletin de Psychologie* 1966 5/254: 276–87; 1967 6-7/255: 388–97; 1967 20-21/260: 1280–305; 1967 23-24/262: 1475–478.
1968a. “Entretien sur la mécanologie.” (TV Interview by J. Le Moyne) In *Révue de synthèse* 2009 130/1: 103–32.

1968b. "Perception et modulation." In *Communication et information. Cours et conférences*, 187–245. Chatou: La Transparence, 2010.
1968–69. "L'invention et le développement des techniques." In *L'invention dans les techniques. Cours et conférences*, 75–226. Paris: Seuil, 2005 (First edition: "Ethique de l'intervention." *Les cahiers du CERFEE* 1992/8-9: 27–41).
1969. "Revue critique" of Oparin, A. I. *L'origine de la vie sur la terre*." *Revue philosophique de la France et de l'Étranger* 1969/46: 71–90.
1969–70. "La perception de longue durée. Plan du cours." *Journal de Psychologie normal et pathologique* 1969/4: 397–419; 1970/2: 153–70; 1970/4: 403–22.
1970a. "Préface" to Rabaudy, C. and Rolland, B. (eds.) *Sophia. Recueil de textes philosophiques pour la classe terminal A avec présentations, notes et questions. Tome I: La connaissance*, 3. Paris: Hatiers.
1970b*. "Mentalité technique." *Revue philosophique de la France et de l'Étranger* "Gilbert Simondon" 2006/3: 343–57 (English translation: De Boever, A. "Technical Mentality." In De Boever, A. et al. (eds.) *Gilbert Simondon. Being and Technology*. Edinburgh University Press, 1–15).
1970c. "Naissance de la technologie." In *Sur la technique*, 131–178. Paris: PUF, 2014.
1970–71. "Formes et niveaux de la communication." Published as "Cours sur la communication." In *Communication et information. Cours et conférences*, 53–155. Chatou: La Transparence, 2010.
1971a. "L'invention dans les techniques." In *L'invention dans les techniques. Cours et conférences*, 227–72. Paris: Seuil, 2005 (First edition: In *Mécanologie. Les cahiers du centre culturel canadien* 1971/2: 41–66).
1971b. "Discussion" following Jones, H. "Implications conceptuelles de la mécanologie." In *Mécanologie. Les cahiers du centre culturel canadien* 1971/2: 29–40.
1971c. "Sans homme, pas de machine: pas d'homme sans machine » Entretien on les sciences et les techniques." (Radio Interview by J. Le Moyne) In Charbonnier, G. *France Culture*, 30 Avril 1971 (Italian transcription by Carrozzini, G. 2008. *Il Protagora* 36/12: 444–75).
1972. "Technique et eschatologie: le devenir des objets techniques." In *Sur la technique*, 331–36. Paris: PUF, 2014.
1974. "La résolution des problèmes." In *L'invention dans les techniques. Cours et conférences*, 305–25. Paris: Seuil, 2005 (abstracts).
1974–75. "L'homme et l'objet." Summary of the course of *Psychologie générale*, cyclostyled text, 32 pp. non numbered.
1976a. "Le relais amplificateur." In *Mécanologie 2. Deuxième colloque la Mécanologie, 21–22 mars 1976. Les cahiers du centre culturel canadien* 1979/4: 135–43.
1976b. "Discussion" following Jones, H. "Machines, être, concrétude." In *Mécanologie 2. Deuxième colloque la Mécanologie, 21–22 mars 1976. Les cahiers du centre culturel canadien* 1979/4: 75–87.
1976c. "Invention et créativité." In *L'invention dans les techniques. Cours et conférences*, 327–43. Paris: Seuil, 2005 (abstracts).

1980. "Art et nature (Maîtrise technique de la nature)." In *Sur la technique*, 179–200. Paris: PUF, 2014.
1982. "Sur la technoesthétique." Draft for a letter to Jacques Derrida, 3rd July 1982 *Les Papiers du Collège International de Philosophie* 1992/12 (English translation: De Boever, A. "On Techno-Aesthetics." *Parrhesia* 2012/14: 1–8). Published with some "Suppléments" as "Réflexions sur la techno-esthétique." In *Sur la technique*, 392–96. Paris: PUF, 2014.
1983a. "Trois perspectives pour une réflexion sur l'éthique et la technique." In *Ethique et technique. Annales de l'Institut de philosophie et sciences morales de l'Université de Bruxelles* 32: 107–18.
1983b. "Sauver l'objet technique." (Interview) *Esprit. Changer la culture et la politique* 76/4: 147–52.
1989. *L'individuation psychique et collective. À la lumière des notions de forme, information, potentiel et métastabilité*. Paris: Aubier.
1995. *L'individu et sa genèse physico-biologique*. Grenoble: Million[2].

Brief Note on the Editorial Vicissitudes of *Individuation*

In 1958 Simondon accomplished his two main works, *L'individuation à la lumière des notions de forme et d'information* [I] and *Du mode d'existence des objets techniques* [MEOT], respectively as a doctoral thesis and a complementary doctoral thesis. While MEOT was published in the same year, *Individuation* was scattered in subsequent editions. In 1964 Simondon published *L'individu et sa genèse physico-biologique* [IGPB] which contains the introduction, conclusion and the first two sections of *Individuation*, with the exception of chapter I. 3. Only in 1989 did he partially re-edit *L'individuation psychique et collective* [IPC] (published posthumously), which contains the conclusive part on psychic and collective individuation, a new introduction made up of the original one plus *Forme, information, potentiels* [FIP], and the unedited *Note complémentaire sur les conséquences de la notion d'individuation* [NC]. In the 1995 edition of IGPB the two programmatic texts *Allagmatique* [A] and *Analyse des critères de l'individualité* [AI] also appeared. Only in 2005 was the original text of *Individuation* published with NC, FIP (without the following debate [FIPD]), the programmatic texts, and the further addition of the unedited *Histoire de la notion d'individu* [HNI], originally supposed to be the second part of the main thesis which Simondon never completed.

Index

A. Bardin, *Epistemology and Political Philosophy in Gilbert Simondon*,
Philosophy of Engineering and Technology 19, DOI 10.1007/978-94-017-9831-0

MIX
Papier aus verantwortungsvollen Quellen
Paper from responsible sources
FSC® C105338

If you have any concerns about our products,
you can contact us on
ProductSafety@springernature.com

In case Publisher is established outside the EU,
the EU authorized representative is:
Springer Nature Customer Service Center GmbH
Europaplatz 3, 69115 Heidelberg, Germany

Printed by Libri Plureos GmbH
in Hamburg, Germany